T0343264

Spectral Imaging of the Atmosphere

This is Volume 82 in the
INTERNATIONAL GEOPHYSICS SERIES

A series of monographs and textbooks
Edited by RENATA DMOWSKA, JAMES R. HOLTON and H. THOMAS ROSSBY

A complete list of books in this series appears at the end of this volume

Spectral Imaging of the Atmosphere

Gordon G. Shepherd

Centre for Research in Earth and Space Science,
York University, Toronto, Canada

An Elsevier Science Imprint

Amsterdam Boston London New York Oxford Paris
San Diego San Francisco Singapore Sydney Tokyo

This book is printed on acid-free paper.

Academic Press
An imprint of Elsevier Science
84 Theobald's Road, London WC1X 8RR, UK
http://www.academicpress.com

Academic Press
An imprint of Elsevier Science
525 B Street, Suite 1900, San Diego, California 92101-4495, USA
http://www.academicpress.com

ISBN 0-12-639481-4

Library of Congress Catalog Number: 2002107237

A catalogue record for this book is available from the British Library

Cover images:
The upper image shows the Upper Atmosphere Research Satellite under preparation for launch, courtesy of the NASA Goddard Space Flight Center UARS Project. The lower data image shows the influence of the tidal wind on the airglow emission at the equator, prepared by Charles McLandress with the support of the WINDII Team.

Typeset by Newgen Imaging Systems (P) Ltd., Chennai, India
Printed and bound by Antony Rowe Ltd, Eastbourne
Transferred to digital print on demand, 2005

02 03 04 05 06 07 MP 9 8 7 6 5 4 3 2 1

DEDICATED TO IRENE AND GEORGE SHEPHERD

in memory of their unfailing love and encouragement

Who has seen the Wind?
Neither you nor I:
But when the trees bow down their heads,
The wind is passing by.

Christina Rossetti (1830–1894)

"Here was the least common denominator of nature, the skeleton requirements simply, of land and sky – Saskatchewan prairie. It lay wide around the town, stretching tan to the far line of the sky, shimmering under the June sun and waiting for the unfailing visitation of wind, gentle at first, barely stroking the long grasses and giving them life; later, a long hot gusting that would lift the black topsoil and pile it in barrow pits along the roads, or in deep banks against the fences."

"High above the prairie, platter-flat, the wind wings on, bereft and wild its lonely song. It ridges drifts and licks their ripples off; it smoothens crests, piles snow against the fences. The tinting green of Northern Lights slowly shades and fades against the prairie nights, dying here, imperceptibly reborn over there."

Who Has Seen the Wind
W.O. Mitchell (1914–1998)

CONTENTS

Preface **xiii**

1 Observing Atmospheric Radiation **1**
1.1 Atmospheric Radiation 1
1.2 Measuring Atmospheric Radiation 8
1.2.1 The Integrated Emission Rate 8
1.2.2 Visible Atmospheric Radiation 10
1.2.3 Thermal Atmospheric Radiation 10
1.2.4 Ultraviolet Atmospheric Radiation 11
1.3 The Scope of Spectral Imaging 12
1.4 One-Dimensional (Vertical) Spatial Information 13
1.5 Two-Dimensional (Horizontal–Vertical) Information 16
1.6 Three-Dimensional Information 18
1.7 Spectral Information 20
1.8 Temporal Information 26
1.9 Preview 28
1.10 Problems 29

2 Spectral Concepts **30**
2.1 Introduction 30
2.2 The Spectral Concept 31
2.3 Formal Statement of the Fourier Transform 33
2.4 Fundamental Properties of the Fourier Integral 35
2.5 Doing a Fourier Integral Without Integration 36
2.6 Building Up a Set of Fourier Transforms 37
2.7 Convolutions and Correlations 38
2.8 The Dirac Delta Function and the Dirac Comb 39
2.8.1 Dirac Delta Function 39
2.8.2 The Dirac Comb 40
2.9 The Discrete Fourier Transform 41
2.10 The Autocorrelation Function and Power Spectral Density 44

2.11 Optical Devices as Linear Dynamical Systems 45
2.12 The Diffraction Grating as a Linear Dynamical System 47
2.13 The Fabry–Perot Etalon as a Linear Dynamical System 51
2.14 Problems 52

3 Instrument Responsivity and Superiority 54
3.1 Responsivity of an Elementary Photometer 54
3.2 The Measurement of Irradiance 57
3.3 Responsivity for Line and Continuum Sources 57
3.4 Photometer Calibration 59
3.5 Generalized Definition of Responsivity 61
3.6 Jacquinot's Definition of Étendue 62
3.6.1 Conservation of Étendue 62
3.6.2 Comparison of Astronomical and Atmospheric Sources 63
3.7 Resolving Power and the Superiority of Spectral Imagers 63
3.7.1 The Photometer Becomes a Spectrometer 63
3.7.2 Dispersion and Resolving Power for a Diffraction Grating Spectrometer 64
3.7.3 Superiority of the Diffraction Grating Spectrometer 65
3.7.4 Comparison of Superiority for the Diffraction Grating and Fabry–Perot Spectrometers 66
3.8 Dispersion, Classification and Nomenclature 66
3.9 Problems 68

4 Imaging Concepts 70
4.1 Elementary Detectors and Noise 70
4.2 Scanning Satellite Imager 72
4.2.1 Overview 72
4.2.2 The ISIS-II Satellite Imagers 73
4.2.3 Dynamics Explorer-1 Imager 74
4.3 Weather Satellite Imagers 76
4.4 Introduction to Array Detectors 80
4.5 The Charge Coupled Device (CCD) Detector 81
4.5.1 Introduction to Semiconductors 81
4.5.2 Method of Operation 82
4.5.3 CCD Readout 84
4.5.4 CCD Characteristics 86
4.5.5 Signal-to-Noise Ratio 87
4.6 Spectral Response and Materials 87
4.7 Considerations Specific to Infrared Array Detectors 88
4.7.1 Background Radiation 88
4.7.2 Infrared Detector Readout 89

4.8 Other Types of Array Detectors 89
4.8.1 Photodiode Arrays 89
4.8.2 Charge Injection Devices 90
4.8.3 Intensifiers 90
4.8.4 Position Sensitive Arrays 91
4.9 Early Array Detector Imagers 92
4.9.1 Elementary Imagers 92
4.9.2 Spectral Imagers 93
4.9.3 The KYOKKO Auroral Imager 95
4.10 CCD Satellite Imagers 95
4.10.1 The Viking Ultra Violet Imager (UVI) 95
4.10.2 The Polar VIS Imager and the IMAGE Satellite 98
4.11 Summary 99
4.12 Problems 100

5 The Fabry–Perot Spectrometer **102**
5.1 Introduction 102
5.2 The Idealized Etalon 103
5.3 The Real Etalon 107
5.4 Elementary Fabry–Perot Spectrometer Configuration 108
5.5 The Spherical Fabry–Perot Spectrometer 109
5.6 Scanning Methods for Fabry–Perot Spectrometers 112
5.7 The Application of Fabry–Perot Spectrometers 114
5.7.1 Multiple Ring Aperture Instruments 114
5.7.2 Axicon System 116
5.7.3 Low Light Level Applications 116
5.7.4 Tandem Fabry–Perot Spectrometers 117
5.7.5 Stabilized Fabry–Perot Spectrometers 119
5.8 Applications of the Fabry–Perot Imager 121
5.8.1 Introduction 121
5.8.2 PRESTO – A Programmable Etalon Spectrometer for Twilight Observations 121
5.8.3 MORTI and SATI: Hybrid Spatial and Spectral Instruments 122
5.8.4 Imaging Low Light Level Application 124
5.8.5 Imaging Winds with an FPS Imager 124
5.8.6 CLIO (Circle to Line Interferometer Optical System) 126
5.9 Problems 127

6 The Michelson Interferometer **129**
6.1 Historical Background 129
6.2 Basic Concept 130
6.3 Spectral Resolution 133
6.4 Field of View 134
6.5 The Real Michelson Interferometer 135
6.6 Sampling the Interferogram 135

6.7 Superiority of the Michelson Interferometer 136
6.8 Scanning Methods for the Ordinary Michelson Interferometer 137
6.8.1 Overview 137
6.8.2 Cube Corner Reflectors 138
6.8.3 Cat's Eye Retro-Reflector 138
6.8.4 The Dynamic Alignment System 139
6.9 Some Atmospheric Applications of the Michelson Interferometer 139
6.10 Field Widening 142
6.11 Problems 149

7 Multiplexers and Modulators **151**
7.1 Spectral Operating Modes 151
7.2 Multiplexers 152
7.2.1 Introduction 152
7.2.2 The Hadamard Spectrometer 152
7.2.3 Grating Spectrometers with Array Detectors 154
7.3 Modulators 154
7.3.1 The SISAM 154
7.3.2 The Birefringent Photometer 156
7.3.3 The Grille Spectrometer 158
7.3.4 The Correlation Spectrometer 160
7.3.5 The Pressure Modulator Radiometer (PMR) 160
7.3.6 Instruments for Dayglow Observations 164
7.4 Problems 166

8 Doppler Michelson Interferometry **168**
8.1 The Measurement of Doppler Temperature 168
8.2 The Measurement of Doppler Wind 172
8.3 Phase Stepping Interferometry 173
8.4 The Wide-Angle Michelson Interferometer 175
8.5 Cube Corner Doppler Michelson Interferometer 176
8.6 Achromatizing a Field-Widened Michelson Interferometer 177
8.7 Thermally Stabilizing a Solid Michelson Interferometer 178
8.8 A Fully Compensated Solid Doppler Michelson Interferometer 179
8.9 Defocusing a Wide-Angle Michelson Interferometer 180
8.10 Polarizing Doppler Michelson Interferometers 181
8.10.1 Introduction 181
8.10.2 PAMI Polarization States 182
8.10.3 The SOHO SOI (Solar Oscillations Investigation) and GONG Instruments 183
8.11 The Phase Quadrature Michelson Interferometer 185
8.11.1 Concept 185
8.11.2 Phase-Shifting with Optical Thin Film Multilayers 185
8.12 Optimized Reflective Wide-Angle Phase-Stepping MI 187
8.13 Problems 189

9 Operational Atmospheric Spectral Imagers **191**
9.1 Introduction 191
9.2 The Wind Imaging Interferometer (WINDII) 191
9.2.1 Fundamental Spaceflight Considerations 191
9.2.2 WINDII Optical System 194
9.2.3 The Michelson Interferometer 195
9.2.4 Interference Filters 196
9.2.5 Detector 197
9.2.6 The WINDII Baffle 197
9.2.7 Calibration 199
9.2.8 Wind Measurement Procedure 200
9.2.9 Examples of Results Obtained 201
9.3 ERWIN: An E-Region Wind Interferometer 207
9.3.1 Introduction 207
9.3.2 Instrument Description 207
9.3.3 The Michelson Interferometer 208
9.3.4 Examples of Measurements 209
9.4 MICADO – Michelson Interferometer for Coordinated Auroral Doppler Observations 211
9.5 The High-Resolution Doppler Imager (HRDI) 213
9.5.1 Introduction 213
9.5.2 Input Optics 213
9.5.3 The Etalons 215
9.5.4 The Detector 216
9.5.5 HRDI Results 216
9.5.6 Comparison of HRDI and WINDII 217
9.6 CLAES: The Cryogenic Limb Array Etalon Spectrometer on UARS 220
9.6.1 Introduction 220
9.6.2 Instrument Design 221
9.6.3 Sample CLAES results 223
9.7 MOPITT – Measurements Of Pollution In The Troposphere 223
9.8 Problems 227

10 Future Atmospheric Spectral Imagers **230**
10.1 The TIMED Doppler Imager (TIDI) 230
10.1.1 TIDI Overview 230
10.1.2 Instrument Description 230
10.1.3 TIDI Data Coverage 233
10.1.4 TIDI Science Measurement Summary 233
10.2 The Mesospheric Imaging Michelson Interferometer (MIMI) 235
10.2.1 Introduction 235
10.2.2 General Description of the Instrument 235
10.2.3 Michelson Interferometer 237
10.2.4 Filter Selection 239

10.2.5 Appearance of the O_2 Lines in the Field of View 239
10.2.6 MIMI Status 239
10.3 The Stratospheric Wind Interferometer for Transport Studies (SWIFT) 240
10.3.1 Introduction and Motivation 240
10.3.2 Concept 241
10.3.3 Instrument Description 244
10.4 The Atmospheric Chemistry Experiment (ACE) 248
10.5 The Michelson Interferometer for Passive Atmospheric Sounding (MIPAS) 251
10.6 Problems 254

11 Grating Spectrometers as Spectral Imagers **255**
11.1 Introduction 255
11.2 Fundamental Aspects of the Diffraction Grating Spectrometer 257
11.3 Selected Airglow Missions Accomplished 258
11.3.1 Mariner 10 Ultraviolet Airglow Experiment 258
11.3.2 The Voyager Mission Ultraviolet Experiment 260
11.3.3 Arizona Imager Spectrograph (AIS) 260
11.3.4 Single-Element Imaging Spectrograph (SEIS) 262
11.3.5 Ground-Based Instruments 264
11.4 Selected Atmospheric Missions Accomplished 266
11.4.1 Total Ozone Mapping Spectrometer (TOMS) 266
11.4.2 Stratospheric Aerosol and Gas Experiment (SAGE) 267
11.4.3 Optical Spectrograph and InfraRed Imaging System (OSIRIS) 267
11.4.4 CRyogenic Infrared Spectrometers and Telescopes for the Atmosphere (CRISTA) 269
11.5 Future Atmospheric Missions using Grating Spectrographs 271
11.5.1 GOMOS and SCIAMACHY on Envisat 271
11.5.2 Ozone Dynamics Ultraviolet Spectrometer (ODUS) 272
11.5.3 Measurements of Aerosol Extinction in the Stratosphere and Troposphere Retrieved by Occultation (MAESTRO) 273
11.6 Spatial Heterodyne Spectroscopy (SHS) 274
11.7 Problems 277

12 Postscript **279**

References **281**

List of Symbols **297**

List of Acronyms and Abbreviations **300**

Author Index **305**

Subject Index **310**

PREFACE

This book is the result of a career spent talking to fellow scientists about atmospheric instruments, and working with colleagues in my own laboratory to conceive, propose, design, build, launch and operate atmospheric spectral imagers. For large space projects this has also involved international collaborations, and working closely with colleagues in industry and in government. The period of data collection is followed by data analysis, validation, software revision and the final scientific analysis, all of which leads to further understanding of the instrument. To all of the individuals involved in these activities I am indebted for the ideas shared over the years. While not all can be named, many are identified in the text but for some I wish to describe their contributions more specifically. In particular I am grateful to my M.Sc. and Ph.D. research supervisors, Donald Hunten and the late Harry Welsh respectively for starting me off in a favourable direction.

Herbert Gush and I worked in the same laboratories on our B.Sc., M.Sc. and Ph.D. degrees. We then went separate ways, he to a PDF in Pierre Jacquinot's Laboratoire Aimé Cotton, and I back to the University of Saskatchewan. Herb made me aware of what was happening in France, and made the contact that led to the visit of Robert Chabbal to Saskatoon in 1958. That visit led to the first Fabry–Perot spectrometers built there, but it was my own visit to this laboratory in 1961 where I met Ové Harang and Pierre Connes working on a field-widened Michelson interferometer that expanded my horizons much further. All of this contributed to the conceptual approach which I still follow.

In the early years at the University of Saskatchewan it was the students who built the instruments, John Nilson, Ted Turgeon, Alan Bens, Leroy Cogger, Steve Peteherych, Ken Paulson, Bill Lake, Ronald Hilliard and Harold Zwick. It was also the students there who led to the first course that I taught on this subject.

Later, at York University, where this volume had its origins, I initially elicited the help of various individuals in writing the original chapters. Rick Gerson wrote the original array detector section of Chapter 4, later extended by Stoyan Sargoytchev, with a contribution from Erik Griffioen. Rudy Wiens wrote the beginning of Chapter 5, with contributions from Bob Peterson and Fadia Bahsoun-Hamade. Bill Gault wrote the orginal version of Chapter 6, little changed from its original form. Bill is the individual with whom I have had the longest association, and with whom I have shared more ideas than any other individual named. Recent course students, Jeffrey Czapla-Myers, David Babcock, Jacob Petersen and Gina Infante contributed to a revised Chapter 7 while former students John Bird and Susan McCall contributed to Chapter 8. Charlie Hersom wrote the first WINDII description with additions by Brian Solheim in Chapter 9, while Stephen Brown contributed to the ERWIN description in the same chapter. The descriptions of the new instruments, MIMI and SWIFT, in Chapter 10 were taken from material by Bill Gault and Reza Mani of York University, William Ward of the University of New Brunswick, Yves Rochon of the Meteorological Service of Canada and Neil Rowlands, Alan Scott and Gary Buttner of EMS Technologies. However, I have made many revisions in the material provided to me and take full responsibility for any misconceptions or errors that may have been introduced in the process.

My latest course students tested all of the problems; Bernard Firanski, Itamar Gabor, Craig Haley, Young-Sook Lee, Guiping Liu, Peyman Rahnama and Peter Ryan. Individual chapters, or parts of chapters, were tested on several colleagues; Ian McDade, Christian von Savigny, John Miller, Neil Rowlands, Leroy Cogger, Gonzalo Hernandez, Bill Gault, Fred Taylor, Brian Solheim, Aidan Roche, Peter Bernath, Herbert Fisher and Abas Sivjee. Again, I am responsible for the remaining deficiencies. Two others to whom I am greatly indebted for many hours of helpful discussion are Paul Hays and Raymond Roble.

Most of all, I am deeply grateful to Marianna Shepherd for her patience, encouragement and support during the many years the work was in progress.

Many of my conversations have been with graduate students, so that is the level to which this work is primarily directed; however, it is planned to be accessible to upper year undergraduates with some knowledge of optics, and the problems are designed with them in mind. It also may be used as a reference work, as the chapters need not be read in sequence, and I hope that the long accumulation of references will prove generally valuable to researchers.

I would like to thank the initial and final Senior Editors, Gioia Ghezzi and Frank Cynar, and the Production Project Manager Sutapas Bhattacharya for their consistent encouragement and support.

1

OBSERVING ATMOSPHERIC RADIATION

1.1 ATMOSPHERIC RADIATION

Winter or summer, day or night, above the Earth or on the ground, one is bathed in atmospheric radiation, comprising ultraviolet, visible and infrared light in the region of the electromagnetic spectrum, lying roughly between 30 nm and 100 μm in wavelength. The sources and processes involved in producing this radiation environment are illustrated in Figure 1.1; in this chapter they are briefly reviewed along with the framework within which this radiation is observed. The ultimate source of the observed radiation is the sun, so it is the logical place to begin. It is the brightest object in the sky, so much so that it cannot be viewed directly by eye. From its effective temperature T of 5780 K the Stefan–Boltzmann law, $E = \sigma_{SB}T^4 (\sigma_{SB} = 5.67 \times 10^{-8}\,\mathrm{W\,m^{-2}\,K^{-4}})$, yields an irradiance $E = 6.33 \times 10^7\,\mathrm{W\,m^{-2}}$. This is the total power radiated from each square metre on the solar surface. The monochromatic solar irradiance E_λ, the irradiance as a function of wavelength λ as approximated by a black body with $T = 5780\,\mathrm{K}$ is shown in Figure 1.2; this is called a spectrum. The integrated irradiance E, just introduced, is the integral over this spectrum. The spectrum of a 300 K blackbody representing the outgoing radiation from the Earth (multiplied by 10^6) is shown in the same figure; this is discussed later. For the wavelength range shown, the solar spectrum is sharply peaked, with the wavelength λ_{max} of the peak given by Wien's displacement law, $T\lambda_{max} = 2897\,\mu\mathrm{m\,K}$, or for $T = 5780\,\mathrm{K}$, $\lambda_{max} = 0.501\,\mu\mathrm{m}$, just as shown in the figure.

The spectra of Figure 1.2 are misleading in terms of the energy distribution because λ is inversely proportional to the photon energy; for this reason it is often preferable to present the spectrum as a function of $(1/\lambda)$, which is the number of waves per unit length, called the wavenumber, σ. In fact it is more complicated than this because wavelengths are normally what is measured, in air, while spectroscopists prefer a universal standard, and so define wavenumbers as in vacuum. The correct conversion thus involves the index of refraction of air, but this minor complication is ignored throughout this work. The spectra of Figure 1.2 are shown as a function of wavenumber in $\mathrm{cm^{-1}}$ in Figure 1.3. On this scale the Earth's

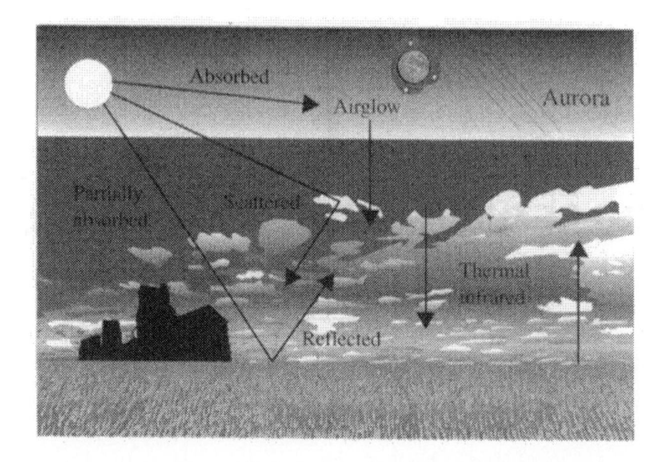

Figure 1.1. Illustrating sky radiation observable through spectral imaging.

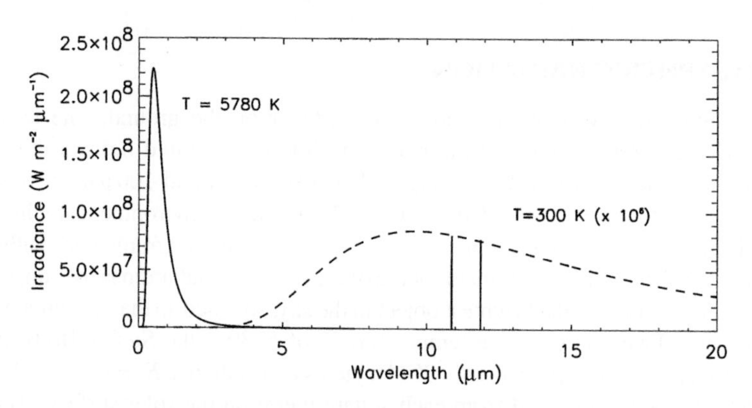

Figure 1.2. The monochromatic irradiance of a 5780 K black body, representative of the sun, and a 300 K black body, representative of the Earth, with the latter multiplied by 10^6. The vertical lines are separated by 1 μm; the interval is centred at 11.37 μm.

spectrum becomes narrow and the solar spectrum is broad which, as noted, reflects more correctly the energy distributions involved.

The monochromatic irradiances shown in Figure 1.2 and Figure 1.3 must specify the spectral interval over which the irradiance corresponds. In Figure 1.2 the interval used is 1 μm, so that the irradiance units are $\mathrm{W\,m^{-2}\,\mu m^{-1}}$. For illustration, a spectral interval of 1 μm centred at 11.37 μm is shown in Figure 1.2, which means that the corresponding value of $7.8 \times 10^7\ \mathrm{W\,m^{-2}}$ is radiated over this wavelength interval. For Figure 1.3 the spectral interval used is $1\ \mathrm{cm^{-1}}$, and for clarity the units are shown as $\mathrm{W\,m^{-2}(cm^{-1})^{-1}}$. While the author's objective is to use SI (System International) units, based on MKS, throughout, the use of $\mathrm{cm^{-1}}$ for spectroscopy is so deeply rooted it cannot normally be avoided, even in this work. Because most spectral imaging instruments use photon detectors it is often

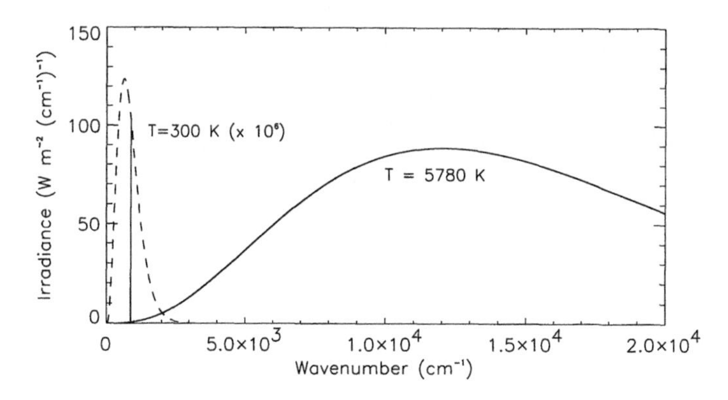

Figure 1.3. The wavenumber spectra corresponding to the wavelength spectra of Figure 1.2. The vertical line at 879 cm^{-1} relates to Figure 1.6.

convenient to use photon units for irradiance. To apply this in a simple way the sun's energy may be imagined to be all in photons of wavelength λ_{max} corresponding to the sharp peak of the solar spectral distribution. From the Planck quantum equation the photon energy is given by $E_{photon} = hc/\lambda = hc\sigma$, where h is Planck's constant, and c the velocity of light; the solar irradiance is then expressed as 1.60×10^{26} photons s^{-1} m^{-2}.

Moving away from the solar disk, the irradiance does not suddenly drop to zero, but decreases gradually because of the solar corona; at about one solar radius from the sun's edge (limb), the irradiance is lower than that of the disk by a factor of about 10^{-6} [Phillips, 1992]. Even close to the sun's disk in the day-time, the coronal light is obscured by the scattered light from the Earth's daytime sky; otherwise the corona would always be visible, not just during solar eclipses. Solar photons are scattered in the Earth's atmosphere by molecules and by aerosols (droplets or particles) as well as by ordinary clouds. The dark blue day-time sky is caused mainly by molecular scattering, called Rayleigh scattering after the third Lord Rayleigh, who showed that the scattered irradiance is proportional to $(1/\lambda)^4$. This factor, when combined with the solar spectrum, produces the beautiful blue colour, because blue is the shortest wavelength detected by eye. This scattered day-sky source of photons is much weaker than the sun but, expressed in green photons as above, is still roughly 10^{20} photons s^{-1} m^{-2}. Thus the coronal irradiance at one solar radius of 1.60×10^{20} photon m^{-2} s^{-1} in the green region is comparable to that of the day-time sky. But since the Rayleigh scattered light is much greater in the blue region of the spectrum, it does overwhelm the overall coronal observation. After the sun has set there is still visible light from the nighttime sky, since it is possible to find one's way around even on a dark night. If the Moon is up, its spectral character is much like that of the sun, but with a surface irradiance that is essentially the same as for the Earth (which is discussed shortly), since it is at about the same distance from the sun. The lunar irradiance is 2×10^{-5} of the solar irradiance, or allowing for 10% lunar reflectance (albedo), 2×10^{-6}. In energy units the lunar irradiance is 130 W m^{-2}, or in photon units, 3×10^{20} photons s^{-1} m^{-2}. This is a factor of three larger than that for the blue sky, consistent with the fact that the moon is visible in the day-time.

In the absence of the Moon, there is still an irradiance of about 2×10^{13} photon s^{-1} m^{-2} in green photons from starlight and zodiacal light. Zodiacal light is sunlight scattered from dust and electrons extending far from the sun, reaching beyond the Earth's orbit. Around the beginning of this century, astronomers began to suspect that there was a non-stellar component in the light of the night sky, on the basis of two clues. First, the radiance of the Milky Way predicted on the basis of star counts was not as great as expected in comparison with other regions of the sky [Newcomb, 1901]. The second clue was that the brightness of the night sky increased as one approached the horizon, independently of the star background [Burns, 1906]; this is called the van Rhijn [1921] effect and it indicates an atmospheric layer for which the viewing path is longer at the horizon than overhead. Spectral analysis (see Chamberlain [1961] for historical details) showed that a significant visual component of this was caused by a single spectral line at 557.7 nm which is now called the atomic oxygen green line airglow. Later the fourth Lord Rayleigh [1930] estimated that this feature contributed 7% of the visual brightness of all of the light of the night sky, including starlight, zodiacal light and terrestrial light. The life of John Strutt, the fourth Baron Rayleigh (Lord is a more general term that includes Baron), is described by G.R. Strutt [1964a], and his roughly 300 publications are listed by C.R. Strutt [1964b]. A dedicated sky observer, Rayleigh recognized in 1921 that the auroral green line existed in the normal night sky radiation. A current estimate of the irradiance contained within this spectral line is 10^{12} photon s^{-1} $m^{-2} = 3.5 \times 10^{-7}$ W m^{-2}. For the sun to produce an irradiance this weak at the Earth it would have to be at a distance of about one light year. Since airglow emission is produced photochemically its irradiance cannot be predicted as with black body radiation; it is said to be non-thermal, but the irradiance can be predicted from a knowledge of atmospheric composition and chemical reaction rates as is discussed later. The aurora, produced by the impact of electrons energized in the magnetosphere, is brighter than the airglow, by a factor anywhere from 5 to 500. These sources are all depicted in Figure 1.1, but for beautiful photographs of the real thing, see Meinel and Meinel [1991], Eather [1980] and Brekke and Egeland [1994] while for more quantitative measurements the reader may consult Roach and Gordon [1973].

In Figure 1.4, images of the entire sky are shown, taken at night from a ground-based observatory using a very sensitive monochromatic all-sky CCD camera (the same type of CCD detector used in a commercial video or digital camera), and a narrowband filter that transmits the hydroxyl airglow and suppresses the starlight background, though stars are still clearly evident in the raw image on the left. The hydroxyl airglow is another airglow emission, from vibration–rotation transitions of the OH radical called the Meinel bands. This image shows strong Van Rhijn horizon brightening around its perimeter. The image on the right is the difference between the image on the left and a later image. This reveals a great deal of structure, indicating the extent of the density perturbations in the thin layer of atmosphere in which the light is produced, between roughly 80 and 105 km above the Earth's surface. The emission is too weak to be detected by a visual observer.

Behind all of this radiation lies the Cosmic Background Radiation (CBR), a remnant of the so-called *big bang* which created the universe that exists now. This is highly redshifted

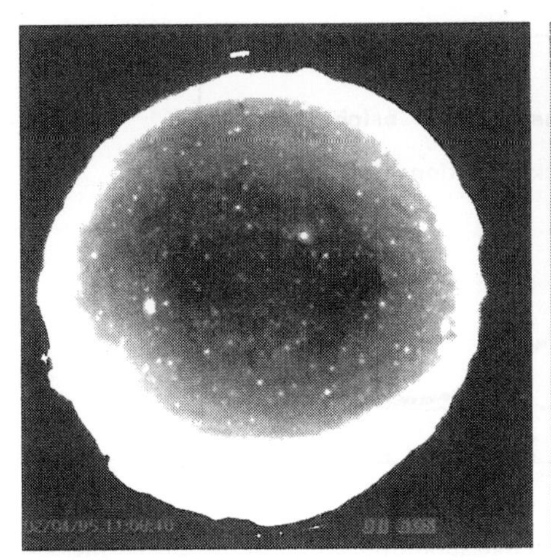
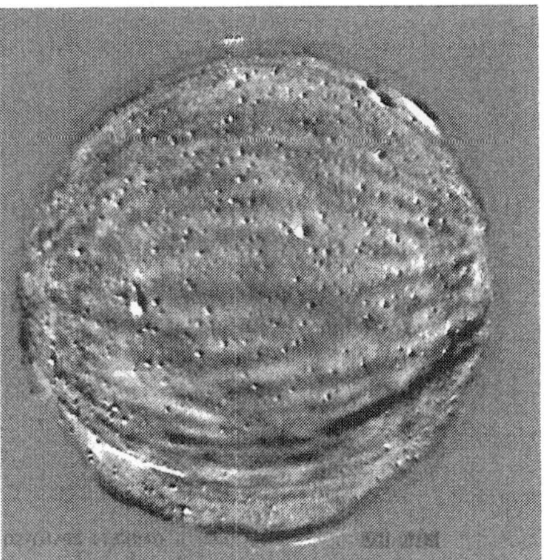

Figure 1.4. A raw hydroxyl airglow image of the entire sky (left), showing Van Rhijn enhancement near the horizon, with wave structures and stars overhead, and (right) an image obtained by taking the difference between images taken at different times, which dramatically shows the gravity wave structure. The images were taken at Albuquerque, New Mexico on Feb. 4, 1995. Courtesy of G.R. Swenson (University of Illinois – Champaign).

radiation which was in equilibrium with matter just as the temperature of the universe fell to a value permitting the formation of neutral hydrogen. According to one theory, the spectrum is Planckian (thermal), the form remaining invariant under the expansion of the universe. The spectrum of this radiation, measured from a rocket with a Michelson interferometer [Gush *et al.*, 1990] is shown in Figure 1.5. The measured spectrum is very close to that of a black body with a temperature of 2.736 K, an important result and a remarkable accomplishment. The units on the vertical axis are $\mathrm{W(cm^2\,str\,cm^{-1})^{-1}}$, which is not monochromatic irradiance, but monochromatic radiance, the power per unit area per unit wavenumber interval radiated into a solid angle of one steradian (str in the figure, but sr throughout this work). It is shown in Section 3.2 that, with certain assumptions, the irradiance is just π times the radiance; the corresponding wavelength-integrated irradiance is $3.14 \times 10^{-6}\,\mathrm{W\,m^{-2}}$, which is comparable to the green line airglow.

It is remarkable that sources as different as the sun, the Earth and the CBR all have spectra which can be fitted over a large fraction of their spectral range by the Planck equation, each corresponding to a single temperature. In an enclosure at a uniform temperature, an equilibrium is established between the walls of the enclosure and the radiation within it. The spectrum is determined by the population of available energy levels, which is determined by the thermal energy available. If the enclosure is pierced with a small opening, the emerging radiation has essentially the same character as that within the enclosure since the loss rate is small compared with the production rate; it is called black body radiation because only a very small fraction of any radiation entering the cavity can escape. The sun is by no means an enclosure, but the photosphere, the relatively thin and opaque layer which gives the

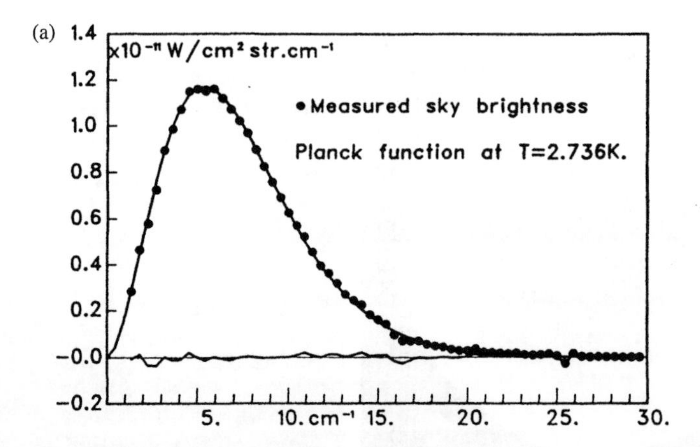

Figure 1.5. Spectrum of the Cosmic Background Radiation over the range 2–30 cm^{-1} measured from a rocket using a Michelson interferometer, with the measurements shown as solid circles and the fitted curve for a temperature of 2.736 K shown as the solid line. After Gush *et al.* [1990].

sun its "solid" appearance is strongly absorbing and so is effectively black. Similarly the Earth's atmosphere is absorbing at certain infrared wavelengths. At wavelengths for which the atmosphere is not fully absorbing the degree of blackness is described by introducing a quantity called *emissivity*, the value by which the Planck equation must be multiplied to obtain the observed radiation. Thus spectral lines that are not fully absorbed are assigned an emissivity that is less than unity.

It has already been mentioned that some radiation is absorbed in the solar chromosphere on its way out from the photosphere; this absorption produces the Fraunhofer lines. Similarly on its way through the Earth's atmosphere some of the solar photons are absorbed, as illustrated in Figure 1.1; those having wavelengths corresponding to absorption lines of atmospheric species. Observing the solar spectrum thus provides a method of detecting the absorbing atmospheric species. Absorption of solar radiation into molecular spectral continuum bands also occurs, dissociating molecular oxygen into its atomic form. In the thermosphere the atomic oxygen recombines, creating the airglow; in the stratosphere the recombination produces ozone. Multiplying the solar irradiance value by the area of the sun (radius $= 6.96 \times 10^8$ m), and spreading this energy over a sphere having a radius equal to that of the Earth's orbit, 1.50×10^{11} m (which is called one Astronomical Unit, or AU), gives the irradiance at the top of the Earth's atmosphere as 1370 W m^{-2}, the quantity known as the *solar constant*. Some authors [Seyrafi and Hovanessian, 1993] reserve the term irradiance for the radiant power falling onto unit area of a surface and, for the radiant power leaving unit area of a surface, use the terms *radiant emittance*, or *radiant exitance*. In the present work this distinction is not made, and the term irradiance is used for both. The term monochromatic is used when appropriate, but usually only when its use may be unclear from the context. The corresponding green photon value is 3.4×10^{21} photon s^{-1} m^{-2}. With this irradiance the sun is not only blindingly bright, but its radiation can be felt, as warmth.

This irradiance warms the earth, bringing it (including the effect of the atmosphere which is considered later) to roughly 300 K. Thus the Earth emits thermal radiation, like the sun, except that since the irradiance goes as T^4, it is much lower than that of the sun, at 459 W m^{-2}. For the Earth's surface (assumed to be at $T_E = 300$ K) the spectral peak is of much longer wavelength, in the infrared, with a broad peak at 9.65 μm as shown in Figure 1.2 or as the sharp peak in Figure 1.3. This infrared radiation is strongly absorbed by the atmosphere because of "infrared active" minor constituents, particularly water vapour, carbon dioxide and ozone. At the peaks of the absorption lines where the absorption is strong the emission can be described by the Planck function at these wavelengths. Both the Earth's surface and the atmosphere exchange this radiation, determining the resulting Earth's surface temperature.

Figure 1.6 shows the monochromatic radiance profile of atmospheric emission from a minor constituent, HNO_3, located at a wavelength of 11.37 μm, from a satellite instrument viewing the Earth's limb, as illustrated in Figure 1.10 so that tangent point altitude of viewing is the vertical coordinate. The profile was taken by the CLAES (Cryogenic Limb Atmospheric Etalon Spectrometer) instrument [Roche *et al.*, 1993] on NASA's Upper Atmosphere Research Satellite (UARS). This is thermal emission because the relevant energy levels of HNO_3 are excited through collisions with the surrounding gas whose constituents have an energy distribution corresponding to the atmospheric temperature. The radiance depends on the species number density, the absorption coefficient and the temperature. The spectral region observed by CLAES is shown by the pair of lines in Figure 1.2 and as a vertical line at 879 cm^{-1} in Figure 1.3. Note that the altitude scale is vertical in Figure 1.6, even though it is the independent variable; because this is the way that atmospheric scientists visualize their data, up being up. Multiplying by π to get

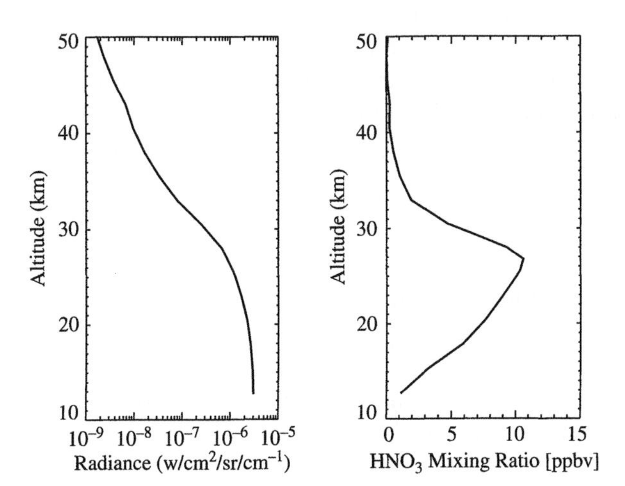

Figure 1.6. Radiance profile of HNO_3 (left), and its inversion to mixing ratio (right) as measured by the CLAES instrument on UARS. Courtesy of A. Roche (Lockheed Research Center). Note that W cm^{-2} sr^{-1} (cm^{-1})$^{-1}$ is written as w/cm^2/sr/cm^{-1}.

the irradiance (see Section 3.2) the peak irradiance value at 15 km tangent height is about $0.1\ \mathrm{W\,m^{-2}}$ for a spectral interval of $1\ \mathrm{cm^{-1}}$, a small fraction of the total infrared radiance, but still much larger than for the airglow. In the right panel these data have been processed to yield an altitude profile of HNO_3 concentration, expressed as a volume fraction of the total atmosphere in PPBV (parts per billion by volume). This illustrates how extremely sensitive such radiance measurements are to the concentrations of infrared emitting molecules and how such radiance measurements may be used to determine species concentrations, and it makes clear the role of minor constituents in generating atmospheric radiation.

But what does an irradiance for the atmosphere mean? It implies that the blue sky or the thermal infrared radiation is regarded as coming from a "material" surface at some distance. While this approach gives the correct irradiance at the ground, something different is needed to describe the radiation emerging from an extended diffuse glowing atmosphere. That quantity, in photon units, is the photon emission rate V, not the emission from a unit area, but from a unit volume of the atmosphere. The *volume emission rate* is the number of photons emitted per second in all directions from a unit volume of one $\mathrm{m^3}$, so that the units of V are photon $\mathrm{s^{-1}\,m^{-3}}$. The challenge in spectral imaging of the atmosphere is to measure this quantity.

The purpose of presenting the range of photon numbers in this section is to set the stage for thinking about observing *atmospheric radiation*, in an environment exposed to a continuous "rain" of photons. This is the basis of *remote sensing*, involving a technology referred to in this book as *spectral imaging*. A ground-based spectroscopic instrument extracts only a tiny fraction of the information available, for example, a photometer with an aperture of 5 cm diameter collects only about 10^{-14} of the photons emitted by an atmospheric volume 100 km distant. A satellite at 600 km may view the Earth's limb, some 2500 km distant, as shown in Figure 1.10, making the fraction collected that much smaller. For the sun, the photon irradiance is usually not a problem, but for radiation emitted from the atmosphere, the ultimate limits of instrument sensitivity must be sought. The question then becomes one of how efficiently these photons are used. The higher the efficiency, the greater the fraction of the available information that is retrieved. In the next section the way in which an instrument collects atmospheric photons is described.

1.2 MEASURING ATMOSPHERIC RADIATION

1.2.1 The Integrated Emission Rate

In observing the atmosphere, the fundamental quantity to be measured is the volume emission rate V just described, the number of photons $\mathrm{s^{-1}}$ emitted from a volume of $1\ \mathrm{m^3}$. Such an elementary volume, located at distance r from a receiver of area A, results in a collected photon number of $dF_{pV} = VA/4\pi r^2$ photon $\mathrm{s^{-1}}$ as illustrated in Figure 1.7(a). If the receiver of area A looks outward into a field of view of solid angle Ω as shown in Figure 1.7(b), then the photon count at the receiver corresponding to a layer of emission 1 m thick and filling the field of view, is dF_{pV} multiplied by the area Ωr^2 of this layer

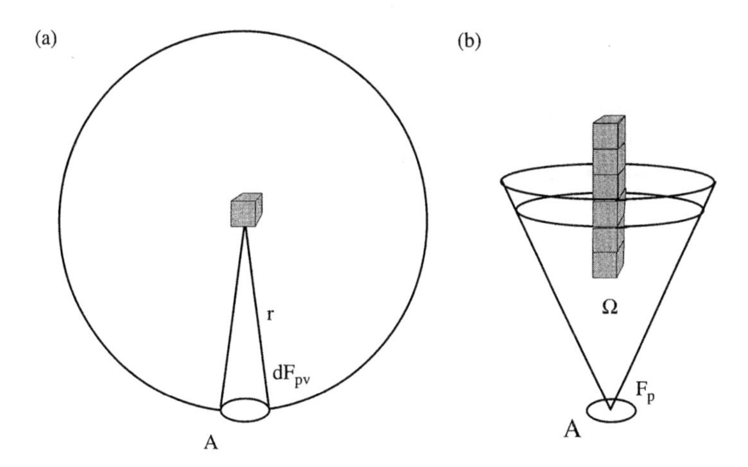

Figure 1.7. Viewing the atmosphere from below; (a) the emission of photons from an elementary volume of 1 m^3, (b) vertical integration along a 1 m^2 column.

intersecting the field of view, or:

$$dF_{pA} = \frac{VA}{4\pi r^2}\Omega r^2 = \frac{A\Omega V}{4\pi}$$

The photon rate F_p, collected by the receiver, is the integral along the line of sight:

$$F_p = \frac{A\Omega}{4\pi}\int_{r_1}^{r_2} V(r)\,dr$$

where r_1 and r_2 correspond to the bottom and top of the emitting layer, and the emission is assumed to be horizontally uniform over the field of view. The integral depends only on the atmosphere (but also on the viewing direction); it corresponds to the number of photons emitted per second from a column of cross-section 1 m^2 oriented along the line of sight, as illustrated for vertical viewing in Figure 1.7(b). Because of the importance of this quantity in aeronomical studies, it was defined as a basic photometric unit by Hunten *et al.* [1956] and called the rayleigh (R), in honour of the fourth Lord Rayleigh, who has already been introduced, a pioneer in nightglow observations and the first to determine the radiance of the 557.7 nm emission in true quantal units [Rayleigh, 1930]. This quantity is denoted by I so that:

$$I = \int_0^\infty V(r)\,dr\ \text{photon s}^{-1}\,\text{m}^{-2}\,\text{column}^{-1} = \frac{1}{10^{10}}\int_0^\infty V(r)\,dr\ \text{rayleigh} \qquad (1.1)$$

and it is called the *apparent emission rate*, or the term *integrated emission rate* is also commonly used. To give reasonable numbers for airglow, the unit was chosen to correspond to the emission of 10^{10} photon s^{-1} m^{-2} column^{-1}. In summary, one rayleigh corresponds to the emission of 10^{10} photon s^{-1} from a column of one square metre in area extending along the line of sight in any direction through the emitting region. While this unit was conceived by airglow investigators, it is useful for any kind of atmospheric emission, or scattering. It also applies to observations from space vehicles, balloons, rockets or satellites. The geometry of the integrating path is different in each case, but the fundamental considerations are the same. The quantity $A\Omega$ depends solely on the instrument; A determines the area of an incoming wavefront captured by the instrument, and Ω determines the angular extent of the ensemble of collected wavefronts. Following Jacquinot [1960], $A\Omega$ is called the étendue in this work and the evaluation of this quantity for different instruments is the focus of much of what follows in succeeding chapters.

1.2.2 Visible Atmospheric Radiation

In the previous section it was assumed that there was no absorption of atmospheric photons along the integrating path. This is usually the case for non-thermal emission, though not always. The reason for this is that many non-thermal airglow emissions are forbidden; that is, they arise from transitions which do not obey spectroscopic selection rules corresponding to electric dipole radiation. Since the absorption transition probability is proportional to that in emission, forbidden spectral lines are very weakly absorbed. However, forbidden transitions of O_2 are absorbed in the atmosphere for long integration paths because molecular oxygen is a major atmospheric constituent. All visible region emissions are non-thermal, because of the energies involved in their excitation.

The night-time atomic oxygen green line airglow emission at 557.7 nm is produced by recombining atomic oxygen atoms. The atomic oxygen is created in the day-time, as indicated earlier, through absorption of solar radiation in the Schumann–Runge continuum. Sidney Chapman proposed that the excitation of the green line proceeded through the three-body process: $O + O + O \rightarrow O(^1S) + O_2$. Here O indicates ground-state atomic oxygen and $O(^1S)$ is the upper state of the 557.7 nm emission. If this reaction proceeds with a rate constant C, and it is assumed that every $O(^1S)$ produced emits a photon, then $C[O]^3 = V_{557.7}$ where $[O]$ denotes the concentration of atomic oxygen. Thus a measurement of $V_{557.7}$ leads to a determination of atomic oxygen concentration, so long as C is known. The Chapman reaction is now known not to be the correct one, though it is equivalent to the currently accepted Barth mechanism [McDade *et al.*, 1986], and it illustrates the diagnostic way in which non-thermal airglow emission may be employed. The fundamental physics of both non-thermal and thermal atmospheric radiation is clearly described by Chamberlain and Hunten [1987].

1.2.3 Thermal Atmospheric Radiation

Where an atmosphere is not truly black it can be considered *gray*. The *gray-body* emission from the atmosphere is given by the Stefan–Boltzmann equation, but multiplied by the emissivity. For a more detailed exposition of this, and atmospheric radiation generally, the work by Salby [1996] may be consulted. By Kirchhoff's law, the emissivity is equal to the absorptivity, so that an atmosphere that emits thermal radiation strongly will also absorb it strongly. Thus absorption cannot be ignored for thermal infrared radiation. As a measure of blackness of the atmosphere it is convenient to introduce the *optical depth* $\tau(z)$, which is the integral of atmospheric density ρ multiplied by the mass absorption coefficient k, integrated through the atmosphere to level z.

$$\tau(z) = \int_z^\infty \rho k \, dz' \tag{1.2}$$

The vertically upwelling irradiance from the atmosphere, emitted between the Earth's surface at optical depth τ_s and the level of interest $\tau(z)$ is given by:

$$E(z) = \int_{\tau(z)}^{\tau_s} B(\tau') e^{\tau(z)-\tau'} d\tau' \tag{1.3}$$

Here B is the black-body radiance emitted from level τ', attenuated through the exponential term by absorption between the τ' level and the level of observation $\tau(z)$ and integrated over τ'.

The principles of infrared atmospheric remote sensing have been thoroughly described by Houghton *et al.* [1984] and more recently by Stephens [1994].

1.2.4 Ultraviolet Atmospheric Radiation

The ultraviolet spectral region is divided into the Far UltraViolet (FUV) region, from 120 to 200 nm, and the Extreme UltraViolet (EUV) from 30 to 120 nm. Below 30 nm is the X-ray UltraViolet (XUV) region. Like visible radiation, ultraviolet spectral emissions are also non-thermal. But unlike visible region emissions, ultraviolet radiation can be strongly absorbed because of the higher photon energies. For example, the atomic oxygen 130.4 nm emission is strongly absorbed in the atomic oxygen of the thermosphere. Thus an analysis of its emission must be carried out using an approach as indicated in Eq. (1.3), where the black-body term B is replaced by a photochemical source function. The principles of ultraviolet remote sensing have been well described by Huffman [1992].

The atmospheric regions relevant to spectral imaging are shown in Figure 1.8, with the approximate altitudes indicated. The boundaries between the regions, the "pauses" are also shown. In the context of the previous discussion, non-thermal emission, both visible and ultraviolet comes mostly from the thermosphere, which has no well-defined upper limit. The upper mesosphere includes non-thermal emission, while the lower mesosphere, stratosphere and troposphere are all characterized by thermal emission. The name *middle atmosphere* is often given to the stratosphere and mesosphere, though not always consistently, and the name upper atmosphere is generally applied to the region above. The troposphere is the region long associated with meteorology, but now the need to study all of these regions jointly is well recognized.

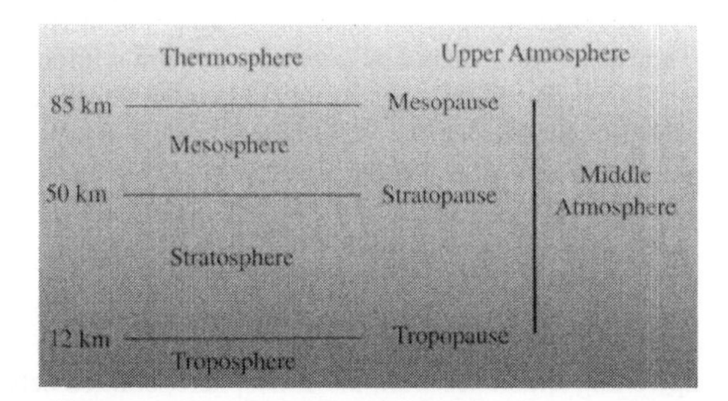

Figure 1.8. The atmospheric regions relevant to spectral imaging.

1.3 THE SCOPE OF SPECTRAL IMAGING

From what has been said so far, it is implied that measurements in three-dimensional geometric space are required, in spectral space and in time. These can be regarded as the three information domains. Consider first geometrical space (x, y, z). As already indicated, the volume emission rate is desired for every cubic metre (or even cubic km) in the atmospheric region of interest, as a function of wavelength and time. To conduct a feasible experiment these requirements have to be reduced, normally by collapsing these dimensions in some way. In Figure 1.4 the volume emission rate is collapsed onto a surface because of integration along the path; this apparent surface can be characterized by its irradiance E. For an optically thin atmosphere, that is, one of small optical depth, the relation between apparent emission rate in rayleigh (I), the irradiance E and the radiance L may be thought about in this way. The atmospheric emission from the 1 m^2 column along the line of sight is assumed to be isotropic, so that the volume emits uniformly over 4π steradians. For the emission into one steradian, the integrated emission rate is divided by 4π so that the radiance, the power radiated from one square meter into a steradian (sr) is given by:

$$L = 10^{10} \frac{I}{4\pi} \text{ photon m}^{-2}\text{ sr}^{-1}\text{ s}^{-1} \tag{1.4}$$

Where there is no height information available in an atmospheric image, one simply maps it onto horizontal coordinates at an assumed altitude. The justification of this for airglow is that the layers often are thin, and if the altitude is known, so much the better. For thermal radiation the emission is concentrated near the level where $\tau = 1$, so it also is somewhat localized. Thus where the three-dimensional information can be collapsed onto two horizontal dimensions, i.e. the layer is thin, a great deal of useful information can be extracted. For many studies, however, the vertical information is required, and a spectral imaging method must be designed to provide that information.

The next step is to consider the spectral information associated with each three-dimensional volume element. Intuitively one thinks of "all" the spectral information as the entire spectrum obtained with extremely high resolution. This is impractical on the one hand, but more importantly is not an efficient way to gather information. The instrument should be designed so as to collect just that information that is required. In other words, before designing an instrument, one begins with the *measurement requirements*, and then seeks to find the instrument that will provide those measurements most efficiently. The "entire" spectrum is so vast, and the ultimate spectral resolution elements so small, that it makes no sense to think of a generalized instrument that does everything. From an atmospheric science perspective, spectral information is not inherently useful in its raw form, but only in terms of derived atmospheric quantities such as species concentration, temperature and wind. Even more complex derived quantities are possible, such as heat or momentum fluxes, or eddy diffusion coefficients. Normally the measurement requirements are expressed in the form of such derived quantities.

Finally, the time domain needs to be considered. This is normally determined by the need to collect an adequate number of photons to determine the desired derived quantity, with the spatial and spectral resolutions required and the required precision. The length of time required for an observation can degrade the measurement, as is the case if the target is moving, or the emission rate of a volume element is otherwise changing with time. A more acute limitation occurs if the measurement platform is a rocket or satellite; a low Earth-orbiting satellite has a speed of about 7 km s^{-1}, so for a spatial resolution of 50 km, say, the measurement must be completed in 7 s. To measure the evolution of a phenomenon in time one needs to consider the scale time of the phenomenon, and be able to make successive measurements at a commensurate rate. Again, the platform will impose its own limitations, as from a satellite a given point on the Earth is generally observed only twice per day, once during the day-time and once at night. Thus for the above example, a specific 50 km $\times$ 50 km element on the earth may be observed during a time of 7 s, and then not observed again under the same conditions for 24 hours.

In the following sections some of the methods used to extract spatial, spectral and temporal information are introduced. It will be seen that this cannot be done fully from the ground, so that space measurements, such as from rockets, must be used. Satellites provide global coverage in addition, which is obviously not available from single or even multiple ground-based sites.

1.4 ONE-DIMENSIONAL (VERTICAL) SPATIAL INFORMATION

It can be seen in Figure 1.4 that the emission rate at the horizon is greater than at the zenith (overhead). This is because the integration path is longer at the horizon; it is only the zenithal emission rate that gives the inherent emission rate in the layer. Emission rates at other zenith angles, ζ, can be approximately corrected by applying a cos ζ correction factor, appropriate to a flat earth. Van Rhijn [1921] was the first to consider this problem in detail, and he realized that close to the horizon it was necessary to consider the airglow as a spherical shell; moreover, he realized that the change of emission rate with zenith angle depended on the altitude of the airglow emission. He recognized that by measuring the emission rate of the airglow layer as a function of angle, from the zenith to the horizon, one could deduce the altitude of the layer from ground-based measurements, an important piece of information in the days before rocket measurements of these altitudes had been made. But scattering effects in the lower atmosphere also have a zenith angle dependence, which if ignored lead to ridiculous results. Later, Roach *et al.* [1958], made careful measurements from an excellent mountain-top site of the atomic oxygen 557.7 nm emission on 12 nights, and obtained altitudes varying from 51 to 136 km, with an average of 100 (the currently accepted value is 97 km). Thus an attractive theoretical concept was proven to have no value, because the information required for an accurate solution was much more complicated to acquire and analyze than was feasible.

The first precise determinations of the altitude and volume emission rates of atmospheric emissions were made with sounding rockets. Figure 1.9 (upper panel) shows the measured vertical viewing integrated emission rate measured from a rocket [Deans and Shepherd, 1978], as a function of altitude as the rocket flew through an aurora above Churchill, Manitoba, Canada. The individual data points and a fitted curve are shown. The emission is from the (0,1) band from the $N_2^+ B^2\Sigma_u^+$ state of the molecular nitrogen ion, the so-called N_2^+ First Negative bands. These data are integrals along the line of sight, as shown in Figure 1.7, and are shown in rayleigh (R). At the lowest altitude, about 80 km, the integrated emission rate above the rocket is about 6000 R (6 kR), which is the total value for the aurora. As the rocket enters the aurora, the integrated emission rate above the rocket decreases, finally reaching 20 R at about 170 km. This value corresponds to the background starlight. Differentiation of this curve according to Eq. (1.1) yields the volume emission rate as shown in the lower panel in the same figure. The N_2^+ emission peaks at about 100 km with a peak volume emission rate of about 3000 photons $cm^{-3}\,s^{-1}$ and falls off above this, corresponding to the absorption of the precipitating auroral electrons in the atmosphere. Taking the peak volume emission rate as 3000 photon $cm^{-3}\,s^{-1}$, and the layer shape as a triangle of half-width 20 km, the estimated integrated emission rate is 3000 photon $cm^{-3}\,s^{-1}$ multiplied by the layer thickness of 2.0×10^6 cm, yielding a value of 6000×10^6 photon $s^{-1}\,cm^{-2}$ column^{-1}, or 6 kR, which is the observed value.

Rocket flights have provided a great deal of information on atmospheric volume emission rates, but they are expensive, and highly localized in space and time. For global observations, satellite measurements are required. The viewing geometry is illustrated in Figure 1.10, where the atmosphere has been divided into four vertical shells, in each of which the volume emission rate is assumed to be constant within a given layer (horizontal homogeneity).The viewing ray passing through the topmost layer yields the volume emission rate for that layer, since the path length is known. For the next layer down, there are two unknowns, the volume emission rates for layers 1 and 2, and since the first is known, the second can be solved for. This inversion process, of deducing the volume emission rates from the integrated emission rate profile, is known as an onion peeling inversion. Because the solution equations are ill-conditioned, some type of constraint is normally applied. There are many inversion methods, which have been described by Rodgers [1976, 1990] and Menke [1989].

This method is illustrated for the WIND Imaging Interferometer (WINDII) instrument on the UARS satellite, for the atomic oxygen green line emission for which the apparent, or integrated emission rates in kilorayleighs (kR) are shown in Figure 1.11(a) (labelled J_1). There is a family of profiles taken at different points along the satellite track, each at a different solar zenith angle. The daytime airglow is produced in part by direct solar EUV

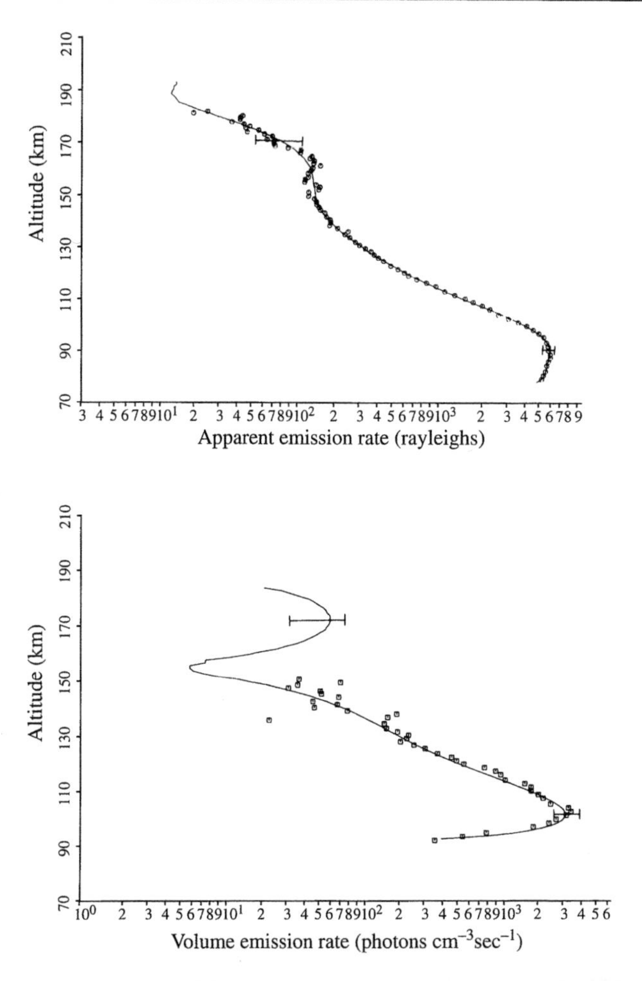

Figure 1.9. Apparent (integrated) emission rate in rayleigh of the N_2^+ 1NG (0,1) 427.8 nm band in aurora measured above the rocket shown as individual data points and the fitted curve (top), and in the bottom panel the volume emission rate given in photon cm^{-3} s^{-1} obtained from the differentiation of the individual data points (squares) and that from the fitted curve. After Deans and Shepherd [1978].

input, which causes the solar zenith angle dependence. In the plot the profiles closest to twilight are on the left, and those closest to noon are on the right. The peak value is about 110 kR, corresponding roughly to a radiance of 10^{-4} W m^{-2}, still a rather small value. The result of inverting these data using a weighted least-squares method [Rochon, 1999] is shown in Figure 1.11(b). These now represent the true volume emission rates for each volume of atmosphere viewed along the track, and show two well-defined peaks. The upper peak near 165 km is caused by absorption of solar UV radiation. The lower peak near

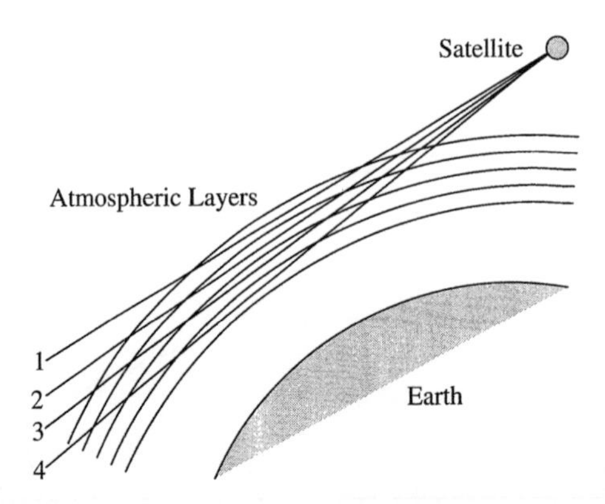

Figure 1.10. Illustrating limb viewing from a satellite and onion-peeling geometry.

100 km is less well understood but includes the recombination of atomic oxygen and likely the absorption of solar hydrogen Lyman-β radiation.

In the thermal infrared region a method of vertical profiling through nadir viewing from a satellite is possible. For a given wavelength, with a specific absorptivity, the radiation coming up from below originates from the vicinity of the region where the optical depth τ is equal to unity. The observed radiance is thus a function of temperature at this altitude level, weighted according to the vertical distribution of the emission, called the weighting function. For a different wavelength, the temperature for a different weighted altitude range can be determined. With a selection of wavelength channels, vertical profiles may be obtained, but with a resolution limited by the spread in the weighting functions. This method is described in detail by Houghton *et al.* [1984]. Higher vertical resolution is possible with limb viewing, but nadir viewing offers the best horizontal resolution.

1.5 TWO-DIMENSIONAL (HORIZONTAL–VERTICAL) INFORMATION

The inversion procedure of the previous section is subject to one strong restriction, namely horizontal homogeneity in each layer. This is not a problem for normal atmospheric measurements, but is definitely a problem where structure is involved, such as for clouds or aurora, but it also applies to more diffuse atmospheric emission where gradients are involved, such as at the Earth's terminator. There is a generalized approach that is capable of solving both of these problems in a generalized way, called tomography.

The mathematical inversion methods of the tomography procedure were first developed for the purpose of medical imaging [Cormack, 1963]. These provided a way to replace the 2-D images of normal x-rays with a depth dimension that allows the 3-D imaging discussed as an objective of atmospheric remote sensing. However, the method is normally applied in "slices"; that is, one begins with a 2-D image of a thin layer of the human body. The "intensity" (or absorption) of a number of parallel x-ray beams passing through the target are measured for a set of equally-spaced angles around the target, covering 360° of revolution.

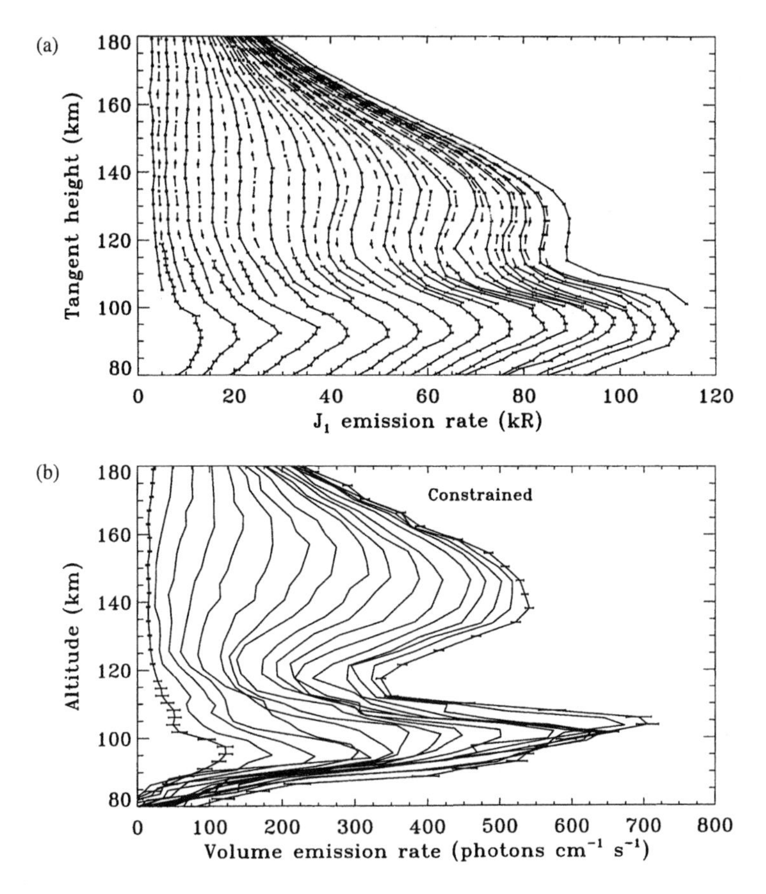

Figure 1.11. (a) Integrated emission rate profiles of atomic oxygen 557.7 nm green line daytime emission observed by WINDII. Note that it is the tangent height that is shown; units are kilo-Rayleigh (kR). The different profiles correspond to different solar depression angles. (b) Family of volume emission rate profiles inverted from the WINDII observations of (a), using a constrained least-squares inversion. The units are photon cm^{-3} s^{-1} (the figure label is incorrect). After Rochon [1999].

These data may be "inverted" to obtain the volume distribution (analogous to the volume emission rate) as has been shown by Vest [1974]. Of course one can obtain as many slices as are required by the physician, and so build up the 3-D information as required. The geometry for a very different application, a rocket flight over an auroral feature is as illustrated in Figure 1.12, and is described by McDade *et al.* [1991]. The rocket, flying over the auroral form had its long axis perpendicular to the trajectory plane, and the photometer viewing perpendicular to this axis, in the trajectory plane, with the rocket spinning about its axis. In this way, repeated scans were taken across the volume of auroral emission, from different directions as the rocket moved along its path. These individual scans comprising the raw data are shown stacked together in Figure 1.13(a) and the inverted volume emission rate

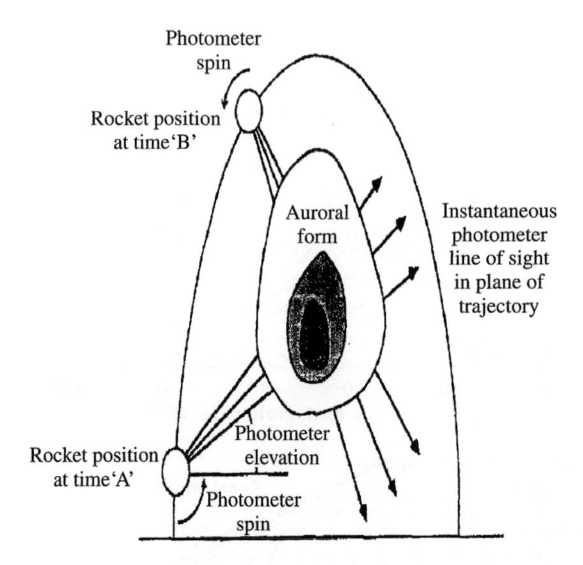

Figure 1.12. Depiction of a rocket auroral tomographic experiment. After McDade and Llewellyn [1991].

distributions are shown in Figure 1.13(b). Thus the volume emission rate is determined in two-dimensional space.

The same technique has been applied to measurements from the Atmospheric Explorer satellite, by Solomon *et al.* [1984, 1985, 1988], for two-dimensional measurements of volume emission rates from the $O(^1D)$ 630.0 nm emission in the aurora.

1.6 THREE-DIMENSIONAL INFORMATION

As already indicated, vertical information can be obtained from a limb-viewing instrument mounted on a satellite, and the associated horizontal information is determined by the track (in latitude and longitude, say) made by the observing footprint on the atmosphere. Typical satellite tracks for one day are shown in Figure 1.14, in a plot of latitude versus longitude. The WINDII, MLS (Microwave Limb Sounder) and CLAES instruments on UARS [Reber *et al.*, 1993] viewed to one side of the orbit; with MLS and CLAES viewing at 90° from the velocity vector and WINDII at 45° and 135° as illustrated in Figure 1.14. With 15 orbits per day, the longitudinal distance between successive orbits is about 2500 km, rather poor horizontal resolution in that direction, even though measurements were made every few hundred km along the track. The HRDI and ISAMS instruments were able to view on both sides of the spacecraft, improving the longitudinal horizontal resolution by a factor of two. The highest resolution obtained to date was with the CRISTA (CRyogenic Infrared Spectrometers and Telescopes for the Atmosphere), flown twice on the space shuttle [Offermann *et al.*, 1999]. By using three fields of view, separated by 18°, the gap between orbits was filled in to a considerable extent, with a longitudinal horizontal resolution of

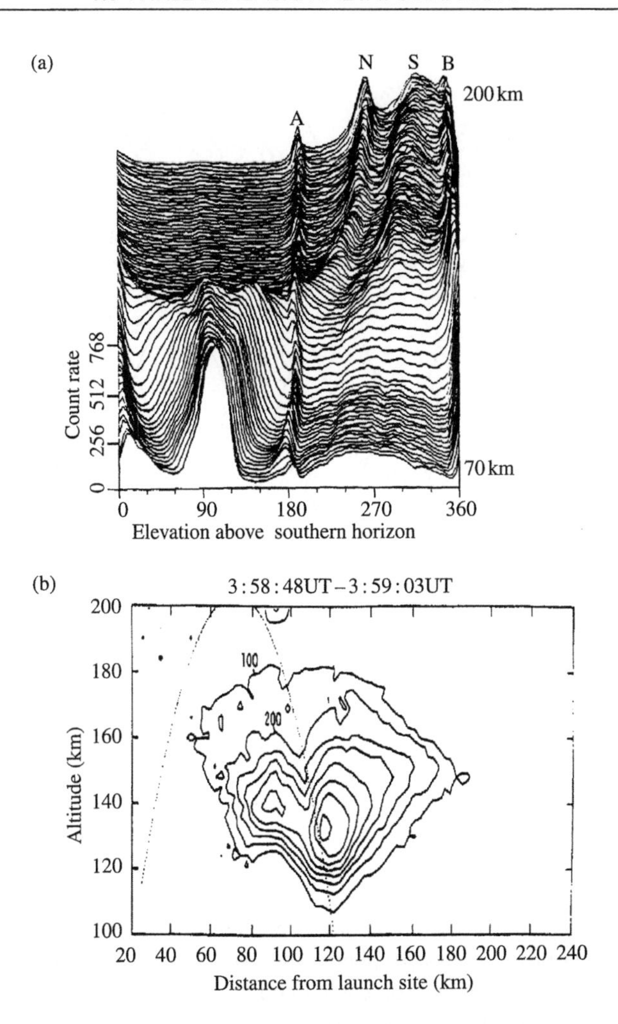

Figure 1.13. (a) Raw auroral scans and (b) tomographically inverted volume emission rates. After McDade *et al.* [1991].

about 650 km. The result is equivalent to taking Figure 1.14 and replacing each orbital track by three. This is an approach for which further progress may be expected in the future.

Satellite measurements thus provide a means of determining atmospheric volume emission rates on a global basis. If such measurements are distributed over the globe then one can say that a three-dimensional image of the global atmosphere has been obtained. However, because of the large distance between successive orbits, the spatial resolution is very poor, and the result is far from the total global image that is potentially achievable. Higher horizontal resolution can be obtained with nadir viewing, as is done with the MOPITT (Measurements Of Pollution In The Troposphere [Drummond *et al.*, 1999])

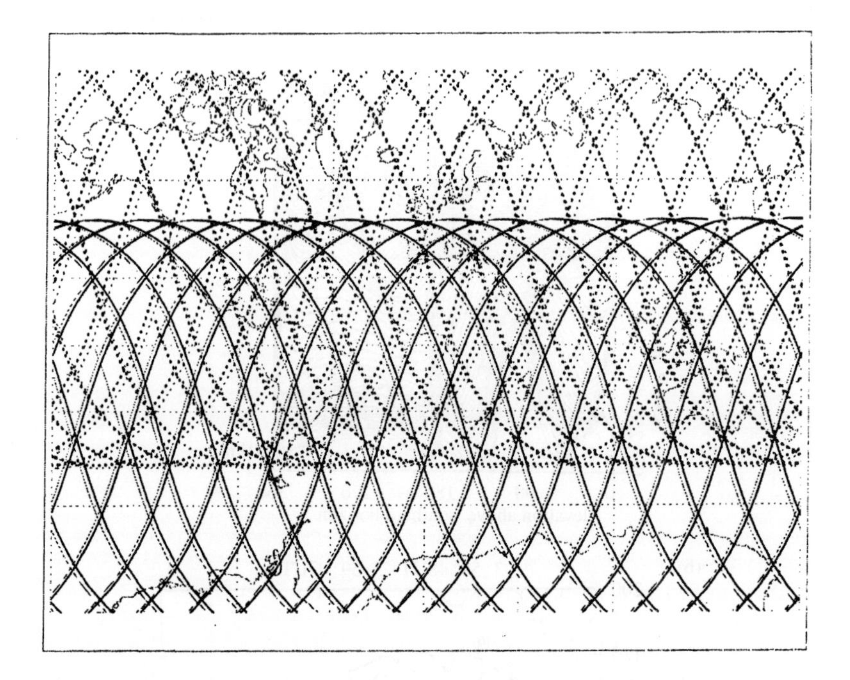

Figure 1.14. Satellite viewing tracks for one day of observation. The tracks are for a satellite of 65° orbit inclination, with two fields of view, one at 45° from the velocity vector, the other at 135°. These pairs of tracks can be seen slightly displaced. The solid tracks correspond to "southward viewing" from the spacecraft, and the dotted tracks to "northward viewing". Courtesy of D.-Y. Wang (CRESTech).

instrument launched on the NASA Terra satellite in December, 1999. This instrument measures CO concentrations in the troposphere, and uses 29 scan positions, transverse to the orbital motion, combined with an orbit cycle that fills in the gaps between orbits during the course of five days. The scan pattern is illustrated in Figure 1.15. Some vertical information is available with nadir viewing as described in Section 1.4, providing three-dimensional information.

1.7 SPECTRAL INFORMATION

The spectral information domain is based, as already discussed, on wavelength or wave number. For wavelength, the classical Ångstrom (Å) unit is avoided, with some regret; and units based on the metre followed; μm, nm and pm (picometre). While σ should be in m^{-1}, the use of the cm^{-1} is so established that it is usually followed here. For each spectral measurement there is a defined spectral range, and a specific number of spectral elements measured. The fundamental spectral information is defined in terms of resolution elements, in which each element contains information that is independent of all the other elements.

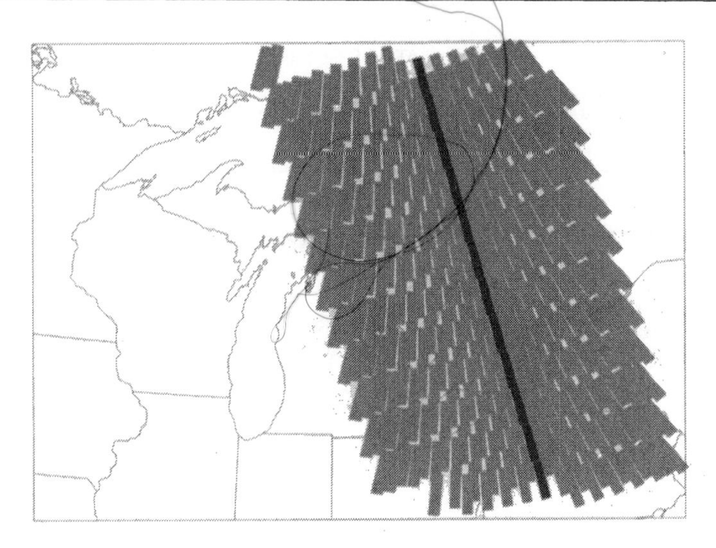

Figure 1.15. Scan pattern for the nadir-viewing MOPITT instrument, for a portion of an orbit near the Great Lakes of North America. Courtesy of J. Drummond (University of Toronto).

The measured spectrum may be oversampled to improve the signal-to-noise ratio. In spectroscopy, the *resolving power*, or resolvance, is traditionally defined as $\mathcal{R} = \lambda/d\lambda$, where $d\lambda$ is the spacing between spectral lines that are just resolvable. In this work, $d\lambda$ is often taken as the width of the instrumental passband function, which is approximately the same.

In Figure 1.16 is shown a measured spectrum of the middle atmosphere covering a comparatively broad spectral range taken with a Fourier transform spectrometer (a Michelson interferometer) called ATMOS (Atmospheric Trace MOlecule Spectroscopy experiment) from the space shuttle [Gunson *et al.*, 1990]. Spectra are shown for different tangent point altitudes of the field of view; 90, 70, 50, 25 and 15 km. Isolated regions of these spectra are expanded in the bottom panels. The excellent resolution of the instrument is matched by the richness and complexity of the spectrum, which is an absorption spectrum taken by viewing the sun through the atmosphere. The goal here is to measure the concentrations of many different minor constituents in the atmosphere, which can be done accurately only if the spectral lines of the different constituents are well resolved. For many applications the spectrum is less complex, and more specialized instruments may be used to advantage.

A non-thermal spectrum of the upper atmosphere taken by the AIS (Arizona Imager Spectrograph) instrument mounted on the space shuttle is shown in Figure 1.17. The instrument [Broadfoot *et al.*, 1992] is a set of five grating imaging spectrographs from which spectra covering the range from 120 nm to 900 nm are shown, for the nightglow (night-time airglow) in Figure 1.17(a), the dayglow (day airglow) in Figure 1.17(b) and the aurora in Figure 1.17(c). Thousands of papers have been written on the origins of these emissions; the descriptions given here are necessarily brief. Beginning with Figure 1.17(a) in the upper panel at 120 nm = 1200 Å the 1216 Å hydrogen Lyman-α line is evident. This is sunlight which is resonantly scattered off atomic hydrogen in the Earth's high atmosphere. Since

Typical ATMOS Spectra

The upper figure shows the decreasing absorption with altitude, while the bottom figures illustrate spectral features used for trace gas profile retrievals. For clarity, vertical scales differ among the displayed spectra.

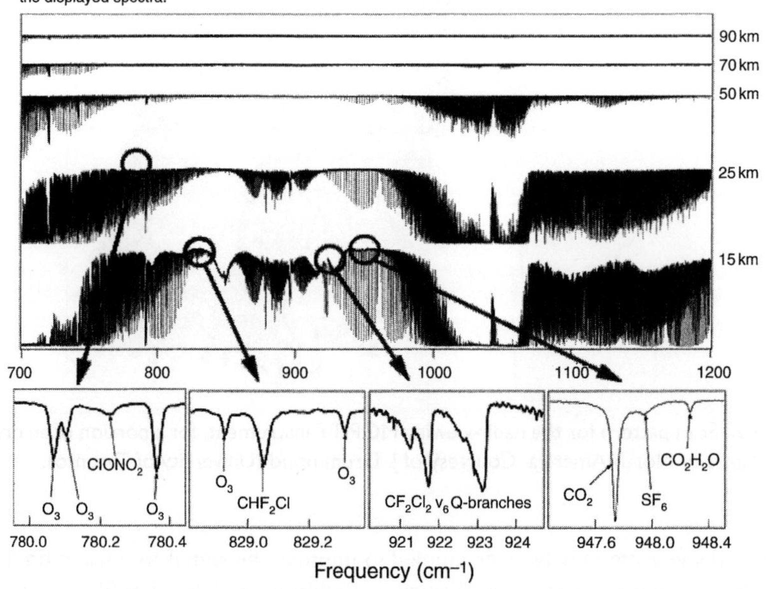

Figure 1.16. An absorption (occultation) spectrum taken by the ATMOS instrument on board the space shuttle. The different curves are for different tangent altitudes as indicated. Courtesy of M.R. Gunson (JPL – California Institute of Technology).

Ly-α terminates in the ground state, a solar photon may be absorbed and then re-emitted. Because the diffraction grating has multiple orders, the 1216 Å line appears again at 2431 Å. Throughout the remainder of the upper panel the spectrum is dominated by O_2 Herzberg and Chamberlain bands, which are produced by recombining atomic oxygen. In the lower panel, the atomic oxygen green line at 5577 Å is prominent, also produced by atomic oxygen recombination. The spectrum at longer wavelengths is dominated by two band systems; the O_2 Atmospheric band is also produced by atomic oxygen recombination and so also are the very many Meinel bands of OH (which are not labelled) though the production is more complicated. These run in sequences, that is the (7,1) band near the green line is a transition from the vibrational level $v' = 7$ to $v'' = 1$, a Δv of 6, which is the same for (8,2) and (9,3). At 9000 Å $\Delta v = 5$. These vibrational levels are populated through the reaction $O_3 + H \rightarrow OH^* + O_2$, which populates 9 vibrational levels indicated as OH^*. Since the ozone is produced by atomic oxygen, this accounts for the influence of atomic oxygen concentration on this emission.

In Figure 1.17(b) the dayglow spectrum has the same features, but they appear weaker because the scale is changed by a factor of 5. Many new features are added because of the

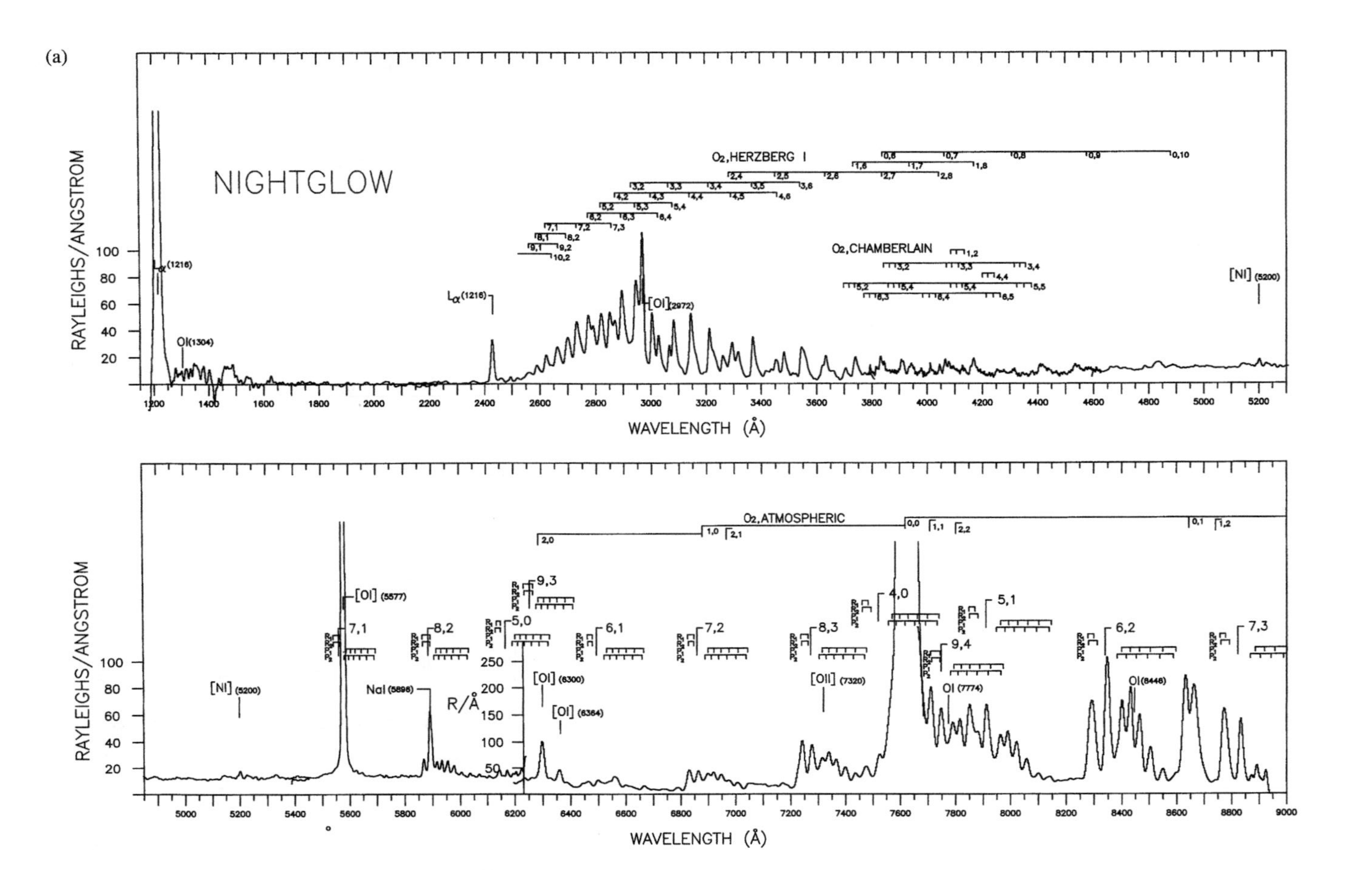

Figure 1.17a. AIS spectrum of the night airglow taken from the space shuttle. Courtesy of the Principal Investigator, A.L. Broadfoot (University of Arizona).

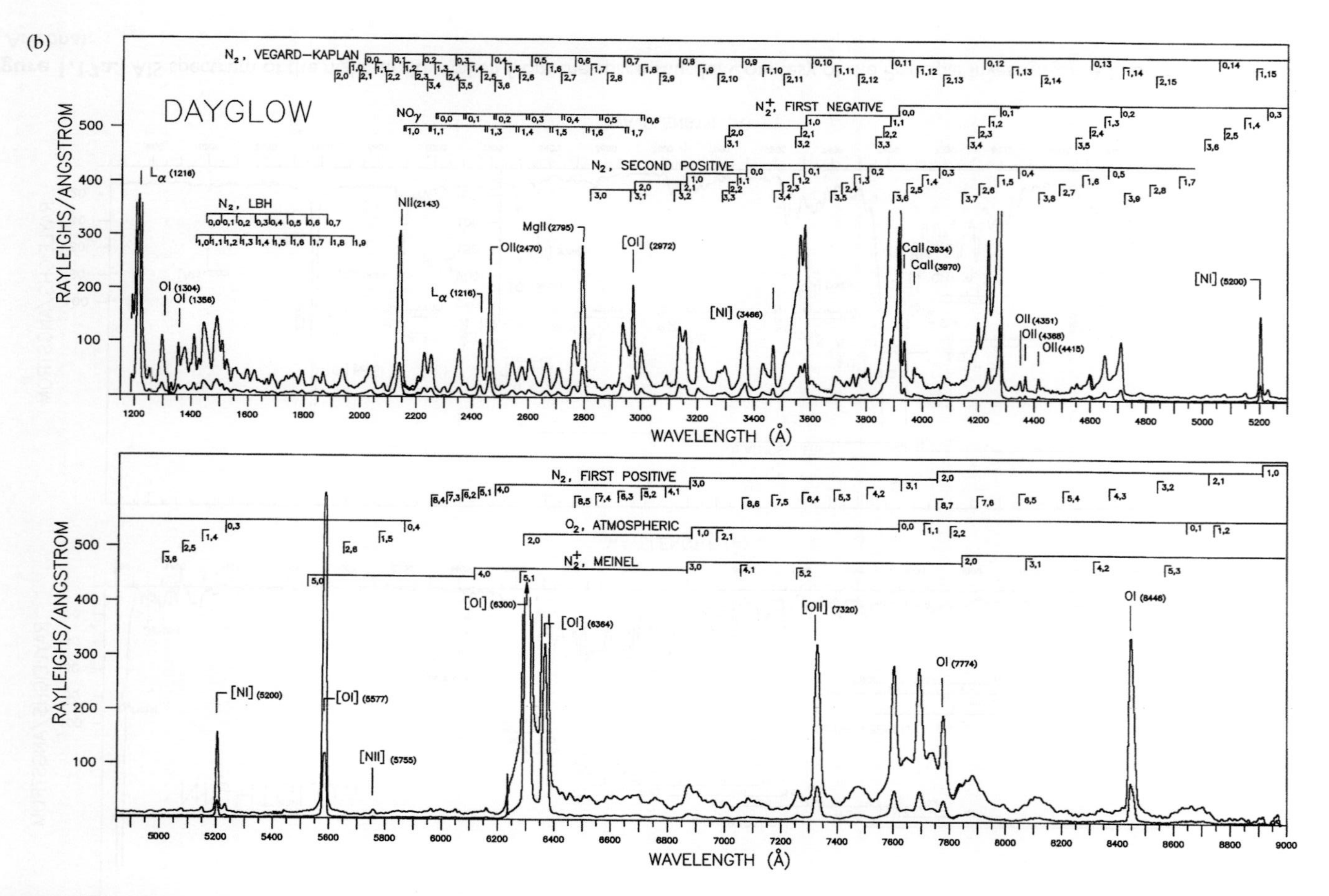

Figure 1.17b. AIS spectrum of the day airglow from 120 to 900 nm taken from the space shuttle. Courtesy of the Principal Investigator, A.L. Broadfoot (University of Arizona).

(c)

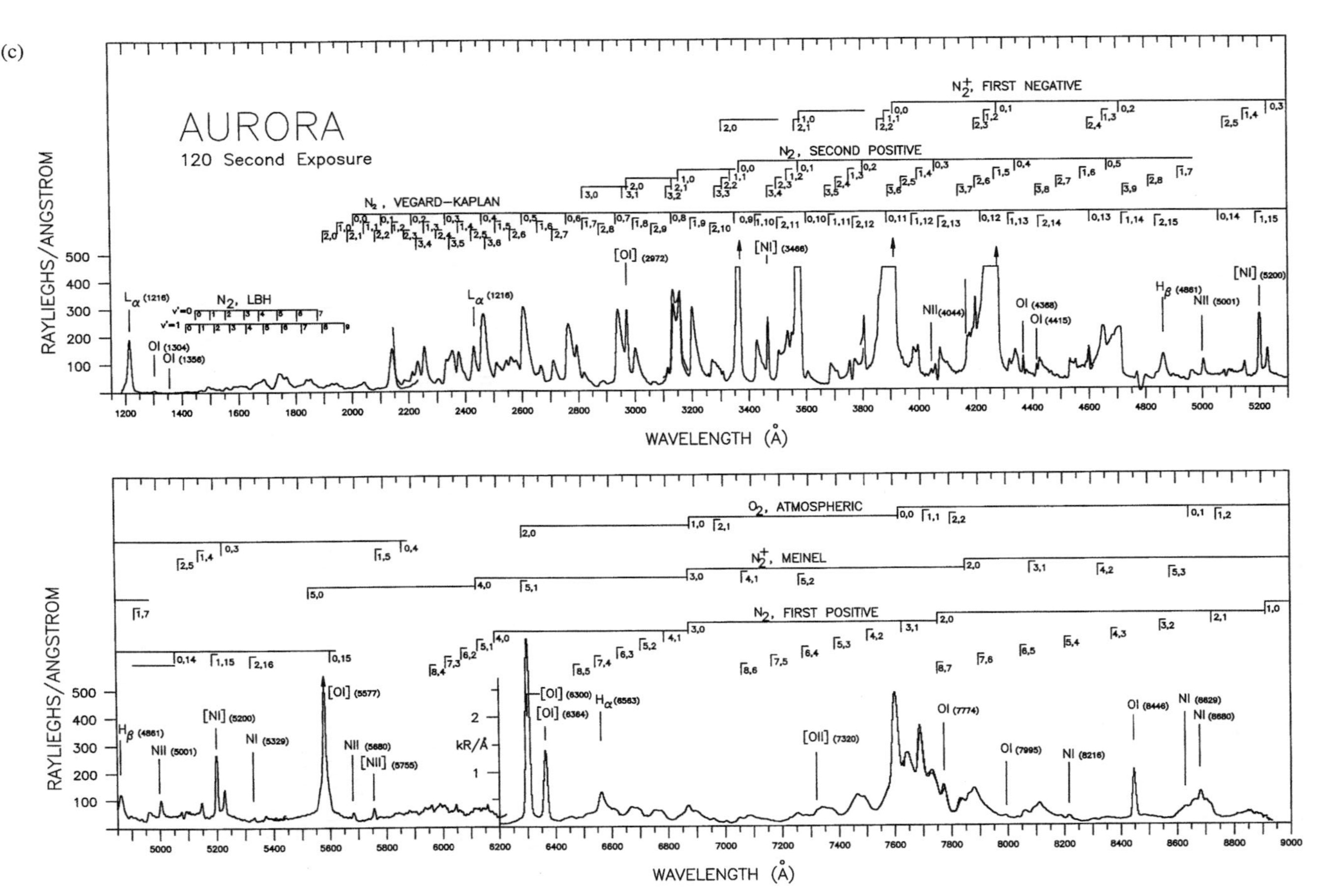

Figure 1.17c. AIS spectrum of the aurora from 120 to 900 nm taken from the space shuttle. Courtesy of the Principal Investigator, A.L. Broadfoot (University of Arizona).

presence of solar EUV radiation which ionizes atmospheric species, producing photoelectrons which excite other species through electron impact. Most of the features in this figure are produced in this way. In particular, the N_2 Vegard–Kaplan, N_2 Second Positive and N_2^+ First Negative band systems appear in the upper panel, and the N_2 First Positive and N_2^+ Meinel band systems in the lower panel. These names are simply historical; positive and negative refer to the polarity of the discharge tube electrodes where the emissions appear, and the other names are of the discoverers. Ionized atomic lines also appear. NII at 2143 Å refers to singly ionized atomic nitrogen; ionized atomic oxygen lines appear at 2470 Å, and 7320 Å and an ionized Mg line at 2795 Å. An allowed atomic oxygen transition appears at 8446 Å. The atomic oxygen red line appears at 6300 Å with a companion at 6364 Å; these are produced in part through electron impact but also through ion reactions in the ionosphere. All the same features appear in the auroral spectrum of Figure 1.17(c) where higher energy auroral electrons of a few keV energy produce a population of low energy secondary electrons through ionizing collisions. The secondary electrons have energies similar to the day-time photoelectrons so produce similar spectra. However, the overall auroral spectrum is different, so that the ratios of spectral features do change. For more detailed explanations the reader may consult Rees [1989] and Vallance Jones [1974].

1.8 TEMPORAL INFORMATION

The final information domain is time. All instruments require a time interval to make a measurement, which is normally chosen to be compatible with the assumption that the optical characteristics do not change during the interval.

For example, the airglow shown in Figure 1.4 was perturbed by atmospheric gravity waves produced near the Earth's surface by winds flowing over mountains, or by storm fronts. They exist through the whole vertical range of the atmosphere and because they transport momentum and energy, they are crucial to an understanding of it. Hines [1974] is the pioneer in this understanding, and Hines and Tarasick [1987] predicted the effects of gravity waves on airglow. Gravity wave periods are typically an hour, and while they can be shorter, the emission rate is not likely to change within time intervals of less than four minutes, the natural buoyancy frequency of the atmosphere. Figure 1.18 shows the typical temporal variation of the integrated emission rate and rotational temperature from the O_2 (Atm) band for gravity waves over a time period of 5 hours. This was obtained by a spectral imager called MORTI (Mesopause Oxygen Rotational Temperature Imager), described by Wiens *et al.* [1991]; these are the same gravity waves as seen in Figure 1.4.

When the assumption that the source remains constant during a measurement is not valid, measurement time and phenomenological time become mixed in a way that may be impossible to separate. For example, if the quantity that is measured over time has a space or spectral dimension, then the latter can become mixed with the time dimension. Single *in situ* satellite instruments cannot separate space and time effects, so the phenomenon is normally assumed to be stable within the measurement interval. But remote sensing instruments do have an advantage here, since the same volume can be viewed repeatedly to observe the real time variations, either to verify the assumption of stability, or to measure phenomenological

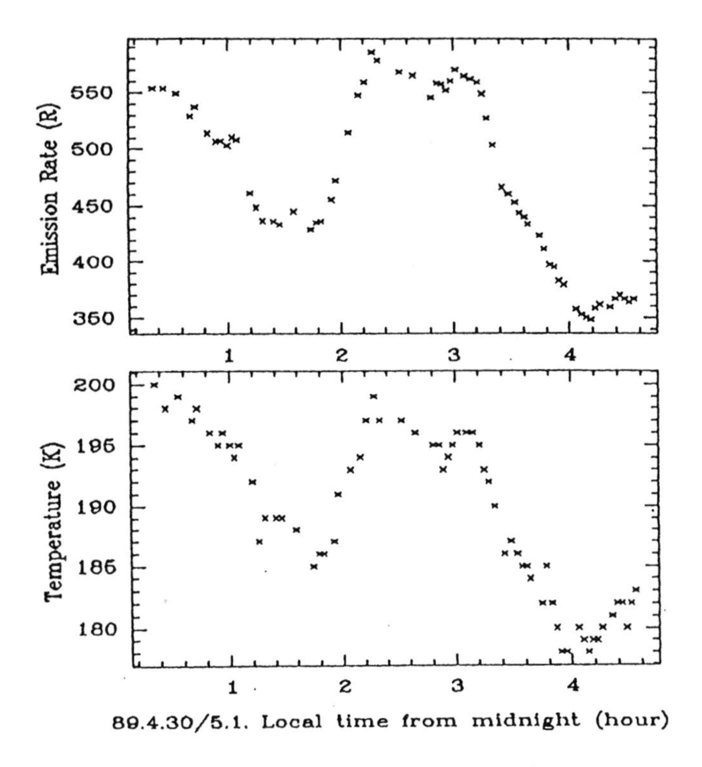

Figure 1.18. Temperature and emission rate perturbations for $O_2(b^1\Sigma)$ airglow emission measured with the MORTI ground-based spectral imager as caused by the passage of a gravity wave. Courtesy of S.P. Zhang (York University).

changes in time. Ground-based instruments are at their most advantageous in providing temporal information. A different kind of situation arises in deriving local time variations from satellite measurements. During the course of a day a satellite is approximately fixed in inertial space; this means that for a given latitude all of the ascending crossings of the orbits shown in Figure 1.14 have the same local time. The precession of the orbit from one day to the next depends on the orbit inclination. A strictly polar orbit has no precession and is fixed in sidereal space; it requires one year to cover all local times. Sun-synchronous orbits, with an inclination of 98°, precess at the same rate as the sun and so maintain perpetually fixed local times, one for the ascending crossings and another for the descending. The UARS satellite, with an inclination of 57°, precesses by 20 minutes per day, and near-equatorial orbits precess much more rapidly. For the UARS satellite it is necessary to acquire data for 36 days in order to cover 24 hours of local time which is necessary for the study of phenomena such as atmospheric tides (in 36 days the orbit moves through 12 hours, but with both ascending and descending orbits the 24 hours are covered). In order to accomplish a local time study of a particular phenomenon, it is necessary to assume that

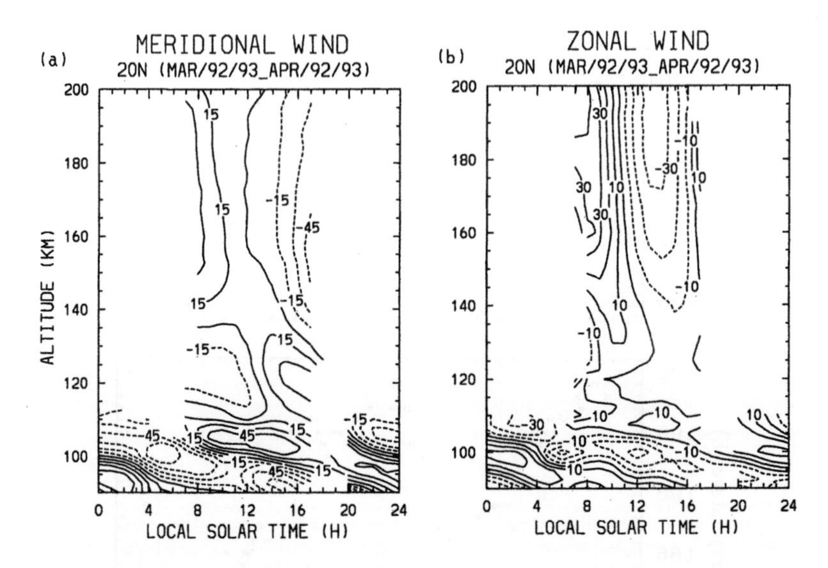

Figure 1.19. Satellite measurements of meridional and zonal winds as a function of altitude, from 90 to 200 km altitude. Measurements were made during the months of March and April in order to cover the 24 hour local time range. Solid contours are positive, i.e. north for meridional, and eastward for zonal, and the units are m s^{-1}. After McLandress *et al.* [1996a].

the phenomenon is stable over that period, which can be checked only through the use of independent measurements.

Figure 1.19 shows the meridional (north–south) and zonal (east–west) winds measured by the WINDII instrument on the UARS spacecraft over the altitude range 90–200 km. In order to cover the full 24 hour range of local times, two months of data were used, March and April. The sloping contours of alternating wind direction correspond to the winds of the diurnal tide, the analysis of which depends on having 24 hours of local time. A fundamental question is whether day-to-day variations invalidate this kind of approach; this is discussed in the paper by McLandress *et al.* [1996a] from which this figure is taken.

1.9 PREVIEW

The following three chapters introduce the basic concepts that form the foundation of this work; spectral concepts; spectrometer responsivity and superiority; and imaging concepts. The next four chapters are devoted to four fundamental classes of instruments, the spectrometer (of Fabry and Perot), the interferometer (of Michelson), multiplexers and modulators, and Doppler Michelson interferometers. Chapter 9 describes selected currently operational instruments and Chapter 10 describes some selected future instruments. Finally, Chapter 11 brings the work to a close by describing the application of the diffraction grating spectrometer to atmospheric observations.

1.10 PROBLEMS

1.1. The Planck equation gives the monochromatic radiance of a black source as a function of λ and T for a wavelength interval $d\lambda$ as follows:

$$B_\lambda(T) = \frac{2hc^2}{\lambda^5\left(e^{hc/\lambda kT} - 1\right)}\, d\lambda \tag{1.5}$$

(a) Using this equation, reproduce the Earth spectrum shown in Figure 1.2. It is recommended that MKS units be used for this calculation.
(b) Convert the Planck equation to wavenumber. Note that $d\lambda$ must be converted to $d\sigma$. Using this equation, reproduce the solar spectrum shown in Figure 1.3.

1.2. For Figure 1.5, do an approximate numerical integration over wavenumber to find the integrated radiance of the Cosmic Background Radiation. Compare this value with the irradiance calculated from the Stefan–Boltzmann law.

1.3. An airglow layer extends from 90 km to 110 km. The volume emission rate is zero at 90 km and increases linearly with increasing altitude to 75×10^6 photons m^{-3} s^{-1} at 100 km, then decreases linearly with increasing altitude to zero at 110 km. A photometer with a circular input 0.1 m in diameter and a field of view of 1° half-angle views the layer at an angle of 45° above the horizon.
(a) Determine the vertically integrated emission rate in rayleigh.
(b) Calculate the vertically viewed radiance of the layer in photon units.
(c) Calculate the vertically viewed radiance of the layer in energy units, for a wavelength of 557.7 nm.
(d) Calculate the photon rate into the instrument.

1.4. Compute the apparent emission rate, in rayleigh, for a satellite view of the emission layer in Problem 1.3, for a tangent height of 100 km. This may be done numerically by dividing the atmosphere into layers 1 km thick, and computing the path length for each layer. Calculate the ratio of this emission rate with the vertically integrated emission rate.

1.5. Derive the Van Rhijn equation for the emission rate $I(\zeta)$ at zenith angle ζ, where $I(0)$ is the overhead emission rate, z is the altitude of the layer, and R is the radius of the earth.

$$\frac{I(\zeta)}{I(0)} = \frac{1}{\sqrt{1 - (R/(R+z)\sin\zeta)^2}}$$

1.6. In Figure 1.17 the integrated emission rate is given in Rayleighs/Angstrom.
(a) The integrated emission rate in R is just the integral in wavelength over the line. Assuming that the observed lineshape for the 8446 Å line in Figure 1.17(b) is triangular, make an estimate of the integrated emission rate in R.
(b) Convert the peak value of R/Å for the 8446 Å line in (a) to W m^{-2} (cm^{-1})$^{-1}$.

2

SPECTRAL CONCEPTS

2.1 INTRODUCTION

It was Newton who first dispersed sunlight into its spectrum of colours, using a prism. This effect involved refraction, the dependence of the velocity of light on the medium, and on its wavelength, but it was not correctly interpreted by him. Newton supported the corpuscular theory of light, according to which the corpuscles were attracted by the medium, thereby having a greater velocity in the medium than in air. Later experiments of the velocity of light in a dense medium showed that the result was the reverse of this. A more correct interpretation of the nature of light was provided by Young, who demonstrated the interference of light waves emerging from two closely-spaced slits. This provided a powerful demonstration of the wave character of light, and a means of determining its wavelength with simple tools, though it was the measurement of the velocity of light in a refractive medium that was considered definitive in establishing its wave nature. Both interference and refraction play an important role in the spectral imagers described in this work. Diffraction, the interference of one part of an obstructed wave front with another, is also important, as is polarization. All of these phenomena reinforce the concept of light as a vibrating transverse electric field, with an accompanying magnetic field.

Atmospheric radiation, on the other hand, can be understood only from the perspective of quantization, and energy levels. The processes of absorption, emission and black body radiation all involve the photon, and discrete energy levels. Remarkably, the two worlds of quantum energy and wavelength are connected with the simple equation $E_{photon} = hc\sigma$, where E_{photon} denotes quantum energy. For this reason there is no problem in observing quantum radiation from the atmosphere with instruments based on a concept of wavelength. Although the optical elements interact with the radiation in a wavelike fashion, the detectors are generally also quantal in nature; fundamentally they count photons.

The original optical identification of the spectrum with colour was very restricted, so that in parallel with the study of optics a much broader interpretation of what is meant by *spectrum* was developed. The concept of spectrum can in fact be applied to any time-varying quantity. The mathematical basis of this approach was developed by Fourier, and interpreted by later investigators. This broader approach strengthens the narrower optical one, and is followed as much as possible herein. Thus this chapter is devoted to a summary of spectral concepts, and their application in understanding spectral imagers.

2.2 THE SPECTRAL CONCEPT

An example of a very different phenomenon that displays a spectrum is that of a vibrating string. Bernoulli showed in 1753 that the solution to the vibrating string problem could be written as a series of sines and cosines. Euler pointed out that this would imply that the same would be true of any arbitrary function, which he held to be impossible. Fourier apparently made the first statement about the expansion of an arbitrary function in 1807, but when his memoirs on heat were submitted to the French Academy in 1811 they were rejected for publication. When Fourier became Secretary of the Academy in 1824, his first act was to publish his memoirs. In his *Theorie Analytique de la Chaleur* of 1822 Fourier was thus still the first to assert that an arbitrary function could be expressed as a series of sines and cosines. The correctness of this assertion continued to be challenged by mathematicians, and only later was accepted as valid. Even then, the applicability of this remarkable theorem was not recognized and it was only after Norbert Wiener's [1933] *The Fourier Integral and Certain of its Applications* was published that the spectral concept associated with this theorem became fully appreciated. Physicists found this intuitive concept easy to accept, and from that point on, the Fourier spectrum began to contribute widely to the analysis of data and of linear dynamical systems. An example of this is the work by Bracewell [1986], a radio astronomer who described spectral concepts in an elegant and highly visual way. The works by Papoulis [1962] and Kanasewich [1981] may also be consulted.

In the context of spectral imaging, the basis of the Fourier approach is that a time-varying signal can be resolved into the sum of a series of sine and cosine signals. For example, the signal in Figure 2.1(a) is the sum of the elementary cosinusoidal signals shown in (b), (c)

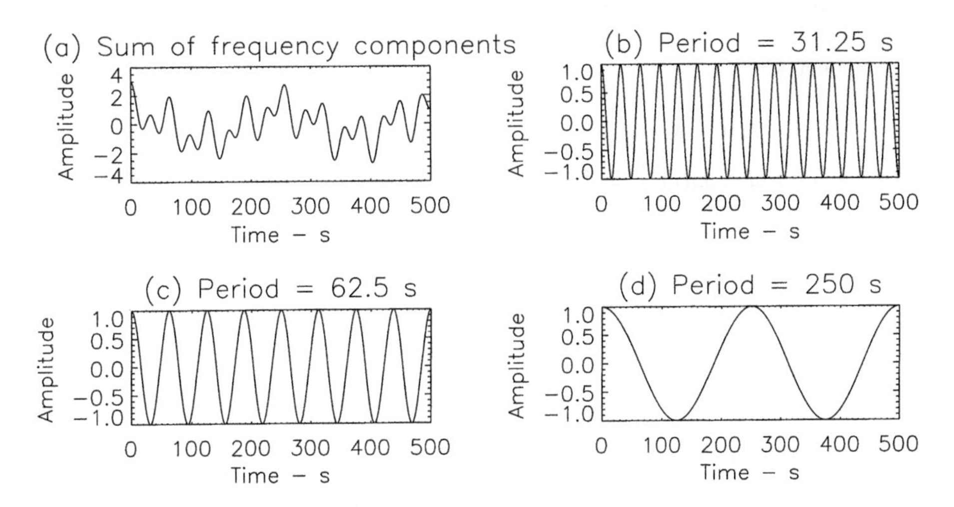

Figure 2.1. (a) The sum of the cosinusoidal signals shown in (b), (c) and (d).

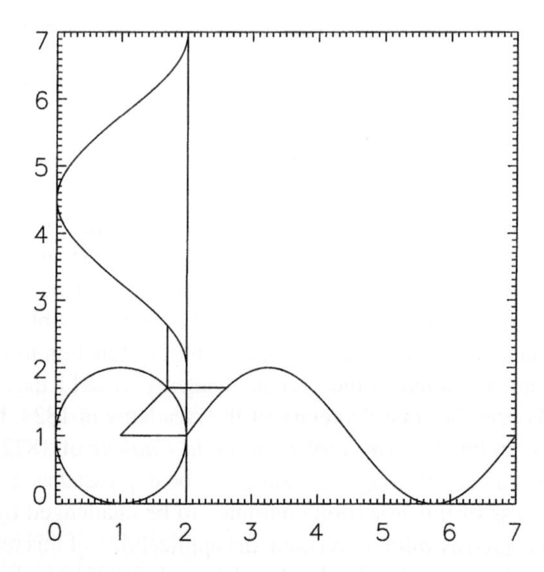

Figure 2.2. Generating cosinusoidal and sinusoidal wave components from a rotating vector.

and (d). Each of (b)–(d) correspond to single periods, or frequencies, of amplitude $A(\nu)$, for which a single frequency ν_o can be represented by a vector of length $A(\nu_o)$ rotating at frequency ν_o as shown in Figure 2.2, where the vector is seen to generate both sinusoidal and cosinusoidal components. The cosinusoid is generated by the horizontal component of the vector so the instantaneous value is given by $x(t) = A(\nu)\cos\theta$ where the rotation angle $\theta = 2\pi\nu_o t$. Similarly, a sinusoidal component $y(t) = A(\nu_o)\sin 2\pi\nu_o t$, is generated from the vertical component as shown in the same figure. Because the circular diagram of Figure 2.2 corresponds to the complex plane, with the vertical axis corresponding to the imaginary $j = \sqrt{-1}$ direction, the instantaneous vector can be written as the sum of the two components:

$$f(t) = A(\nu_o)[\cos 2\pi\nu_o t + j\sin 2\pi\nu_o t] = A(\nu_o)e^{j2\pi\nu_o t} \tag{2.1}$$

Thus a *spectral element*, called by spectroscopists a *spectral line*, can be represented by a vertical line of unit amplitude in the spectrum, say at ν_o on the ν axis, as shown in Figure 2.3(a). In fact to represent a real signal it requires two lines, one at ν_o and the other at $-\nu_o$ as shown in the figure. The idea of a negative frequency seems strange, but it simply corresponds to a vector rotating in the opposite sense (clockwise) to the vector of positive frequency. As shown in Figure 2.3(c), if the vectors start together on the positive horizontal axis then the vector sum of the two is an oscillating horizontal vector of amplitude 2, that is, the real quantity $2\cos 2\pi\nu_o t$. To generate a vertical (imaginary) oscillation in the same way it is necessary to flip the negative frequency vector by 180° so as to locate the sum vector on the vertical axis as shown in Figure 2.3(d). This flipping of the vector by 180° to the negative direction is represented in Figure 2.3(b) by inverting its amplitude. Thus the pair of inverted lines of unit amplitude yields $2j\sin 2\pi\nu_o t$.

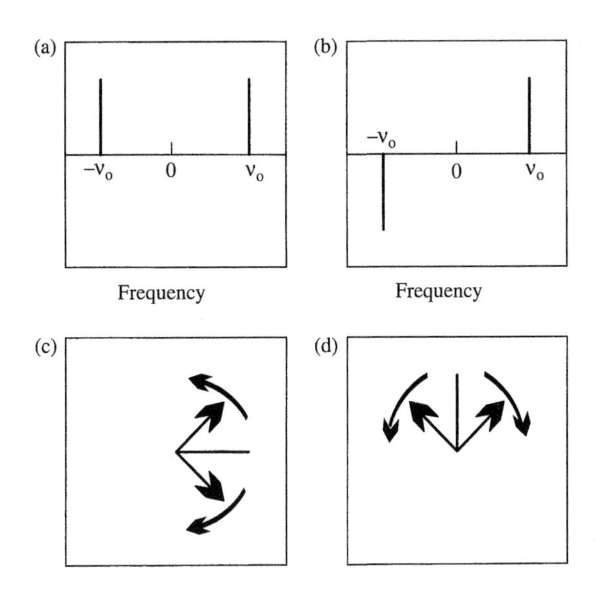

Figure 2.3. Generating cosine and sine waves from counter-rotating vectors.

In mathematical terms this is obvious as the counterclockwise vector is represented by $e^{j2\pi\nu t}$ and the clockwise one by $e^{-j2\pi\nu t}$ with a sum that is just $2\cos 2\pi\nu t$, and similarly $e^{j2\pi\nu t} - e^{-j2\pi\nu t} = 2j \sin 2\pi\nu t$.

2.3 FORMAL STATEMENT OF THE FOURIER TRANSFORM

The Fourier Transform is the means by which signals are transformed from the temporal domain to the frequency domain, and vice versa. It takes different forms, depending on the choice of variable. For example, frequency ν or angular frequency $\omega = 2\pi\nu$ may be used. Here it is stated in the form that uses ν, giving a symmetrical pair of equations. In the previous section the spectrum was comprised of discrete values, but here it is represented as a continuous function of frequency.

$$F(\nu) = \int_{-\infty}^{\infty} f(t)e^{-j2\pi\nu t}\,dt \qquad f(t) = \int_{-\infty}^{\infty} F(\nu)e^{j2\pi\nu t}\,d\nu \tag{2.2}$$

This remarkable transform pair depends on the orthogonality of the sine and cosine functions, which can be briefly outlined as follows. The average value of a cosine function over an integral number of periods is zero. Now consider the integral of the product of two cosines over an interval T which corresponds to integral multiples of the periods of both cosines. That is, $T = uT_1 = u/\nu_1 = vT_2 = v/\nu_2$ where u and v are integers and the integral is:

$$\int_0^T \cos(2\pi\nu_1 t)\cos(2\pi\nu_2 t)\,dt \tag{2.3}$$

Since the cosines can be expressed as exponentials, the integral involves products of the form:

$$e^{j2\pi\nu_1 t} \times e^{j2\pi\nu_2 t} = e^{j2\pi(\nu_1+\nu_2)t} \tag{2.4}$$

where $\nu_1 + \nu_2$ is a new frequency denoted ν', corresponding to a period T'. Further,

$$T' = \frac{1}{\nu_1 + \nu_2} = \frac{1}{u/T + v/T} = \frac{T}{u+v}$$

so that $T = (u+v)T'$. Now if the above integral is taken over the interval T', the result is zero because T' corresponds to one period of the new frequency. But since T is an integral multiple of T', the integral over T is also zero. This is true for the product of any two frequencies that are different from one another. However, when $\nu_1 = \nu_2$ the cosine becomes squared, and does not integrate to zero. Thus when a signal that is a superposition of different frequencies is multiplied by a wave of a single frequency, and integrated, the result is a measure of the spectral content contained in the signal at that multiplying frequency. In Eqs. (2.2), $f(t)$ is a time-varying signal comprising many different frequencies; when multiplied by $e^{-2\pi\nu_o t}$ and integrated, the result is $F(\nu_o)$, the spectral value of the wave component of frequency ν_o. The reverse is also true in Eqs. (2.2) where the spectrum $F(\nu)$ is multiplied by the corresponding wave component, $e^{j2\pi\nu t_1}$ to obtain the value of $f(t)$ at t. This is the basis of the Fourier transform equations.

The result is that information given as a function of one variable can be transformed to information as a function of another variable, and vice versa. Here the first variable, t, has been given the meaning of time, and the other variable ν the meaning of frequency, which has the dimensions of reciprocal time. Thus what is stated is that the information contained in a signal that varies with time, $f(t)$, is entirely equivalent to the information contained in its spectrum, $F(\nu)$. A convenient way of expressing this is to say that $f(t)$ and $F(\nu)$ are members of a Fourier Transform (FT) pair. It is conventional to use an upper case symbol for the spectral function, and a lower case symbol of the same letter for the temporal function.

For physical applications relevant to optical instruments, the signal $f(t)$ is always real, but because of the complex exponential, $F(\nu)$ is normally complex. The complex spectrum may be described in one of the two following ways:

$$F(\nu) = R(\nu) + jX(\nu) \tag{2.5}$$

$$F(\nu) = A(\nu)e^{j\phi(\nu)} = A(\nu)\cos\phi(\nu) + jA(\nu)\sin\phi(\nu) \tag{2.6}$$

such that $R(\nu)$ is the real part of the spectrum and $X(\nu)$ the imaginary part, or alternatively $A(\nu)$ is the amplitude of the spectrum and $\phi(\nu)$ its phase. If $\phi(\nu) = 0$ then $X(\nu) = 0$ and the spectrum is purely real, while if $\phi(\nu) = 90°$ then $R(\nu) = 0$ and the spectrum is purely imaginary. The phase is the angle between the single vector of amplitude $A(\nu)$ at time zero, and the positive horizontal axis, and is a way of formalizing the "starting angles of the rotating vectors" discussed in relation to Figure 2.3. The instantaneous value of the spectral component of frequency ν_o involves both t and the phase and is given by:

$$f(t) = A(\nu_o)e^{j(2\pi\nu_o t + \phi(\nu_o))} \tag{2.7}$$

It is clear that $R(\nu) = A(\nu)\cos\phi(\nu)$ and $X(\nu) = A(\nu)\sin\phi(\nu)$. Thus (2.2) is really two transforms, a cosine transform (c) and a sine transform (s), as follows:

$$F_c(\nu) = \int_{-\infty}^{\infty} f(t)\cos(2\pi\nu t)\,dt \tag{2.8}$$

$$F_s(\nu) = -j\int_{-\infty}^{\infty} f(t)\sin(2\pi\nu t)\,dt \tag{2.9}$$

This leads to two important properties. If $f(t)$ is odd in $t, f(t) = -f(-t)$ and $F_c(\nu) = 0$ (since the cosine is even in t), so that the spectrum $F(\nu)$ is purely imaginary. If $f(t)$ is even in $t, f(t) = f(-t)$ and $F_s(\nu) = 0$ so that the spectrum $F(\nu)$ is purely real. Thus $F_c(\nu) = R(\nu)$ and $F_s(\nu) = jX(\nu)$.

2.4 FUNDAMENTAL PROPERTIES OF THE FOURIER INTEGRAL

The following properties of the Fourier integral allow the manipulation of spectra in a way that is helpful to the understanding of spectral imagers and lead to an easy visualization of spectral relationships. They are not proven here, but will be understood through their application. The symbol $\Leftrightarrow$ connects the members of a Fourier transform pair, so that in all cases $f(t) \Leftrightarrow F(\nu)$.

Properties of the Fourier Transform

1. Linearity: If $F_1(\nu) \Leftrightarrow f_1(t)$, and $F_2(\nu) \Leftrightarrow f_2(t)$, then $[a_1 f_1(t) + a_2 f_2(t)] \Leftrightarrow [a_1 F_1(\nu) + a_2 F_2(\nu)]$
2. Inversion: $F(\pm t) \Leftrightarrow f(\mp\nu)$
3. Multiplication of t by a constant: $f(at) \Leftrightarrow (1/|a|)F(\nu/a)$
4. Time shifting: $f(t - t_o) \Leftrightarrow F(\nu)e^{-j2\pi t_o \nu} = A(\nu)e^{j[\phi(\nu) - 2\pi t_o \nu]}$
5. Frequency shifting: $F(\nu - \nu_o) \Leftrightarrow f(t)e^{j2\pi\nu_o t}$
6. Frequency differentiation: $d^n F(\nu)/d\nu^n \Leftrightarrow (-j\pi t)^n f(t)$
7. Time differentiation: $d^n f(t)/dt^n \Leftrightarrow (j2\pi\nu)^n F(\nu)$
8. Convolution theorem: If $f_1(t) \Leftrightarrow F_1(\nu), f_2(t) \Leftrightarrow F_2(\nu), F(\nu) \Leftrightarrow f(t)$, and $f(\tau) = \int_{-\infty}^{\infty} f_1(t)f_2(\tau - t)\,dt$, then $F(\nu) = F_1(\nu) \times F_2(\nu)$
9. Dirac comb: $\delta(t - uT) \Leftrightarrow \delta[\nu - (u/T)]$

In the following sections, examples are given of the application of these properties. The first of these shows that, using these properties, a Fourier integral may be evaluated without carrying out a mathematical integration.

2.5 DOING A FOURIER INTEGRAL WITHOUT INTEGRATION

In the example given in Section 2.2, $f(t)$ was a sinusoidal pattern, and $F(\nu)$ consisted of two spectral lines, but because of the symmetry of the Fourier transform it could just as well be the other way around. The two lines may be a pair of inverted pulses in the time domain, separated by T (here T does not denote period, but simply a time interval) as shown in Figure 2.4(a). In this figure the time-varying signals are on the left, and their FTs (spectra) on the right. The derivative of the rectangular function in (c) is the pair of inverted pulses in (a) so (c) $= f(t)$ and (a) $= df(t)/dt$. The FT of (a) is $-2j[\sin 2\pi\nu(T/2)] = -2j\sin\pi\nu T$ from Section 2.2 so (b) $= -2j\sin\pi\nu T$. The FT of (c) is (d) $= F(\nu)$. But, by property 7, the FT of the derivative is $-j2\pi\nu F(\nu)$ (the negative sign is needed as in property 6 because time and frequency have been inverted here) and this must also be equal to (b). These two expressions for (b) may be equated to yield $j2\pi\nu F(\nu) = 2j\sin\pi\nu T$ so that:

$$F(\nu) = \frac{\sin\pi\nu T}{\pi\nu} = T\frac{\sin\pi\nu T}{\pi\nu T} = T\,\mathrm{sinc}(\nu T) \tag{2.10}$$

which can of course be derived more formally from the Fourier transform equations above. This relation defines the sinc function, sinc(x). Note that, as ν approaches 0, $\sin\pi\nu T$ may

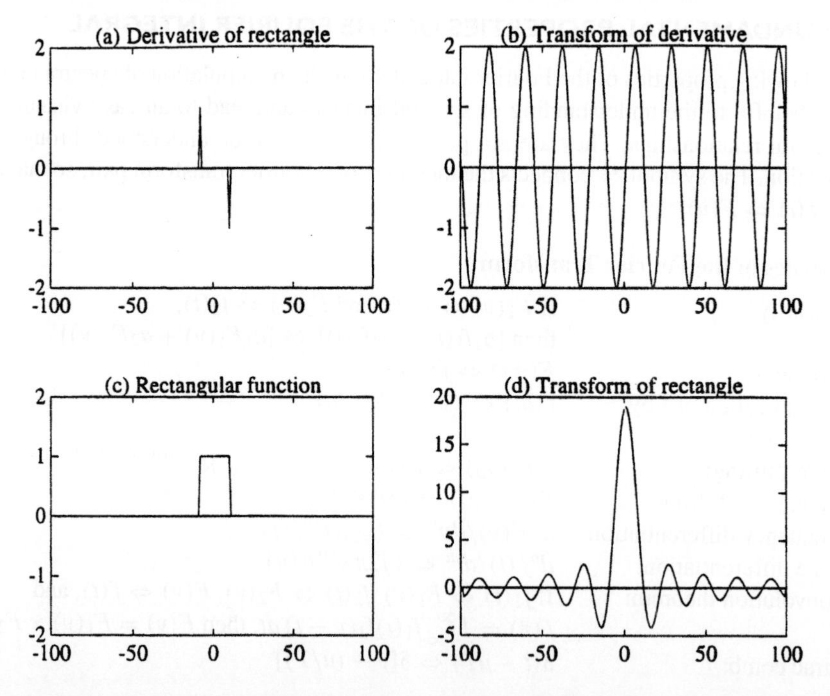

Figure 2.4. Deriving the Fourier Transform of a rectangular function. The impulses in (a) are shown with unit height although they have infinite height but a finite area of unity. The time-varying functions are on the left, and their FTs are on the right.

be replaced by $\pi \nu T$ so that $F(0) = T$, a value of 20 for this example, which is confirmed in Figure 2.4(d).

2.6 BUILDING UP A SET OF FOURIER TRANSFORMS

Rectangular functions are important in understanding spectral imagers. Consider a rectangular time pulse of width T centred at the origin. From the previous section it has a spectrum given by $F(\nu) = T\text{sinc}(\nu T)$. Now if the pulse is shifted to the left by t_o and Property 4 is applied, the Fourier transform is given by:

$$F_a(\nu) = T\text{sinc}(\nu T) \times e^{-j2\pi t_o \nu} \tag{2.11}$$

If the pulse is shifted by the same amount to the right, the following Fourier transform is obtained:

$$F_b(\nu) = T\text{sinc}(\nu T) \times e^{j2\pi t_o \nu} \tag{2.12}$$

From Property 1 (Linearity) the FT of a double pulse can be obtained from $F_a(\nu) + F_b(\nu)$:

$$F(\nu) = T\text{sinc}(\nu T)[e^{-j2\pi t_o \nu} + e^{j2\pi t_o \nu}] = T\text{sinc}(\nu T) \times 2\cos(2\pi t_o \nu) \tag{2.13}$$

This is shown in Figure 2.5 for three different values of T as compared to t_o.

As T becomes vanishingly small with respect to t_o, the Fourier transform approaches $2 \times \cos(2\pi t_o \nu)$. In other words, the FT of a pair of spikes of unit height is a cosine function of amplitude two. This confirms again the simple picture presented in Section 2.2.

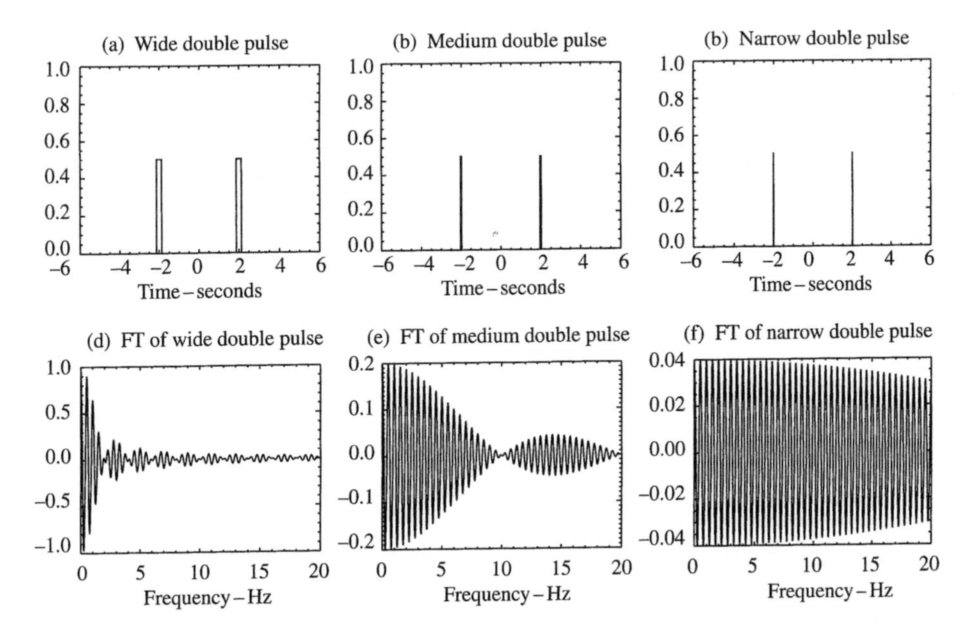

Figure 2.5. The Fourier transforms of three double pulses.

2.7 CONVOLUTIONS AND CORRELATIONS

Property 8, the convolution theorem, is particularly useful. The equation defining convolution, as given in the statement of Property 8, is shown below along with a compact form of notation introduced by Bracewell [1986].

$$\int_{-\infty}^{\infty} f_1(t) f_2(\tau - t)\, dt = f_1(t) \star f_2(t) \tag{2.14}$$

Property 8 states that the FT of the convolution of two functions is just the product of the FTs of those two functions. This allows building up new transforms from known ones. For example, suppose $f_1(t)$ and $f_2(t)$ are identical rectangular pulses of width T as shown in Figure 2.6(a). The FT of either rectangle is shown in Figure 2.6(c); from earlier it is simply $T\,\mathrm{sinc}(\nu T)$. The convolution of the two rectangles of width T is a triangle of base $2T$. This result is obtained by reflecting one rectangle, and sliding it over the other; for each separation value of τ, the product is integrated over t, forming a triangle as a function of τ as shown in Figure 2.6(b). The sliding together of the rectangles generates the "up" slope of the triangle and the moving away generates the down slope. The Fourier transform of this triangle is the product of the FTs of the two rectangles, that is, the square of the FT of a rectangle:

$$F(\nu) = T^2\,\mathrm{sinc}^2(\nu T) \tag{2.15}$$

as shown in Figure 2.6(d). Thus using these basic properties of the FT, a body of transforms can be built up. This non-mathematical approach to developing FT relationships is not to avoid performing integrations; it gives greater insight into the meaning of the FT operations which, as shall be seen, are also operations performed by spectral imaging instruments.

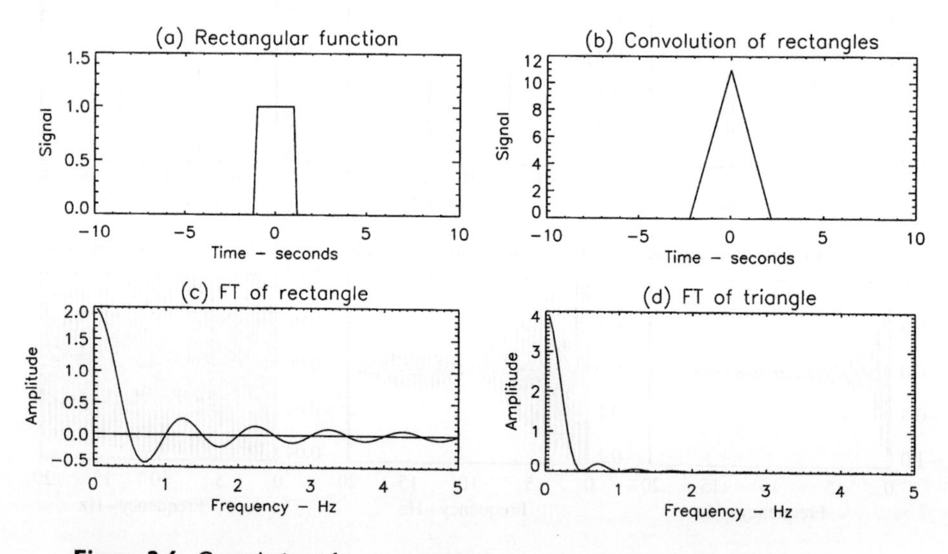

Figure 2.6. Convolution of two rectangles into a triangle, and its Fourier Transform.

A function closely related to that in Eq. (2.14) is the cross-correlation function, which is given below, using Bracewell's [1986] notation.

$$\int_{-\infty}^{\infty} f_1(t)f_2(t-\tau)\,dt = \int_{-\infty}^{\infty} f_1(t+\tau)f_2(t)\,dt = f_1(t) \star f_2(t) \qquad (2.16)$$

In the argument of f_2 in the first integral, t and τ are interchanged with respect to the convolution. Thus the convolution and cross-correlation functions give the same result for symmetrical functions, but for asymmetrical functions the results are different, as one of the functions is reflected before multiplication by the other. The convolution of a function with itself is called a self-convolution, and the cross-correlation of a function with itself is called an autocorrelation function; the latter is discussed later. The relationship has been described by Chamberlain [1979] as: $f \ast f(-) = f \star f$, where the $(-)$ indicates a change of sign of the variable.

2.8 THE DIRAC DELTA FUNCTION AND THE DIRAC COMB

2.8.1 Dirac Delta Function

An impulse, or pulse, is intuitively easy to understand as a very narrow function of finite height. But how finite? Chamberlain [1979] described it as a pulse of sufficiently narrow width that the system cannot distinguish between it and an even shorter pulse. Dirac introduced the delta function $\delta(t)$, which he referred to as an improper function, as one that was everywhere zero except at $t = 0$, and for which the integral over time was unity.

$$\int_{-\infty}^{\infty} \delta(t)\,dt = 1 \qquad (2.17)$$

This definition allows mathematical operations to be carried out, multiplication by another function, for example.

$$\int_{-\infty}^{\infty} \delta(t-t_o)f(t)\,dt = f(t_o) \qquad (2.18)$$

Since the delta function is zero everywhere except at t_o, the result of the multiplication is simply the value of $f(t)$ at t_o. The integration is needed to deal with the infinite height and infinitesmal width of the impulse, but the result is as though $f(t)$ had been simply multiplied by a pulse of unit height located at t_o, which is what is done in the numerical examples that follow. Thus the multiplication is shown without the integration, which is nevertheless implied.

The FT of a delta function at t_o is given by:

$$F(\nu) = \int_{-\infty}^{\infty} \delta(t-t_o)e^{-j2\pi\nu t}\,dt = e^{-j2\pi\nu t_o} = \cos(2\pi\nu t_o) - j\sin(2\pi\nu t_o) \qquad (2.19)$$

Since these are just the real and imaginary amplitude vectors, Dirac delta functions can be used to construct sine and cosine functions, as described earlier. If $t_o = 0$, then $F(\nu) = 1$; the spectrum is constant, i.e. uniform over the spectrum, or "white".

2.8.2 The Dirac Comb

The final, and perhaps most remarkable property, Property 9, concerns the Dirac comb, a series of equally-spaced Dirac delta functions, a mathematical way of describing the *lines* or *impulses* referred to earlier. For a comb of spacing Δt its Fourier transform is also a Dirac comb, of spacing $(1/\Delta t)$. An example of a comb in t is shown in Figure 2.7(a) with a Δt of 0.5 s. From an applications point of view the comb has two important uses; sampling and replication. In Figure 2.7(c) the result of multiplying a triangular function (Figure 2.7(b)) by a Dirac comb (Figure 2.7(a)) is shown; the continuous triangular function has been effectively sampled, or digitized, with a sampling interval of 0.5 s. This is a procedure that needs to be applied to any continuous function for which the FT is to be numerically calculated, as is described in the next section. The FT of the continuous function is shown in Figure 2.7(e); it is of the form $T^2\text{sinc}^2(\nu T)$. The FT of the original comb is shown in Figure 2.7(d) – it too is a comb but of spacing $1/\Delta t = 2\,\text{Hz}$. Finally, since the sampled triangle of Figure 2.7(c) is the product of (a) and (b), the FT of (c) is the convolution of (d) and (e), a Dirac comb convolved with $T^2\text{sinc}^2(\nu T)$. When a single Dirac delta function is convolved or correlated with another function as that of Figure 2.7(e), it replicates the original function; thus convolution with a comb generates a replicated sequence of the original function, as shown in Figure 2.7(f). This result is extremely significant in that the sampled function has the same spectrum as that of the continuous function. That is, sampling a continuous function does not result in an incorrect spectrum when the FT is calculated; the only consequence of sampling is that the spectrum of the function has been

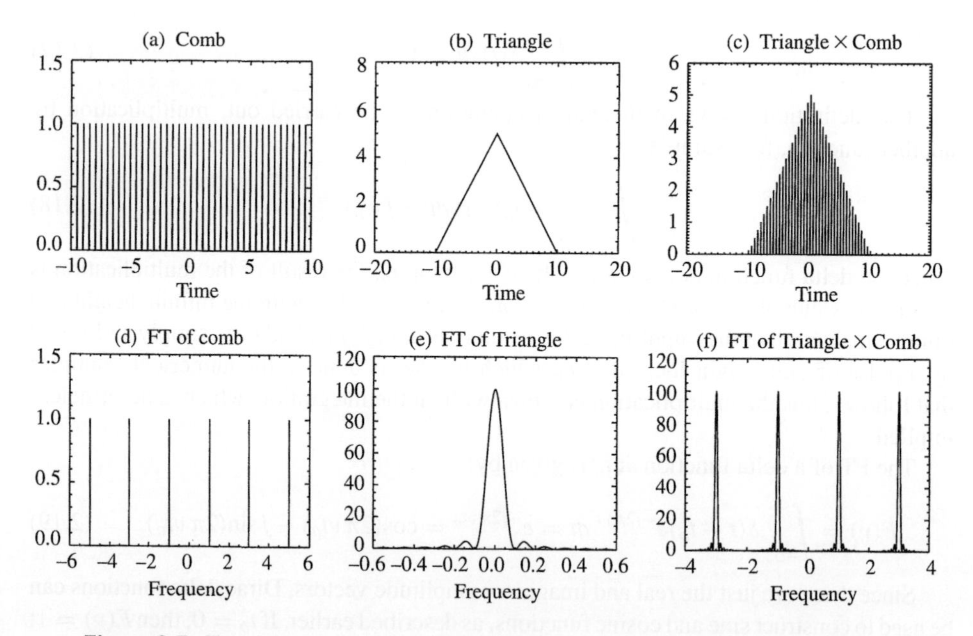

Figure 2.7. Transforms involving the Dirac comb, illustrating sampling and replication.

replicated over and over again (it has multiple orders). Thus one can compute correct spectra from sampled continuous functions. However, to ensure that the spectrum is correct, the different "orders" of the spectrum must not overlap, otherwise one order will contaminate the other. The frequency separation between orders, $1/\Delta t$, must be twice the highest frequency ν_{max} present in the spectrum, so that $1/\Delta t = 2\nu_{max}$, or $T_{max} = 2\Delta t$, where $T_{max} = 1/\nu_{max}$. This means that the sampling interval Δt must be such that two samples are obtained for each wave period T_{max} of the highest frequency present in the spectrum. This statement is known as Shannon's sampling theorem. It is remarkable that such a large part of the signal may be "thrown away" and yet the true spectral character is retained, but this just reflects the fact that for a band-limited signal, the amount of information is finite.

The second application mentioned above was replication. The convolution of the Dirac comb in Figure 2.7(d) with the $T^2\,\mathrm{sinc}^2(\nu T)$ function of (e) generates the replicated function of (c). Thus convolution with a Dirac comb is a means of replication. There will often be uses for replicating a function in this way.

2.9 THE DISCRETE FOURIER TRANSFORM

The continuous (or integral) Fourier Transform presented in the previous section is useful for the conceptual idealizations considered there. When it comes to working with real data, however, the Fourier transform is normally computed numerically. Physical data are often in analog or continuous temporal form. In order to input them to a computer they must be made discrete; they are said to be digitized. This may be done by an analog-to-digital converter, or it may be done by scaling a chart record by hand, but in either case, it is necessary to select a starting point in time, an ending point, and a sampling interval. If N samples are collected, it is convenient to number them from 0 to $N-1$, identified by the running number k. The time variable t is thus given by $t = k\Delta t$. The lowest frequency that can be calculated in the spectrum is that corresponding to fitting one wavelength into $N\Delta t$; this corresponds to $\nu_1 = 1/(N\Delta t)$. Higher frequencies can then be represented by the running index r, such that $\nu = r\nu_1 = r/(N\Delta t)$. Substituting these values for ν and t into Eqs. (2.2), and using summation in place of integration, the Discrete Fourier Transform (DFT) is obtained in the following form:

$$F_r = \sum_{k=0}^{N-1} f_k e^{-j2\pi(rk/N)} \tag{2.20}$$

$$f_k = \frac{1}{N}\sum_{k=0}^{N-1} F_r e^{j2\pi(rk/N)} \tag{2.21}$$

There are several considerations to keep in mind when using the DFT, as follows. The Fourier transform is defined over a range from minus to plus infinity while the DFT is restricted to a finite interval for which the values are all positive. It is implicit therefore that the true $f(t)$ is periodic with a period equal to the record length $N\Delta t$. Since the user is

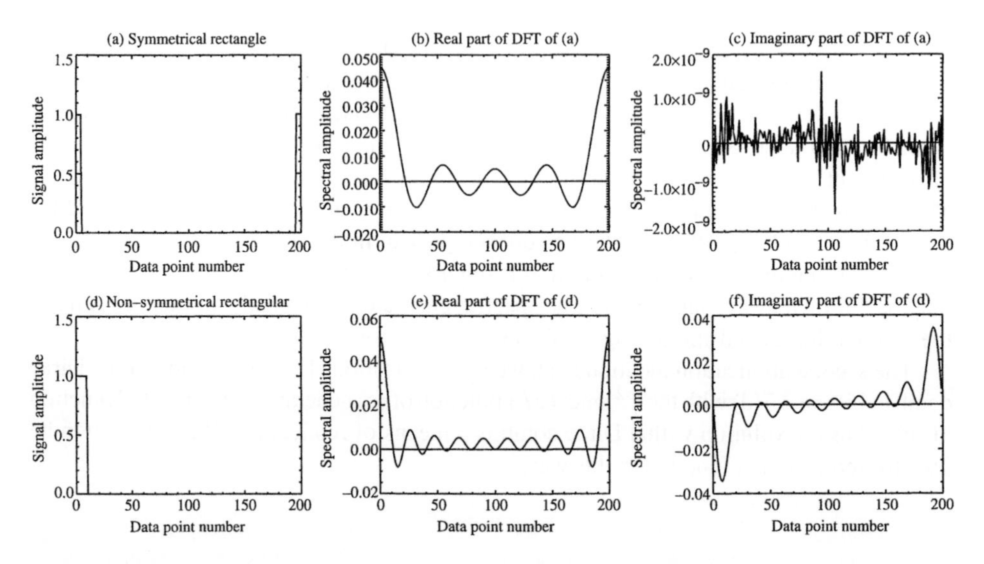

Figure 2.8. Illustrating the application of the DFT as applied to a symmetrical rectangle (wrapped around), and a non-symmetrical rectangle that is not wrapped. The former gives a purely real spectrum while the latter gives a complex spectrum.

concerned only with what happens inside the measurement range, this assumption doesn't present a problem, but from a more practical point of view it means that the data "wrap around" from the end of the record to the beginning. For example, if the intent is to create a rectangular $f(t)$, centred at zero so that it is an even function and its DFT will be purely real, it would be constructed as shown in Figure 2.8(a), with the "negative" portion of the rectangle wrapped around to the right-hand end of the plot. Note that data point $N+1$ is just a replication of point 0. The real part of the spectrum shown in Figure 2.8(b) is symmetrical, while the imaginary part shown in Figure 2.8(c) is essentially zero (about 10^{-7} of the real part). In Figure 2.8(d) a rectangle of the same width is shown, but only on one side of zero (non-symmetrical). The real part of its DFT is shown in Figure 2.8(e), which is different from (b). The imaginary part is shown in Figure 2.8(f); it is not zero, it is about the same magnitude as the real part and is asymmetrical about the middle of the plot. In other words, the spectrum of the non-symmetrical rectangle is complex.

The lowest frequency computed is the one that has a period equal to the record length. This means that the record length determines the frequency resolution in the computed spectrum and this clearly guides the user in the choice of record length. The sampling interval determines the highest frequency in the computed spectrum, according to the Shannon criterion of two samples per wave for the highest frequency wave present. The sampling thus defines what is called the Nyquist, or folding frequency $\nu_N = 1/(2\Delta t)$, which is the ν_{max} introduced earlier. The significance of this frequency is that the spectrum may be considered to be folded back on itself as shown in Figure 2.9. Both axes in (a) are frequency, but the horizontal one has been folded back upon itself three times. All the

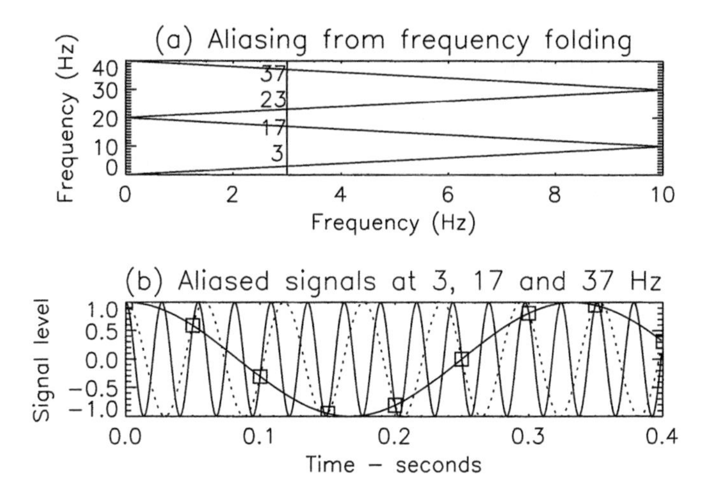

Figure 2.9. Illustrating aliasing in the DFT. The squares mark the times at which the samples are taken, every 0.05 s corresponding to a folding frequency of 10 Hz. All three waves intersect at these locations, meaning that they are aliases of one another.

frequencies on a given vertical line in Figure 2.9(a) are said to be aliases of one another since they all yield identical data sets when sampled. To demonstrate the truth of this, in Figure 2.9(b) the waveforms for 3, 17 and 37 Hz are shown. The folding frequency (Figure 2.9(a)) is 10 Hz, so the sampling interval is 0.05 seconds. In Figure 2.9(b) it is seen that all three waveforms cross at 0.0, 0.05, 0.10 etc. seconds. Thus the sampled waveforms are identical. Since the spectrum is computed only up to the Nyquist frequency, any information present above this frequency is aliased back onto the computed spectrum. In other words, one must ensure that the sampling interval is at least small enough to ensure that there is no information present in the signal above the folding frequency. This might be done by inserting an analog filter in the system before the signal reaches the digitizer, or a numerical filter could be used. When this condition is satisfied, the computed spectrum is identical to the true spectrum, and there is no loss of information caused by the sampling process.

Finally, while it is not really necessary here to prove the Fourier theorem given at the beginning of this chapter, it is of interest to demonstrate it for the DFT, for a particular value of N which here is taken to be $N = 4$. Suppose that the signal f_m consists of four values, f_0, f_1, f_2 and f_3. The spectrum is calculated from Eq. (2.20) and then the signal is re-calculated from Eq. (2.21), yielding f_k. If the theorem is correct, $f_k = f_m$. Following this procedure, f_k is given by the double summation resulting from inserting into (2.20) into (2.21).

$$f_k = \sum_{r=0}^{N-1} \sum_{m=o}^{N-1} f_m e^{-j2\pi(rm/N)} \times e^{j2\pi(rk/N)} = \sum_{r=0}^{N-1} \sum_{m=o}^{N-1} f_m e^{-j2\pi r((m-k)/N)} \qquad (2.22)$$

Table 2.1. Demonstration of the Discrete Fourier Transform

	$r=0$	$r=1$	$r=2$	$r=3$	Sum
$m=0$	f_0	$f_0e^{j\pi}$	$f_0e^{j2\pi}$	$f_0e^{j3\pi}$	0
$m=1$	f_1	$f_1e^{j\pi/2}$	$f_1e^{j\pi}$	$f_1e^{j3\pi/2}$	0
$m=2$	f_2	f_2	f_2	f_2	$4f_2$
$m=3$	f_3	$f_3e^{-j\pi/2}$	$f_3e^{-j\pi}$	$f_3e^{-j3\pi/2}$	0

The 4×4 terms obtained in the double summation are shown in Table 2.1: but since one value of k is computed at a time, the value $k = 2$ is chosen for this illustration, and the corresponding values are shown in the table. The "vectors" along each row sum to zero, except for the row corresponding to $m = 2$. Allowing for the $1/N$ term of Eq. (2.21), the entire table sums to f_2, which is as expected. The orthogonality of the sine and cosine function is manifested by the symmetry of the rotating vector additions in this representation.

This representation of the Fourier components as vectors led to the realization that numerical Fourier transforms need not be calculated simply as series of sine and cosine functions using computer library function routines. Using the vector approach, and taking advantage of the fact that the angles appear repetitively, allows the calculation to be speeded up enormously. This method, called the Fast Fourier Transform (FFT), was conceived in the 1960s, and had a major impact on the practical use of the FT, especially as computers were so slow at that time. Nowadays the problem of computer power is not a major factor, but the FFT allows FTs to be calculated in real time, and has made possible the development of FFT chips. From this point on, the terms FT, DFT and FFT are used interchangeably, depending on the context.

2.10 THE AUTOCORRELATION FUNCTION AND POWER SPECTRAL DENSITY

The autocorrelation function $R(\tau)$ of a signal $f(t)$ is obtained from the definition of the cross-correlation function given in Eq. (2.16) by making the two functions the same:

$$R(\tau) = \frac{1}{2T}\int_{-T}^{T} f(t)f(t+\tau)\,dt \tag{2.23}$$

For a real signal of finite length, it is physically not possible to carry out the integration from minus infinity to plus infinity so instead an average over the measurement time of length $2T$ is taken; τ is the lag time between the two locations of the same signal. This quantity gives a measure of the coherence at lag τ, and reveals periodicities in the signal. The FT of

$R(t) = f(t) \star f(t)$ is given as follows:

$$\int_{-\infty}^{\infty} [f(t) \star f(t)] e^{j2\pi\nu t}\, dt = \int_{-\infty}^{\infty} d\tau \int_{-\infty}^{\infty} dt\, f(\tau) f(\tau + t) e^{j2\pi\nu t} \tag{2.24}$$

$$= \int_{-\infty}^{\infty} d\tau \int_{-\infty}^{\infty} dt\, f(\tau) f(\tau + t) e^{j2\pi\nu(t+\tau-\tau)} \tag{2.25}$$

$$= \int_{-\infty}^{\infty} d\tau f(\tau) e^{-j2\pi\nu\tau} \int_{-\infty}^{\infty} dt\, f(\tau + t) e^{j2\pi\nu(t+\tau)} \tag{2.26}$$

Thus

$$\mathrm{FT}[f(t) \star f(t)] = F^*(\nu)\, F(\nu) = |F(\nu)|^2 = S(\nu) \tag{2.27}$$

This is known as the Wiener–Khinchin theorem, which states that the FT of the autocorrelation function of the signal $f(t)$ is the modulus squared of the FT of the $f(t)$, a quantity denoted $S(\nu)$ and called the *power spectral density* of the signal or, in optical terms, simply the spectrum. This is done by analogy in the sense that if $f(t)$ is a time-varying voltage, or amplitude, then its square is power. Since $R(\tau)$ is even in (τ), $S(\nu)$ is always real, but in general $F(\nu)$ is complex, so the complex conjugate is used.

2.11 OPTICAL DEVICES AS LINEAR DYNAMICAL SYSTEMS

With optical devices it is convenient to change variables; since the frequency is not measured directly by conventional optical methods, ν is replaced by the wavenumber σ where $\nu = c\sigma$. In place of time t which is also difficult to measure directly, the distance that light travels in vacuum in that time, $\Delta = ct$, is used instead. The result is that νt is replaced by $(c\sigma)(\Delta/c) = \sigma\Delta$ where Δ is called the *optical path difference* (OPD), with σ in m^{-1} and Δ in m. However, as long as σ and Δ have reciprocal units it doesn't matter what they are as the units will cancel; it can be convenient to use cm^{-1} as is done by spectroscopists. In any case the result is a very convenient form of the FT equations for optical use, namely:

$$F(\sigma) = \int_{-\infty}^{\infty} f(\Delta) e^{-j2\pi\sigma\Delta}\, d\Delta \tag{2.28}$$

$$f(\Delta) = \int_{-\infty}^{\infty} F(\sigma) e^{j2\pi\sigma\Delta}\, d\sigma \tag{2.29}$$

The linear dynamical system may be mechanical, electrical or optical in nature. The only requirements are that it have an input, an output, and that it be linear. Let the input be $f(\Delta)$ and the output $y(\Delta)$. The Fourier transform of $f(\Delta)$ provides the spectrum of the input.

$$F(\sigma) = \int_{-\infty}^{\infty} f(\Delta) e^{-j2\pi\sigma\Delta}\, d\Delta \tag{2.30}$$

A real system will not respond equally to all frequencies; the function which describes how the amplitude and phase of an input wave are modified by the system, is a multiplicative quantity $H(\sigma)$, that is called its frequency response function. In order to modify both the amplitude and phase it must in general be a complex function of frequency. The output has a spectrum modified from that of the input by this function, so that the output spectral amplitude for frequency σ' is:

$$H(\sigma')F(\sigma')e^{j2\pi\sigma'\Delta} \tag{2.31}$$

The output signal is then a superposition of all frequency components, calculated as the integral over σ:

$$y(\Delta) = \int_{-\infty}^{\infty} H(\sigma)F(\sigma)e^{j2\pi\sigma\Delta}\,d\sigma = \int_{-\infty}^{\infty} Y(\sigma)e^{j2\pi\sigma\Delta}\,d\sigma \tag{2.32}$$

where $Y(\sigma)$ is the output spectrum, and is given by:

$$Y(\sigma) = H(\sigma)F(\sigma) \tag{2.33}$$

and it is seen that $y(\Delta)$ and $Y(\sigma)$ are Fourier transform pairs. In other words, the influence of the system on the signal can be computed from the spectrum if the frequency response function is known. $H(\sigma)$ can be thought of as a filter, with a transmittance that is frequency dependent as is the phase shift, creating the output spectrum, $Y(\sigma)$.

The influence of the system can also be described in the time domain. If the FT is taken of both sides of Eq. (2.33) the result is:

$$y(\Delta) = h(\Delta) \star f(\Delta) \tag{2.34}$$

where $\star$ denotes cross-correlation, and $h(\Delta)$ is the FT of $H(\sigma)$. To visualize the nature of $h(\Delta)$, consider the case where $f(\Delta) = \delta(\Delta)$, a Dirac delta function. Then since the cross-correlation of a delta function with another function simply replicates the latter function:

$$y(\Delta) = h(\Delta) \star \delta(\Delta) = h(\Delta) \tag{2.35}$$

For this particular case, the output $y(\Delta)$ is just $h(\Delta)$ which is thus the output of the system for an impulsive input; $h(\Delta)$ is called the *impulse response function* or sometimes the *weighting function* of the system. In summary, there are two ways to measure the frequency response of a system, a stereo system, for example. One can input a sinusoidal signal, and measure the ratio of the output to the input, and the phase difference between output and input, as one scans the sinusoid through all frequencies, providing $H(\sigma)$ directly. The other way is to input a pulse to the system, measure the output as a function of time, which is $h(\Delta)$ from the above, and then proceed in one of two ways. The first is to calculate the autocorrelation function $h(t) \star h(t)$ and take its FT which from (2.27) is equal to $|H(\sigma)|^2 = W(\sigma)$, where $W(\sigma)$ is the instrument passband function. The second is to calculate the FT of $h(\Delta)$, yielding $H(\sigma)$, and then compute $W(\sigma) = H(\sigma)^*H(\sigma) = |H(\sigma)|^2$. In the next section the instrument passband function, or simply the passband, of the diffraction grating spectrometer is computed using the autocorrelation method, and then the passband of the Fabry–Perot spectrometer is found by the second method, taking the FT of $h(\Delta)$.

The quantities $f(\Delta)$ and $y(\Delta)$ can be thought of as mechanical displacements, or electrical voltages, or in the case of an optical system, as the electric field. Optical detectors cannot follow the individual oscillations of the electric field, and so do not measure $y(\Delta)$, but $\overline{y(\Delta)^2}$, the average power. However, with interferometers it is possible to introduce time delays in the electric field before combining them, allowing the measurement of quantities such as $\overline{f(\Delta)f(\Delta+\Delta_o)}$, which is seen to be related to the autocorrelation function. This suggests another approach to obtaining the spectrum which is investigated shortly.

2.12 THE DIFFRACTION GRATING AS A LINEAR DYNAMICAL SYSTEM

The impulse response method has been used by a number of optical practitioners; one example is the analysis by Stoner [1966]. The diffraction grating is used as the first example of an optical device as a linear dynamical system, by using the impulsive response approach. The diffraction grating is made up of N rulings, with the grating tilted so that for an impulsive plane wave input pulse, each ruling returns a pulse with a separation in Δ, the optical path difference, that is successively shifted with respect to the previous pulse. For an impulsive input, then the output is a Dirac comb that is not infinite, but contains just N pulses. The output may be considered to be the product of a rectangular function with a Dirac comb as shown in Figure 2.10. For a path difference d between successive rulings, the path difference width of the rectangular function is $(N-1)d \approx Nd$.

As indicated above, the autocorrelation method is used in this analysis, making use of the Dirac comb, as $\delta(\Delta - vd)$, for which each impulse has a value of infinity for $\Delta = vd$ and is equal to zero for other values of Δ. The autocorrelation function for the grating spectrometer is then:

$$R(\tau) = \int_{-\infty}^{\infty} \sum_{v=0}^{N-1} \sum_{u=0}^{N-1} \delta(\Delta - vd)\delta(\Delta - \tau - ud)\, d\Delta \tag{2.36}$$

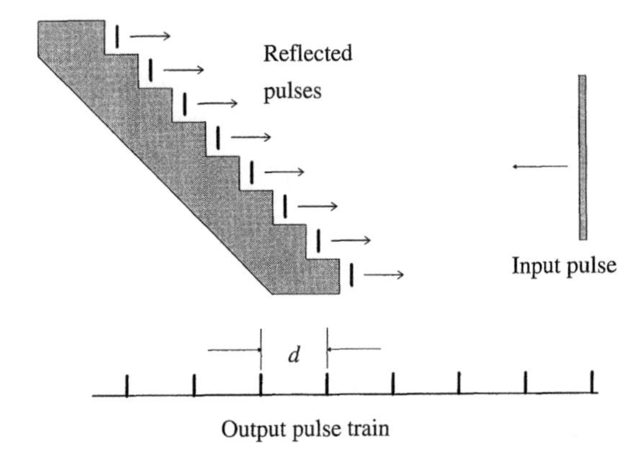

Figure 2.10. Impulse response of a diffraction grating. The input pulse becomes eight small pulses after reflection from the grating, generating a pulse train with a spacing of d, the optical path difference.

for which the integral gives a contribution only when the impulses of the two delta functions fall upon one another, which occurs when τ is a multiple of d, which can be written $\tau = (v - u)d$. After integrating over Δ, the result is a Dirac comb in τ.

$$R(\tau) = \sum_{v=0}^{N-1}\sum_{u=0}^{N-1} \delta[\tau - (v - u)d] \tag{2.37}$$

A new problem arises above with the improper delta function, in that one impulse of infinite height is being multiplied by another, also of infinite height; the integral of the product is not unity, but infinity. Hernandez [1986a] has offered a solution by defining the delta function as:

$$\int_{-\infty}^{\infty} |f(t)|^2 \, dt = 1 \tag{2.38}$$

The FT of the autocorrelation function is $W(\sigma)$, given by:

$$W(\sigma) = \int_{-\infty}^{\infty} \sum_{v=0}^{N-1}\sum_{u=0}^{N-1} \delta[\tau - (v - u)d] e^{-j2\pi\sigma\tau} \, d\tau \tag{2.39}$$

$$W(\sigma) = \sum_{v=0}^{N-1}\sum_{u=0}^{N-1} \int_{-\infty}^{\infty} \delta[\tau - (v - u)d] e^{-j2\pi\sigma\tau} \, d\tau \tag{2.40}$$

There are contributions to the integral only where $\tau = (v - u)d$ so that:

$$W(\sigma) = \sum_{v=0}^{N-1}\sum_{u=0}^{N-1} e^{-j2\pi\sigma(v-u)d} = \sum_{v=0}^{N-1} e^{-j2\pi\sigma vd} \sum_{u=0}^{N-1} e^{j2\pi\sigma ud} \tag{2.41}$$

$$W(\sigma) = \frac{1 - e^{-j2\pi\sigma Nd}}{1 - e^{-j2\pi\sigma d}} \times \frac{1 - e^{j2\pi\sigma Nd}}{1 - e^{j2\pi\sigma d}} \tag{2.42}$$

where the last term is expressed as the product of the sums of two geometric progressions in which a is the first term, n the number of terms, r the common ratio and the sum s is:

$$s = a\frac{1 - r^n}{1 - r}$$

Further, using the trigonometric identities that $e^{j\theta} + e^{-j\theta} = 2\cos\theta$, and $\cos 2\theta = 1 - 2\sin^2\theta$, the result is:

$$W(\sigma) = \frac{2 - 2\cos(2\pi\sigma Nd)}{2 - 2\cos(2\pi\sigma d)} = \frac{\sin^2 \pi\sigma Nd}{\sin^2 \pi\sigma d} \tag{2.43}$$

This passband function is plotted in Figure 2.11, where $N = 10\,000$, $d = 5.0 \times 10^{-5}$ cm which is one wavelength for light of 500 nm, or 20 000 cm^{-1}. Figure 2.11(a) shows the denominator over a wavenumber range of 0 to 40 000 cm^{-1} (the numerator is too

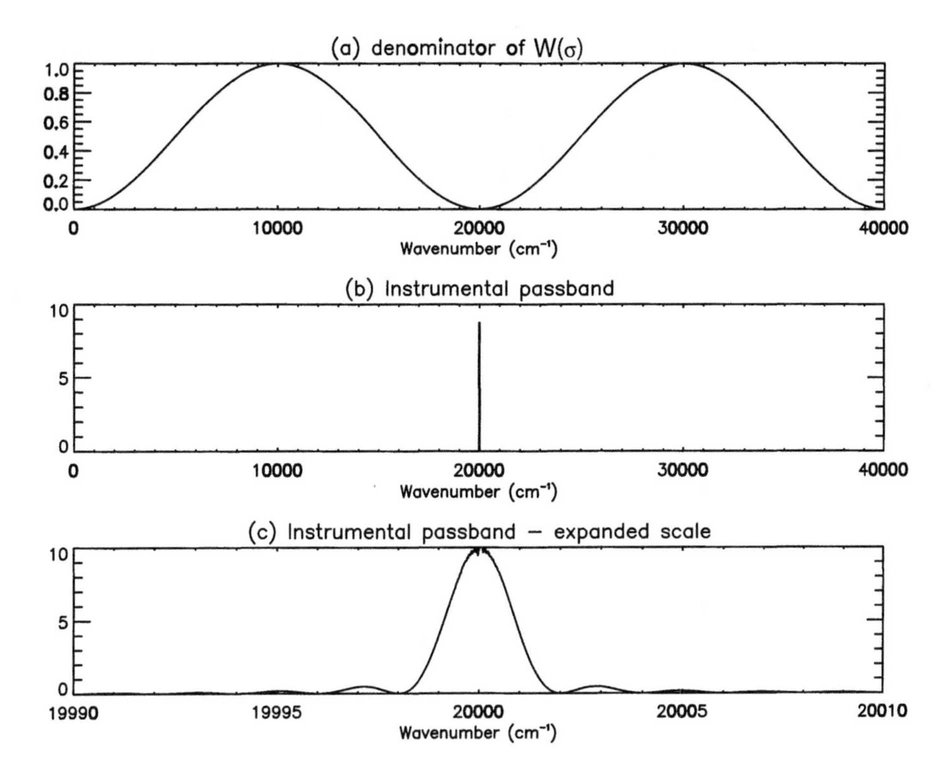

Figure 2.11. For a diffraction grating spectrometer: (a) the denominator of the passband function; (b) the passband function over 0–40 000 cm^{-1}, and (c) the passband function on an expanded scale.

rapidly varying to be shown on this scale). Figure 2.11(b) shows $W(\sigma)$ on the same scale, appearing simply as a spike; the same spike occurs at 0 and $40\,000\,\text{cm}^{-1}$ corresponding to $\sin^2 \pi\sigma d = 0$. $W(\sigma)$ is plotted again in Figure 2.11(c) on an expanded scale of $20\,\text{cm}^{-1}$ around $20\,000\,\text{cm}^{-1}$ where, as noted earlier, it corresponds to the FT of a triangle. There are some perturbations near the peak which are caused numerically through the division by near-zero.

To characterize the passband width one can note that the denominator changes very slowly with σ and can be considered constant over the passband width. The numerator can be written $\sin^2 \pi Nd(\sigma - \sigma_o)$, where σ_o corresponds to the passband peak, and the first zero occurs where $\pi Nd(\sigma - \sigma_o) = \pi$. Thus remembering that $\sigma = 1/d$, the spectral bandwidth is given by $(\sigma - \sigma_o) = \sigma/N$, and the resolving power $\mathcal{R} = \sigma/(\sigma - \sigma_o) = N$, the number of rulings on the grating.

The impulse response approach allows further insights about diffraction gratings. For example, one can readily visualize the results of making certain modifications to the diffraction grating. If the grating grooves were sinusoidal rather than step-like in shape, the exiting wave would be a sinusoidal wave train, instead of a train of pulses. This result can be thought of as a rectangular function multiplied by an infinite sinusoid. The Fourier transform is that of a rectangle convolved with a double pulse. That is, there are no multiple orders with this grating.

One can also see that the shape of $W(\sigma)$, the spectral passband, can be changed by changing the shape of the envelope of the pulse train. For example, if a diamond shaped mask is placed over the grating, with the diagonal axis parallel to the rulings, the output pulse train has a triangular envelope. The shape of the spectral passband is then:

$$W(\sigma) = \frac{\sin^4 \pi\sigma N d}{\sin^4 \pi\sigma d} \tag{2.44}$$

This passband has smaller sidelobes than that for the normal grating, but the bandwidth of the primary passband is wider; the preferable choice depends on whether one wishes to separate closely spaced lines of equal radiance, or to detect weak lines in the presence of nearby strong ones. One can tailor the mask to the desired situation; a Gaussian mask, for example, gives a Gaussian transform with no side lobes at all. This process is called apodization, and has been described by Jacquinot [1950].

Finally, to complete this discussion of the diffraction grating, it is interesting to look in a slightly different way at how it processes the input data. The output train of N pulses can be thought of as a sampling of N points on the electric field signal $\mathcal{E}(\Delta)$, from which the squared sum is presented to the detector. This is illustrated in Figure 2.12 and written:

$$S(\sigma) = a\upsilon \left[\sum_{\upsilon=0}^{N-1} \delta(\Delta - \upsilon d)\mathcal{E}(\Delta) \right]^2 \tag{2.45}$$

This quantity is a measure of the power in the spectrum for $\sigma = 1/d$. As the grating is turned to uniformly change d, the spectrum is scanned. Thus the diffraction grating may

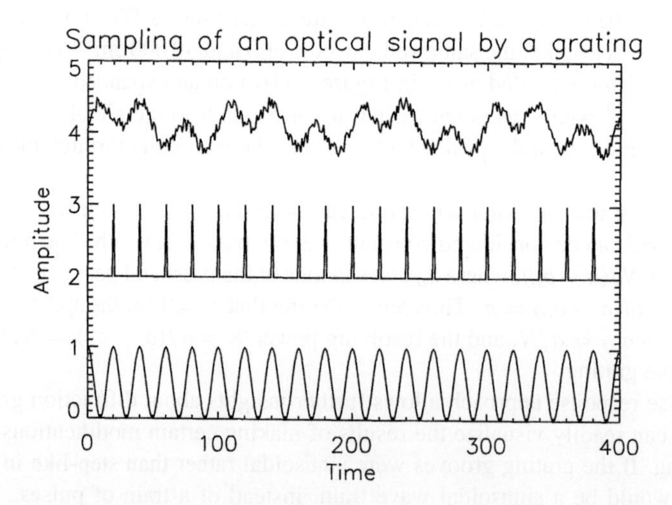

Figure 2.12. Illustrating the sampling of the optical electric field by a diffraction grating.

be thought of as an analog Fourier series calculator. It operates with great efficiency, but restricted flexibility. Interferometers for which the computation must be done numerically are discussed later. These add a certain cost, but provide much greater flexibility.

2.13 THE FABRY–PEROT ETALON AS A LINEAR DYNAMICAL SYSTEM

The basis of the FPS is the etalon, a pair of plane parallel surfaces, each coated with a material of radiance reflectance r_a, a radiance transmittance t_a and separated by a material of refractive index n and thickness t. Light entering the etalon cavity is reflected back and forth across the gap several times with some light emerging on each reflection as shown in Figure 5.1. The impulse response function is a sequence of pulses as shown in Figure 2.13, equally spaced and decaying in amplitude. The time delay between adjacent pulses can be expressed in terms of the optical path difference (OPD), denoted Δ in general, but with the value d for the difference in Δ between adjacent pulses, so that $d = 2nt$ for normal incidence (the general case is considered in Chapter 5). If the input $f(\Delta)$ consists of a single impulse in amplitude, the output is a succession of output pulses, each reduced in amplitude from the previous one by a factor of r_a for two reflections since $\sqrt{r_a}$ is the amplitude reflectance at each surface. In Figure 2.13 the values of $r_a = 0.8$ and $d = 5.0$ cm are used. The envelope of the output pulse train is represented by the equation for the mth pulse where the radiance transmittance of t_a through two surfaces is included:

$$y_{env}(\Delta) = t_a r_a^m = t_a r_a^{\Delta/d} \tag{2.46}$$

The impulse response function $h(\Delta)$ is the product of this envelope with a Dirac comb of spacing d, written as follows. The problem of infinite functions arises again here, and the

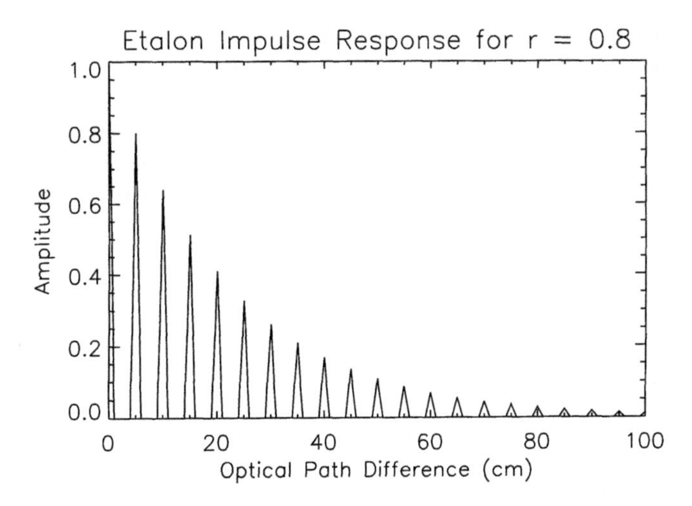

Figure 2.13. Impulse response function of an etalon for $r = 0.8$ and $d = 5$ cm.

reader is referred to Hernandez [1986a] for the solution. Here we simply multiply by unity for the delta function.

$$h(\Delta) = \sum_{v=0}^{\infty} t_a r_a^{\Delta/d} \delta(\Delta - vd) \tag{2.47}$$

$$W(\sigma) = H^*(\sigma)H(\sigma) = t_a \sum_{u=0}^{\infty} r^u e^{-j2\pi\sigma ud} t_a \sum_{v=0}^{\infty} r^v e^{j2\pi\sigma vd} \tag{2.48}$$

since contributions occur only when $\Delta = ud$ or vd. This is just the product of two geometric progressions. since the common ratio is less than unity, the sum of each progression is just the first term divided by (1 minus the common ratio). Thus, using in addition the identity $\cos 2x = 1 - 2\sin^2 x$, the result is:

$$W(\sigma) = \frac{t_a}{1 - r_a e^{-2\pi j\sigma d}} \times \frac{t_a}{1 - r_a e^{2\pi j\sigma d}} = \frac{t_a^2}{1 + r_a^2 - 2r_a \cos(2\pi\sigma d)} \tag{2.49}$$

$$W(\sigma) = \frac{t_a/(1 - r_a)^2}{1 + (4r_a/(1 - r_a)^2)\sin^2 \pi\sigma d} \tag{2.50}$$

This is known as the Airy function, after G.B. Airy who published it in 1833, long before Fabry and Perot constructed their first instrument. This derivation, which he applied to parallel surfaces such as a piece of glass, was reprinted in his book, *On the undulatory theory of optics* [Airy, 1877].

This function, described in more detail in Chapter 5, has repeated passband peaks located where $\sin^2 \pi\sigma d = 0$. For an etalon cavity of refractive index n, $d = 2nt$ so that successive passband peaks may be identified with their interference order m, where $2nt\sigma = m$, and m is an integer.

Equations (2.43) and (2.50) describe the instrumental passband functions of two important spectrometers, the diffraction grating spectrometer and the Fabry–Perot spectrometer, which is an important initial step in understanding spectral imagers. But more than this is needed. The next chapter explains how the associated spectral resolution determines the output signal of these instruments to a given input radiance, a quantity called the responsivity. The term superiority is also introduced, providing a way of comparing different instruments.

2.14 PROBLEMS

2.1. Consider two counter-rotating vectors that both make an angle of 30° with the positive x axis at time zero.

(a) Express these vectors in the form $Ae^{j(2\pi\nu t+\phi)}$, and expand into sines and cosines of $2\pi\nu t$ and ϕ to obtain: $2A[\cos(\pi/6)\cos 2\pi\nu t + j\sin(\pi/6)\cos 2\pi\nu t]$.

(b) Show that this is the representation of an oscillation along a diameter of a circle of radius A centred at the origin, and making an angle of $+30°$ with the positive x axis.

2.2. Let $f_1(t) = 5.0 \sin(14\pi t) + 10.0 \cos(5\pi t)$ where t is in seconds, and let $f_2(t) = 12$ for $0 \leq t \leq 2.0$ seconds, and $= 0$ otherwise.
(a) Compute and plot the FFTs (using a software package) of $f_1(t)$ and $f_2(t)$ over a range of ten seconds, and a suggested value of 200 data points. Note that the FFT will return both a real and an imaginary spectrum. Make the following calculations for both the real and imaginary spectra.
(b) Add these spectra to form $F(\nu) = F_1(\nu) + F_2(\nu)$.
(c) Compute $F^+(\nu)$ as the FFT of $[f_1(t) + f_2(t)]$. By comparing $F(\nu)$ and $F^+(\nu)$ demonstrate the Linearity Property 1 (of Section 2.4).

2.3. For 200 data points, create the Gaussian function $f_1(t) = \exp(-(x-100)^2/4)$ and over 20 data points the function $f_2(t) = \exp(-(x-10)^2/16)$.
(a) Using a software package, convolve these two functions to form $f(t)$ (in order to slide one function over the other, one must be much narrower in dimensions, which is why it has been set up that way). Demonstrate the convolution theorem with these two functions.
(b) From the numerical results of (a) show that if a Gaussian of half-width a is convolved with a Gaussian of half-width b, the result is a Gaussian with half-width $c = \sqrt{a^2 + b^2}$.
(c) Show analytically that the result of (b) is true.

2.4. If a signal record of length 100 seconds is sampled at 20 samples s^{-1}, and the spectrum is computed:
(a) determine four aliases of 27 Hz in that spectrum;
(b) determine the smallest (non-zero) and largest frequencies in the computed spectrum.

2.5. A linear dynamical system has an impulse response function $h(t) = \exp(-0.5\,t)$ where t is in seconds and $h(t)$ is defined only for positive t. Calculate and plot the instrument passband function $W(\nu) = H(\nu)H^*(\nu)$ for this device.

2.6. The normal measure of instrumental passband width is the full-width at half-maximum (FWHM). Determine the FWHM for a diffraction grating.

3

INSTRUMENT RESPONSIVITY AND SUPERIORITY

3.1 RESPONSIVITY OF AN ELEMENTARY PHOTOMETER

An "atmospheric" photometer is a device which measures, for some defined spectral band, the integrated photon emission rate from the volume of atmosphere lying within the solid angle of its field of view. Imagers are considered later but, for now, the photometer can be thought of as one picture element (pixel) of a spectral imager. The elements of the photometer as shown in Figure 3.1 are a spectral passband defining element (shown here as a filter), a lens (normally called an objective lens) that defines an entrance aperture of area A, called the aperture stop, a field stop that determines the field of view in front and subtends a solid angle of Ω at this lens, and a detector that converts the photons into electrons, creating an electron signal, or electric current. The detector receives all the light transmitted by the aperture stop and field stop combination, giving a photon arrival rate F_p at the detector. From Sections 1.2 and 1.3 this can be written:

$$F_p = \frac{A\Omega}{4\pi}\int_0^\infty V(r)\,dr = \frac{A\Omega}{4\pi}I = A\Omega L \tag{3.1}$$

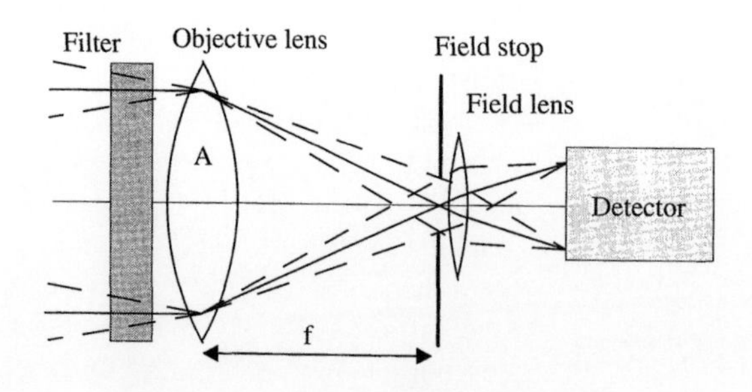

Figure 3.1. The elementary photometer.

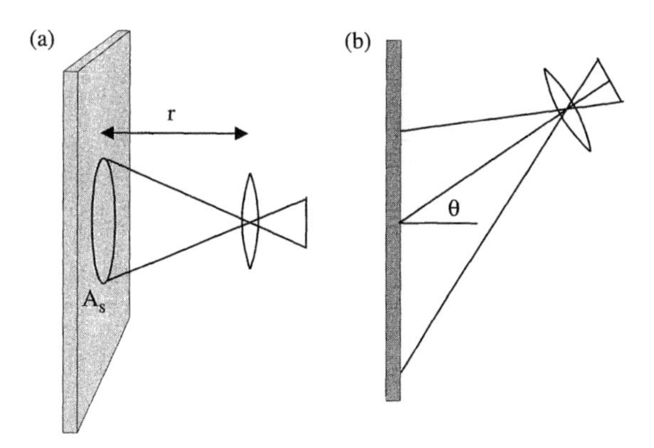

Figure 3.2. (a) Collection of light from an extended source; (b) viewing an extended source from an angle θ.

that is, $F_p = LA\Omega$, for an atmosphere that is assumed to be uniform across the field of view, and is characterized as a "surface" of radiance L. This is illustrated in Figure 3.2(a) where an extended source of radiance L photons $\mathrm{s^{-1}\,m^{-2}\,sr^{-1}}$ is located at a distance r from the photometer. The area A_s of the source, viewed by the photometer on the atmosphere, is the image of the field stop. The photon count rate corresponding to the emission from A_s measured at the detector is then just $F_p = LA_sA/r^2$, but since $\Omega = A_s/r^2$ this is, as has been given above, $F_p = LA\Omega$, and in particular is independent of the source distance r. Allowing for an optical system transmittance of t_s the count rate is given by $F_p = LA\Omega t_s$.

This elementary photometer may be improved by adding a "field" lens after the field stop, to image the aperture stop onto the photomultiplier. The reason this is important is because, without a field lens, light from different parts of the source falls on different parts of the detector. Although the viewed volume is assumed to be of uniform radiance, a real volume may have some structure and since the detector response is not entirely uniform across its surface, the resulting signal will mix both of these effects together, generating a misleading result. For example, a movement of the structure could generate an apparent time-varying signal, which would incorrectly represent the true situation. The field lens eliminates this possibility by imaging the objective lens on the detector, producing a calibration (as is discussed later) that is fully valid with respect to the underlying assumptions.

These elementary ideas are clearly expressed in the design of the Visible Airglow Experiment [Hays *et al.*, 1973] which was flown on the Atmosphere Explorer series of satellites launched between 1973 and 1975. There are two channels, a narrow-angle and a wide-angle channel, optically independent except that they share the same filter wheel as shown in Figure 3.3. The objective lens and the field stop are readily defined in the figure. The "field lens" which images the objective lens onto the photomultiplier is split into two elements that provide a collimated beam between them that accommodates the interference filter. Most of the volume of the instrument is in the baffles, for which some explanation is required. Although in principle the lens field-stop combination determines the only light rays that reach the detector, there are in practice many ways for unwanted light to be detected. Light

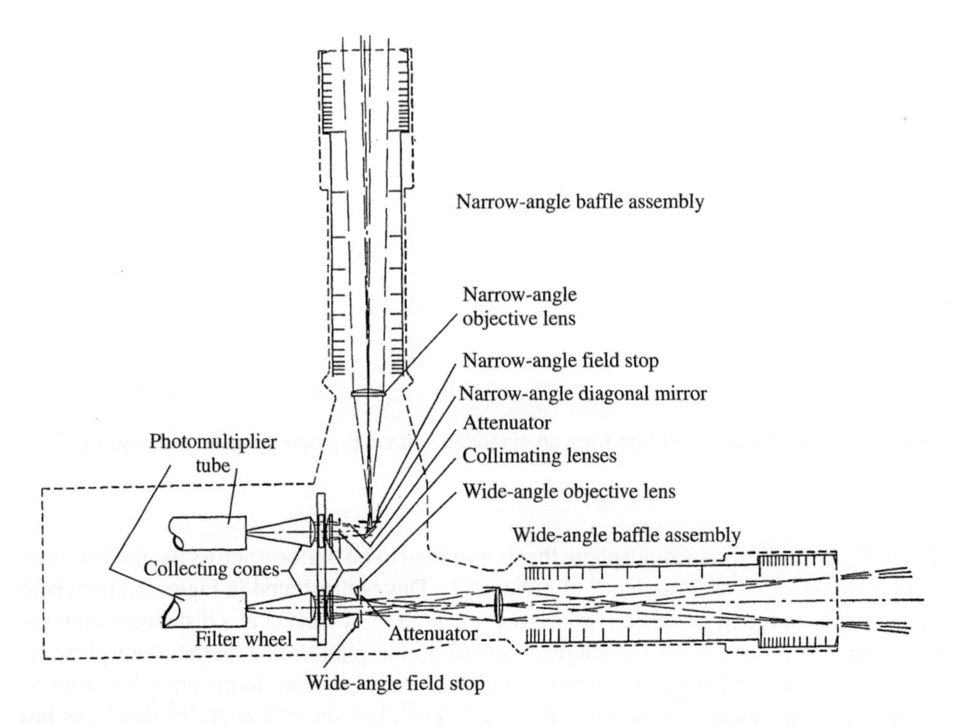

Figure 3.3. The Visible Airglow Experiment (VAE) on the Atmosphere Explorer satellite. There are two channels, a narrow-angle channel, directed vertically in this diagram, but normally at the limb while in use, and a wide-angle channel, directed horizontally in the diagram but at the zenith during observations. See text for details. After Hays *et al.* [1973].

originating outside the field of view may strike the field stop (called a stop because it is supposed to stop the light) and be reflected onto the walls of the instrument and eventually reach the detector. Light from far off the axis will strike the walls and may after reflections reach the detector. Finally, unwanted light from outside the field of view that illuminates the objective lens may be scattered off dust or other irregularities on that lens and thus enter the field of view. A good baffle can eliminate most of this unwanted light by keeping "out-of-field" light off the objective lens. In Figure 3.3 it is evident that both baffles have two sections, called an "inner baffle" and an "outer baffle". The objective lens cannot "see" the outer baffle, although the inner baffles do and the lens cannot see the walls of the inner baffle. This means that the latter can contribute stray light only after multiple reflections.

The responsivity of a spectral imaging system is defined as the ratio of the output to the input, the latter being the source radiance L. The output signal is taken to be the detector electron rate F_e in electrons s^{-1}, where the photons are converted to electrons with a quantum efficiency q ($q = F_e/F_p$ in this case) so that:

$$F_e = LA\Omega t_s q \tag{3.2}$$

and the responsivity is given by:

$$R = \frac{F_e}{L} = \frac{LA\Omega t_s q}{L} = A\Omega t_s q \quad (3.3)$$

which is dependent only on the photometer characteristics. The quantum efficiency $q = q(\lambda)$ is strongly wavelength dependent, while t_s normally varies slowly with wavelength.

3.2 THE MEASUREMENT OF IRRADIANCE

In the configuration described so far, the photometer is viewed perpendicular to the planar source, as in Figure 3.2(a), but to consider irradiance it is necessary to consider other angles as well, as illustrated in Figure 3.2(b) with a photometer viewing the source at angle θ from the normal to the surface. The surface is said to be *isotropic* if the surface is equally "bright" when viewed from different directions by eye, or with the photometer. Such surfaces are said to be *Lambertian*. While the microphysics is complex, many surfaces are closely Lambertian. For example, a piece of paper illuminated by the sun appears equally bright to the eye if the paper is rotated to different angles. In the same way the sun appears as a disk of uniform radiance even though it is a sphere, so that the angle between the viewing direction and the normal to the solar surface changes as one's point of observation moves across the sun. However, the sun is not completely Lambertian, as there is some darkening at its limb.

It has already been shown that the signal measured by the photometer is independent of the distance to the source, and for a Lambertian source it is independent of the angle θ. Since the photometer views an area larger by a factor of $\sec\theta$ than for normal viewing, this means that the energy emitted in this direction by a unit area in the plane must be smaller by a factor of $\cos\theta$. In other words it is the projected area perpendicular to the viewing direction which determines the radiance and the detected signal. The irradiance is the integral of the radiance over the hemisphere and is given by:

$$E = \int_0^{\pi/2} 2\pi L \sin\theta \cos\theta \, d\theta = \pi L \quad (3.4)$$

where $2\pi \sin\theta$ is the element of solid angle and $\cos\theta$ is the area factor discussed above. Thus for a Lambertian surface, the irradiance is just π times the radiance.

On the other hand, an airglow layer is not isotropic because the integrated emission rate I is a function of zenith angle, increasing as $\sec\theta$ because of the longer integrating path, expressed in terms of the van Rhijn factor. In general, the situation can be complex. In remote sensing of the Earth's surface, the bi-directional reflectance must be considered, i.e., the reflectance as a function of both incident and reflected angles in three-dimensional space.

3.3 RESPONSIVITY FOR LINE AND CONTINUUM SOURCES

Although the responsivity as defined above is independent of the source radiance, a complete consideration of responsivity must take into account the spectral character of the source, and the nature of the spectral response function of the instrument. In what follows, a spectral passband function is introduced and the two most frequently encountered situations are considered; the first is that of a continuum source of radiation, while the second is the responsivity for a spectral line source that is narrow with respect to the instrumental passband function. In viewing a continuum source, the monochromatic radiance is the radiance per unit spectral interval and is written $L(\lambda)$ in photons $s^{-1}\,m^{-2}\,sr^{-1}\,nm^{-1}$, and the filter is assigned a relative transmittance $t_f(\lambda - \lambda_o)$, where λ_o is the wavelength of the filter passband peak. The meaning of relative is that the transmittance is normalized to unity at the passband peak, so that the absolute transmittance there is included as part of the overall system transmittance $t_s(\lambda) = t_s t_f(\lambda - \lambda_o)$, where t_s is taken as constant over the narrow spectral range involved. It does vary slowly with wavelength over a broad spectral range, involving the optical elements for the entire instrument, including lenses, mirrors or windows. The quantum efficiency $q = q(\lambda)$ is wavelength dependent, but is usually regarded as constant over the spectral range covered by the photometer and so is taken outside the integral below. However, both t_s and $q(\lambda)$ can be kept inside the integral where appropriate. The electron signal F_{eC} for this continuum source when this is not done is given by:

$$F_{eC} = A\Omega t_s q \int_0^{\infty} L(\lambda) t_f(\lambda - \lambda_o)\, d\lambda \tag{3.5}$$

This is illustrated in Figure 3.4, which displays a portion of the spectrum lying between 500 and 510 nm, where $L(\lambda)$ is represented by a flat continuum spectrum and $t_f(\lambda - \lambda_o)$ is shown as a Gaussian filter with a full width at half-maximum of 2 nm and $\lambda_o = 503$ nm.

If $L(\lambda) = L(\lambda_o)$ is constant over the bandwidth of the filter, as is the case with a narrow band photometer, the radiance can also be taken outside the integral: the responsivity can

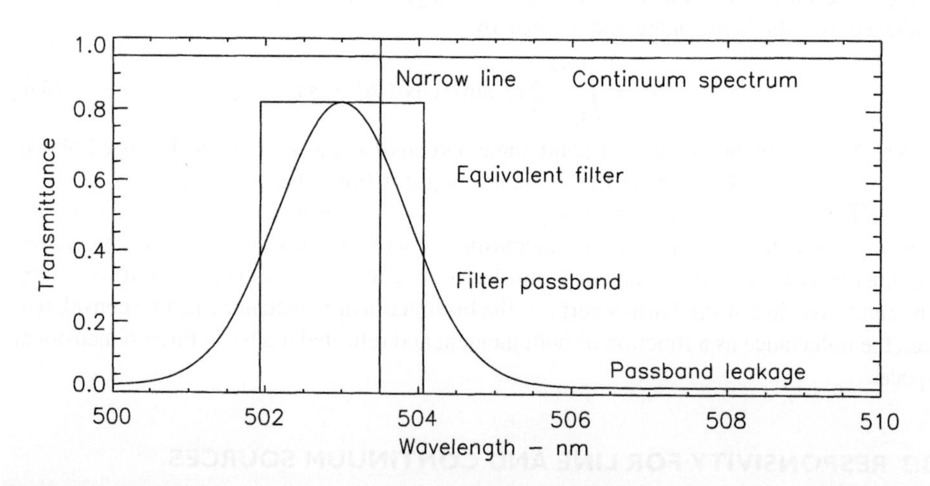

Figure 3.4. Showing a continuum spectrum, a narrow line spectrum, a filter passband with (exaggerated) passband leakage and the equivalent rectangular filter.

be written as follows, which is dependent only on the photometer characteristics.

$$F_{eC} = A\Omega t_s q L(\lambda_o) \int t_f(\lambda - \lambda_o)\, d\lambda = L(\lambda_o) A\Omega t_s q a, \tag{3.6}$$

where

$$a = \int t_f\, (\lambda - \lambda_o)\, d\lambda \tag{3.7}$$

The equivalent width of the filter, a, may be thought of as the width of a rectangular filter transmitting the same radiance as the actual filter, where both the actual and equivalent filters have a peak transmittance normalized to unity. The equivalent filter passband is also shown in Figure 3.4.

The responsivity is then:

$$R_C(\lambda_o) = \frac{F_{eC}}{L(\lambda_o)} = A\Omega t_s q \int t_f(\lambda - \lambda_o)\, d\lambda = A\Omega t_s q a \tag{3.8}$$

The units of continuum responsivity are: electron photon^{-1} m^2 sr nm.

The second case is where the source is a narrow spectral line at λ_N that is represented in Figure 3.4 as a vertical line at 503.5 nm with a radiance $L(\lambda_N)$ photon s^{-1} m^{-2} sr^{-1} because all the energy is within the line. The line is narrow with respect to the filter bandwidth, yielding an electron count rate F_{eN} at the detector given by:

$$F_{eN} = L(\lambda_N) A\Omega t_s q t_f(\lambda_N) \tag{3.9}$$

where $t_f(\lambda_N)$ is the relative filter transmittance at the wavelength of the line, allowing for the possibility that the line may not be located at the filter passband peak, even though that is normally the desired condition. The responsivity is then given by:

$$R_N(\lambda_N) = A\Omega t_s q t_f(\lambda_N) \tag{3.10}$$

and the units are electron photon^{-1} m^2 sr.

One can use the above formulation in designing an optical system to meet requirements, but in general it is both difficult and unnecessary to measure A, Ω, t_s and q to the accuracy required. The responsivity used for actual atmospheric measurements is determined by calibration as described in the following section.

3.4 PHOTOMETER CALIBRATION

In most airglow observations one measures the integrated emission rates of line emissions, but to obtain a value in rayleigh, a calibration must be performed, and all absolute calibration sources have continuous spectra. The ideal calibration source is a black body, whose radiance is determined simply as a function of its temperature. Tungsten sources are an approximation to this, and have the practical advantage of being both inexpensive and having an output that is stable in time. Such secondary sources, calibrated against black body standards, are commercially available. Normally these sources are employed by illuminating a white screen of known reflectance that scatters light uniformly in all directions, i.e. a Lambertian surface. In practice one proceeds as follows. A standard calibration of known $L_S(\lambda_o)$ is first observed, obtaining a signal given by:

$$F_s = R_C(\lambda_o)L_S(\lambda_o) \tag{3.11}$$

The emission line is then observed, yielding a signal given by:

$$F_N = R_N L_N \tag{3.12}$$

Combining Eqs. (3.8), (3.10), (3.11) and (3.12), one obtains for the radiance L_N of the observed line:

$$L_N = \frac{F_N}{R_N}\frac{R_C(\lambda_o)}{F_S}L_S(\lambda_o) = \frac{F_N}{F_S}\frac{a}{t_f(\lambda_N)}L_S(\lambda_o) \tag{3.13}$$

Note that the A, Ω, t_s and q cancel, since they are common to both responsivities. Note also that the transmittance profile $t_f(\lambda - \lambda_o)$ (needed to determine a) and the transmittance $t_f(\lambda_N)$ at the emission line wavelength need be measured only in a relative sense with respect to one another, since the absolute multiplier is absorbed in t_s. In principle, this procedure is straightforward, but the accuracy depends critically on the measurement of the filter passband shape near λ_o, as well as possible leakage outside the passband. Leakage outside the passband is illustrated in an exaggerated way in Figure 3.4; typical transmittances outside the passband for a good filter are 10^{-4}. Because $t_f(\lambda - \lambda_o)$ is in general a function of incident angle and because Ω implies a range of off-axis angles, the responsivities are really weighted averages over solid angle for a particular instrument. To ensure the same weighting for the calibration and measurement the instrumental passband shape must be measured with the actual photometer to be used for atmospheric measurement.

This is done as illustrated in Figure 3.5 by placing a diffuse screen in front of the photometer, illuminated with light exiting from a scanning spectrometer that has a continuous source in front of it. The entrance slit of the spectrometer imposes a rectangular function on the spectrum, as shown in the figure. The exit slit convolves this with its own rectangular

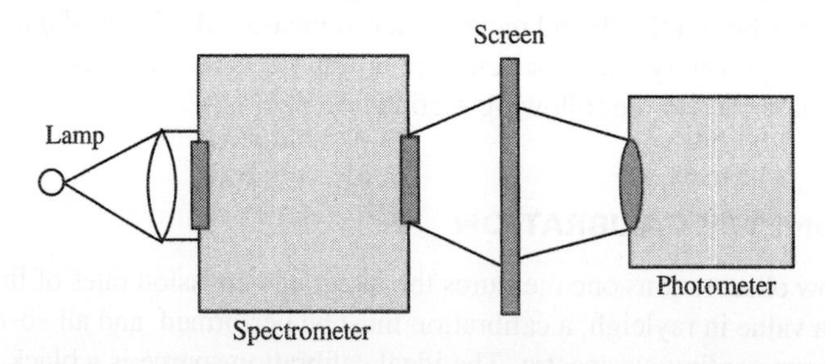

Figure 3.5. Illustrating the measurement of the instrumental passband of a spectrometer with a tungsten lamp and a scanning spectrometer. In this way the inherent filter function and the range of off-axis angles inside the photometer are properly taken into account.

function, creating a triangular instrumental passband for the entire instrument. When the spectrometer is scanned, the photometer detector records the convolution of the spectrometer instrumental passband with that of the photometer (the filter profile). In practice one ensures that the spectrometer passband width is narrow enough not to broaden the measured filter profile; otherwise a numerical deconvolution calculation is required. However, the main point of Figure 3.5 is to illustrate that by filling the field of view of the instrument with calibration light, the true instrument passband is measured, including not only the inherent filter passband, but any broadening effects introduced by the range of off-axis angles inside the photometer. By following the procedure outlined above, large fields of view can be used with no loss of calibration accuracy.

In observing thermal continuum emission from the atmosphere it is the continuum responsivity that is relevant. Here the terms involving the filter passband cancel out so that the continuum radiance $L_C(\lambda_o)$ is given by:

$$L_C(\lambda_o) = \frac{F_C}{F_S} L_S(\lambda_o) \tag{3.14}$$

3.5 GENERALIZED DEFINITION OF RESPONSIVITY

For a scanning spectrometer, an approach can be taken that is somewhat different from that for a photometer having a fixed passband. For most applications, grating spectrometers are not used at the limit of their resolving power, but at much lower resolution, in order to achieve the desired responsivity. In such cases the slit width of the spectrometer (expressed as a spectral width) may be much wider than the spectral line observed. Imagine a scanning spectrometer with an entrance and an exit slit that is scanned by rotating the grating. Here, for a specific wavelength, the exit slit is effectively scanned over the entrance slit, constituting the convolution of two rectangular functions and yielding a triangular passband as already discussed.

The spectrum of a narrow spectral line will thus appear as a triangle, having a full width at half maximum (FWHM) of a. The peak height of the triangle is given by the line responsivity which is $R_N = A\Omega t_s q$, since the peak transmittance is unity, so that the peak signal is $F_N = L_N R_N$. But if the output signal is considered to be not the peak value, but the area of the recorded spectral line, this output is then $F_N a = L_N R_N a$, which from Eqs. (3.8) and (3.10) is equal to $L_N R_C$, for $t_f = 1$. With this procedure, $R_N a$ is equal to R_C. Thus, by changing the definition of output from signal amplitude to signal area, the continuum responsivity may be used, avoiding the need for two responsivities as was described above for the simple photometer; the generalized expression for responsivity is then:

$$R = A\Omega t_s q a \tag{3.15}$$

For a grating spectrometer that records an extended spectrum, as in Figure 1.17, the instrumental passband width is displayed in the spectrum itself, for all spectral lines whose widths are much less than the passband width, which is true for all lines in this spectrum. Thus the measurement of the integral over λ is the same as multiplying the peak value (in R/Å in this case) by the equivalent width a, giving a value in rayleigh. For a photometer, whose passband is fixed, the value of a must be independently measured, as has been described.

3.6 JACQUINOT'S DEFINITION OF ÉTENDUE

3.6.1 Conservation of Étendue

From what has been presented, a key element in the responsivity is the product $A\Omega$ as first identified and named by Jacquinot [1960], the étendue. In the description of the elementary photometer it was implied that Ω was selected to be appropriate for the chosen target. But since R is proportional to Ω, one wants to make Ω as large as possible. There are two problems with a large Ω. The first is that the filter passband may be broadened more than is desired; consideration of this aspect is deferred to a later chapter. The second problem is that Ω may then be larger than the desired field of view, but this problem can be solved with a transfer telescope in front of the photometer, as shown in Figure 3.6. It can be seen that $d = i_1 f_1 = i_2 f_2$, where d is the radius of the field stop, where the i's are off-axis angles and further, that $D_1/f_1 = D_2/f_2$, where the Ds are the lens diameters and the small-angle approximation is used throughout. It follows that $\theta_1 D_1 = \theta_2 D_2$ or, in terms of area and solid angle,

$$A_1\Omega_1 = A_2\Omega_2 \tag{3.16}$$

This is a statement of conservation of étendue, which really arises from conservation of energy flux, or more fundamentally from the Second Law of Thermodynamics, since if this law could be broken, the Second Law would be violated also. In practice it means that each spectroscopic system has an étendue determined by its inherent characteristics. The area of the filter A_2 is determined largely by available technology, which in turn is reflected in its cost and the solid angle Ω_2 is determined by the nature of the dispersing element and the spectral passband width the user can tolerate (this will be made quantitative later on). The user can choose any field of view that is wanted by appropriate transfer lenses as

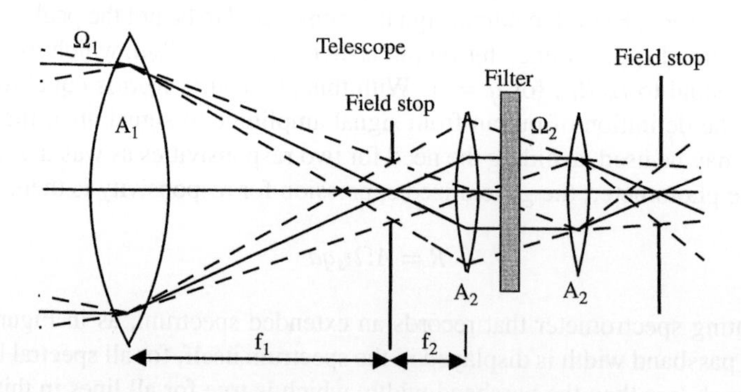

Figure 3.6. Transfer lens and conservation of étendue.

shown, without changing the étendue of the system. If a very small field of view is wanted, then this can be achieved, although the lens area may become large, to the point where it might dominate the cost of the system. Conversely, the user cannot increase the étendue of a system just by employing large fore-optics; one is limited by the filter or whatever the dispersing element may be.

The telescope employed above is called an *afocal* system, because it does not have a focal length. Parallel rays that enter correspond to parallel rays that emerge, and their off-axis angles are related through Eq. (3.16). Afocal telescopes are widely used in spectral imaging instruments.

3.6.2 Comparison of Astronomical and Atmospheric Sources

The idea of the narrow field may be extended further. Atmospheric physicists and astronomers use similar spectroscopic instruments, so they begin with the same inherent étendue. For atmospheric observations from the ground, a field of view of 10° may be fine, or perhaps 1° if structure is involved. For limb-viewing, one arc minute may be required but in any case the decision is up to the observer; it is to some extent a matter of subjective choice since as has been explained, the responsivity of the instrument doesn't depend on it. For an astronomer it might seem a matter of choice too, except that the desired field of view is small, but for a truly point object, a star, the argument is completely different. For a point source the photon flux collected depends *only* on the area of the collecting telescope, and it turns out that even for very large telescopes the associated $A\Omega$ is not larger than can be coupled to the same type of spectroscopic devices as used by atmospheric scientists. This is why astronomers build large telescopes a few meters across and why atmospheric scientists can make useful scientific observations with instruments having telescope lenses a few centimetres in diameter.

There is another class of light collectors, called *non-imaging concentrators* [Welford and Winston, 1978], which are used as solar collectors for power generation. Such devices can be used for the concentration of light onto (non-array) detectors, but they cannot be used to advantage for the spectral imaging systems of concern here.

3.7 RESOLVING POWER AND THE SUPERIORITY OF SPECTRAL IMAGERS

3.7.1 The Photometer Becomes a Spectrometer

In Chapter 2, expressions were obtained for the instrument passband functions of the diffraction grating and Fabry–Perot spectrometers. In the preceding portion of this chapter it was shown how the étendue $A\Omega$ determines the responsivity of the instrument. Without an input telescope this Ω determines the field of view, but it was also shown that an input telescope could be used to obtain any desired field of view, independently of the Ω inside the instrument. What then fundamentally determines the solid angle Ω? It is in fact the instrument passband width $d\lambda$, and the relationship between Ω and $d\lambda$ is explored in the remainder of this chapter.

3.7.2 Dispersion and Resolving Power for a Diffraction Grating Spectrometer

The key to the way in which Ω and $d\lambda$ are related concerns the way in which wavelength varies with changes in the angle of the ray passing through the device. This characteristic is known as dispersion, defined by $di/d\lambda$ where i is the angle of incidence into the spectrometer. If the wavelength varies linearly with the off-axis angle of the ray, the instrument is said to have linear dispersion; that is, $di/d\lambda$ is constant, or nearly so. The dispersion is critically important because, for a given spectral passband $d\lambda$, the dispersion determines the angular range di of the light that is accepted by the instrument. In one sense a perfect instrument would have infinite dispersion in that all light passing through it would have the same wavelength, $d\lambda = 0$, and a solid angle of 2π steradians could be collected. But how then could a spectrum be extracted? Some other variable would have to be changed. Looking at it from a different perspective, the changing wavelength with angle can be used to extract light of different wavelengths without internal scanning. There are many ways of doing this as shall be seen.

Spectral information may be acquired by sliding a spectral window (a filter) over a certain spectral range. Such a scanning filter is normally called a *spectrometer*; it has a finite spectral width, and scans the spectrum in a way that the wavelength varies more or less linearly with time. If the output signal from the detector is plotted versus time, the result is a close approximation to the original spectrum. It isn't the true spectrum, partly because of noise but also because of the finite spectral width of the instrument which is convolved with the spectrum. The important point about a spectrometer is that the detector receives only the light in the spectral element selected at that time. The spectrometer has an appealing simplicity because its output so closely resembles the spectrum, but the classical instrument with a single detector has an important drawback too, in that all of the light contained in the spectrum that falls outside the spectral passband at any instant of time is being wasted; it is blocked inside the instrument and does not reach the detector.

Because the dispersion $di/d\lambda$ is roughly fixed for a given instrument the responsivity is reduced as the spectral bandwidth is decreased. To incorporate this concept quantitatively the spectral resolution is defined in terms of *resolving power* $\mathcal{R}$, the ratio of the wavelength to the spectral passband width, namely $\mathcal{R} = \lambda/d\lambda$. In the next section this concept is applied to the diffraction grating spectrometer as introduced in Chapter 2. The resolving power of the grating in Chapter 2 was the ultimate resolving power determined by the grating alone. In the section following the resolving power is determined by the width of the slit.

To make the photometer of Figure 3.1 into a spectrometer, the filter element is replaced with a diffraction grating as shown in Figure 2.10, the field stop becomes the exit slit, and an adjacent slit is provided as the entrance slit, where the input radiance enters. The geometry of these two slits, the lens, and the angle of the diffraction grating determine the incident and diffracted angles with respect to the grating normal. Arrangements for actual instruments are described in Chapter 11.

3.7.3 Superiority of the Diffraction Grating Spectrometer

The defining equation for a grating spectrometer is:

$$m\lambda = t(\sin i_1 + \sin i_2) \tag{3.17}$$

where m is the order of interference, t is the distance between adjacent rulings of the grating, and i_1 and i_2 are the angles of incidence and diffraction on and off the grating respectively. If i_1 and i_2 are on the same side of the normal then both angles are positive. If they are on opposite sides of the normal then i_2 is negative. For a so-called Littrow arrangement, as described in Chapter 2, where the incoming and outgoing beams are roughly anti-parallel along the same path and the wavelength is changed by rotating the grating, the grating equation can be written from Eq. (3.17) as:

$$m\lambda = 2t \sin i \tag{3.18}$$

$$m\frac{d\lambda}{di} = 2t \cos i \tag{3.19}$$

The reciprocal of the dispersion, $d\lambda/di$ appears here, from which:

$$di = \frac{m\, d\lambda}{2t \cos i} = \frac{m}{2t \cos i} \times \frac{d\lambda}{\lambda} \times \frac{2t \sin i}{m} = \frac{\tan i}{\mathcal{R}} \tag{3.20}$$

where $\mathcal{R}$ is the resolving power. Now the grating looks into a solid angle Ω which is slit-like, with an angular width di and height β. Hence the product of the solid angle and the resolving power can be written:

$$\Omega\mathcal{R} = \beta \tan i \tag{3.21}$$

Thus the quantity $\Omega\mathcal{R}$ is a constant for this particular instrument. Increasing Ω in order to increase the responsivity, reduces the resolving power. This has a simple interpretation for the diffraction grating spectrometer; halving the width of the slit decreases Ω by a factor of two, and increases $\mathcal{R}$ by the same factor. Jacquinot [1954, 1960] recognized that the quantity $R\mathcal{R}$ was a valid way of comparing the performance of different instruments, which he called the *luminosity resolving power product*; since he used the term luminosity in place of responsivity. Others such as Hernandez [1986a] have continued the use of this expression, using the acronym LRP, but the luminosity $A\Omega t_s$ is arbitrary in that the area A is defined only for a specific instrument, and the same is true of the transmissivity (responsivity is $A\Omega t_s q$ which includes the quantum efficiency). When Jacquinot compared two different instruments he assumed the same luminosity and transmissivity for both. To avoid this ambiguity but at the same time to retain his profound discovery, it is suggested that the name *superiority* be used for $S = \Omega\mathcal{R}$, since it is a dimensionless quantity whose absolute value does determine the superiority of one ideal spectral imager with respect to another. Indeed, Jacquinot [1960] used the term superiority in comparing different instruments, although he did not give the term a specific definition. Thus the superiority of the grating spectrometer, S_G is given simply by $S_G = \beta \tan i$. Where attention is paid to using long slits, reasonable values are $\beta = 0.2$ rad and $\tan i = 0.57$, giving a value $S_G = 0.11$. If an instrumental passband width of 1 nm is desired at 500 nm then $\mathcal{R} = 500$ and $\Omega = 2.2 \times 10^{-4}$ sr. Since the slit length is 0.2 rad the corresponding slit width is 1.1×10^{-3} rad, or 0.063°.

3.7.4 Comparison of Superiority for the Diffraction Grating and Fabry–Perot Spectrometers

In Section 2.13 it was shown that for normally incident light on a Fabry–Perot etalon, the wavenumber of a passband peak for normally incident light is given by $2nt\sigma_o = m$, where m is the order of interference. In Chapter 5 it is shown that for off-axis angle θ inside the etalon the wavenumber of the peak changes to $2nt\sigma \cos\theta = m$. Combining these two equations and writing the approximation that $\cos\theta = 1 - \theta^2/2$, the result is:

$$\sigma - \sigma_o = \sigma\theta^2/2 \tag{3.22}$$

Further approximating Snell's law as $i = n\theta$, where i is the angle of incidence outside the etalon, the variation of wavenumber with incident angle is given by:

$$\sigma - \sigma_o = \sigma i^2/2n^2 \tag{3.23}$$

Since an etalon has circular symmetry, the corresponding solid angle is a cone of angle i. In the more general case, the solid angle corresponding to a spectral width $\sigma_1 - \sigma_2$ may be obtained from Eq. (3.23) where σ_1 and σ_2 are approximated as σ_o, with the result:

$$\sigma_1 - \sigma_2 = \sigma_o\frac{i_1^2}{2n^2} - \sigma_o\frac{i_2^2}{2n^2} = \sigma_o\frac{i_1 + i_2}{2n^2} \times (i_1 - i_2) = \sigma_o\frac{\Omega}{2\pi n^2} \tag{3.24}$$

where Ω is the solid angle of a hollow cone of angular width $i_1 - i_2$ and circumference $\pi(i_1 + i_2)$.

Recalling the definition of resolving power: $\mathcal{R} = \sigma/d\sigma$, and writing $\sigma_1 - \sigma_2 = d\sigma$, the superiority $\Omega\mathcal{R}$ of the planar FPS may be expressed as:

$$S_{FPS} = \Omega\mathcal{R} = \Omega\frac{\sigma}{d\sigma} = 2\pi n^2 \tag{3.25}$$

a remarkably simple result, especially for the most common case of the air-spaced etalon, for which $n = 1$ and the superiority is simply 2π. The air-spaced Fabry–Perot spectrometer can thus be compared to the grating spectrometer having the same parameters as in the previous section, as follows:

$$\frac{S_{FPS}}{S_G} = \frac{2\pi}{\beta \tan i} = \frac{2\pi}{0.2 \times 0.57} = 55 \tag{3.26}$$

This is a substantial advantage, and it was this which provided the incentive for the development of these interferometric devices. For solid etalons there is an additional factor of n^2 so that the gain is more than 100; this is what gives interference filter photometers such large responsivities.

3.8 DISPERSION, CLASSIFICATION AND NOMENCLATURE

It was indicated above from Eq. (3.19) that the diffraction grating spectrometer has linear dispersion, while for the Fabry–Perot etalon the wavelength shift varies as the square of the off-axis angle, as shown by Eq. (3.23). These devices may be called quadratic, and there are others of still higher order.

The spectrometers that have been discussed use just one of three modes of spectral observation. This mode consists of scanning a spectral passband (a filter) over a certain spectral range. Only one spectral element is recorded at a time. In the second mode, multiplexing, all of the light entering the device reaches the detector all of the time, but the different wavelengths are coded so that they may be distinguished from one another, and so extracted from the output signal. Fourier transformation is the most common method of coding, but Hadamard codes have also been used. The third device is the modulator in which all of the light from all of the wavelengths reaches the detector at all times, but one spectral passband is modulated so that it can be distinguished from the others. Thus it operates somewhat like a spectrometer in that no de-coding process is required.

Within this framework, in which the instrument operating modes are described as spectrometers, multiplexers or modulators, all spectroscopic devices can be now be ordered into a two-dimensional array as shown in Table 3.1, where the names of a number of instruments have been classified according to operating mode and order of dispersion. All of these instruments will be encountered in subsequent chapters.

In this table WAMI stands for Wide Angle Michelson Interferometer and SISAM for Spectromètre Interférentiel a Sélection par l'Amplitude de Modulation. As each device is explored in succeeding chapters, its superiority is evaluated. Roughly speaking, the least superior devices are found in the upper left corner of the table, and the most superior are in the lower right.

From an historical perspective, the main devices listed in the table: prisms, gratings, Fabry–Perot etalons and Michelson interferometers, were all discovered before the turn of the century. It was Pierre Jacquinot's Laboratoire Aimé Cotton in the Centre National de la Recherche Scientifique (CNRS) in Paris that re-invented these devices as photoelectric instruments after the Second World War [Jacquinot, 1948, 1954]. That caused a kind of revolution in interferometric spectroscopy, and while many of the new ideas such as the photoelectric Fabry–Perot spectrometer, the spherical Fabry–Perot spectrometer and the SISAM came from this laboratory, these developments triggered a worldwide burst of activity in interferometric spectroscopy. More recently, a new revolution is under way, caused by the advent of imaging detectors, and the full implementation of that is now being exploited.

A few final words about terminology are in order, as there is no existing uniform usage. As noted earlier Jacquinot [1960] introduced the word "étendue" to describe $A\Omega$, and this is

Table 3.1. Classification of optical spectroscopic devices

Order of dispersion	Spectrometer	Multiplexer	Modulator
Linear	Prism, Grating	Hadamard	Grille
Quadratic	Fabry–Perot	Michelson	SISAM
Higher order	Spherical Fabry–Perot	Scanning WAMI	WAMI

Table 3.2. Summary of nomenclature relevant to spectral imaging

Quantity	$A\Omega t_s q$	$A\Omega$	$A\Omega t_s q\mathcal{R}$	$\Omega\mathcal{R}$
French	Luminosity	Étendue	Luminosity resolving power product	
English	Light gathering power	Light grasp		
American	Throughput	Geometrical factor	LRP	
Recommended	Responsivity	Étendue		Superiority

sometimes translated as "light grasp". In energetic particle detector technology, $A\Omega$ is often called the "geometric factor". Jacquinot used the word "luminosity" to describe $A\Omega t_s$ which is often referred to as "throughput" in the English language. The "responsivity" as defined above is a radiance response, as suggested by Shepherd [1974]. The terms étendue and responsivity are used in this work, and in addition the term superiority has been introduced. The situation is summarized in Table 3.2.

3.9 PROBLEMS

3.1. A photometer has an interference filter with a Gaussian shape, and a half-width (full width between the points of half the maximum transmittance) of 1.2 nm. The centre wavelength is 558.0 nm. It views a calibrated white light screen of radiance 2.3×10^{13} photon $m^{-2}\ s^{-1}\ sr^{-1}\ nm^{-1}$ and records a signal of 5.4×10^3 el s^{-1}. Viewing the sky vertically at night, the photometer yields a signal of 32 el s^{-1} for the 557.7 nm atomic oxygen green airglow line.
(a) Determine the equivalent width of the filter.
(b) Calculate the vertically integrated emission rate of the night sky green line for this signal, in rayleigh.

3.2. The photometer of Problem 3.1 has an input aperture of 5.0 cm diameter, and an input field of view (full angle) of 10.0°. Its lens has a focal length of 15 cm. It is desired that the field of view be reduced to 1°.
(a) What is the diameter of the telescope lens required, and its focal length?
(b) For the same airglow conditions as in Problem 3.1, determine the photometer signal with the reduced field of view.

3.3. A tungsten lamp, which may be approximated by a point source, radiates 1.5 W nm^{-1} in some defined wavelength band in all directions, and illuminates a white Lambertian surface having a reflectance of 0.87 at an angle of 45° from the normal to the surface and from a distance of 1.8 m.
(a) Calculate the irradiance onto the screen.
(b) Calculate the radiance detected by a photometer viewing the surface at 45° on the other side of the normal to the surface.
(c) Calculate the irradiance associated with the scattered light from the screen.

3.4. A diffraction grating is used at a wavelength of 450 nm, an incident angle of 45° and an angular slit length of 0.15 rad.

(a) If an instrumental passband width of 0.1 nm is required, determine the slit width in radians.

(b) The grating has dimensions of 10 cm × 10 cm. Using its projected area and the calculated Ω, a system transmittance of 0.1 and a quantum efficiency of 0.05 calculate the count rate when viewing the daytime sky having a radiance of 10^{13} photons s^{-1} m^{-2} sr^{-1} $(cm^{-1})^{-1}$.

3.5. Given what has been presented in this chapter, the only way to increase the superiority of a diffraction grating spectrometer is through an increase in $\tan i$, which means increasing the physical size of the grating as well. Consider a grating angle of 75°. Such a configuration is called an echelle, and is normally used at high order.

(a) Calculate the gain in superiority over that provided by the instrument in Problem 3.4.

(b) For a typical diffraction grating ruled at 600 lines mm^{-1}, determine the order for a wavelength of 450 nm, for light returned approximately in the opposite direction to the incident light. For that order, calculate the angle of the refracted light with the normal to the grating.

(c) Calculate the dimensions of the diffraction grating in order that the projected area is the same as for Problem 3.4.

3.6. (a) If an air-spaced Fabry–Perot spectrometer is to be used with a spectral passband width of 0.1 nm as in Problem 3.4, calculate the angular radius of its field of view.

(b) Calculate the ratio of this solid angle to that of the grating spectrometer of Problem 3.1.

(c) Confirm the result of (b) using Eq. (3.26) with the relevant parameters.

(d) Repeat (b) and (c) for the case where the etalon is of glass with a refractive index of 1.5.

4

IMAGING CONCEPTS

4.1 ELEMENTARY DETECTORS AND NOISE

In the previous chapters, it was shown how to construct an elementary photometer and how to generate a spectral passband with a diffraction grating or a Fabry–Perot spectrometer. A formulation was presented for the calculation of the output signal of such an instrument for a given input radiance but this did not provide any information about the precision of the measurement, which depends on the noise. Detectors and noise are discussed in this chapter, first for a single photomultiplier detector and then for visible and infrared array detectors. After describing the relevant detectors, imaging concepts are presented and applied to selected imaging systems, mostly satellite instruments.

Photomultipliers operate on the basis of photoemission from surfaces. Einstein received his Nobel prize for recognizing that the energy of a photon striking a surface could be absorbed in a single electron, and if the photon energy is greater than the amount of energy required to remove the electron from the surface (called the work function), then the electron, called a photoelectron, appears in the vacuum environment inside the device, with the excess as kinetic energy. This electron is attracted to a positively charged dynode, from which a photoelectron current is drawn. For most atmospheric observations the current is very small, too small for the electronic vacuum tube amplifiers available at the time when these photoelectric devices were introduced. So the *photomultiplier* was conceived which introduces a "gain" factor in the electron current by using multiple dynodes. The photoelectron is accelerated in an electric field before striking a dynode, and thus has enough energy to remove two or three electrons from the dynode surface. These released electrons are similarly accelerated to a second dynode, where each of the electrons releases two or three more. After about ten stages of amplification, many thousands of electrons are available at the final dynode – the anode, for each initial photon, providing a readily measurable, though still small, current.

The surface from which the photoelectrons are released is called the photocathode. While a metallic surface can provide photoelectrons, the work function is so large that only ultraviolet photons have enough energy for their removal. Thus the cathode is coated with some material having a lower work function, allowing the detection of light in the visible region, and even extending into the near infrared. However, the photoelectron removal process is not 100% efficient, and while the photocathode materials are selected for the highest possible efficiency, it is difficult to achieve an efficiency of more than about 15%. For the near infrared the *quantum efficiency* may be only a few percent.

Photomultipliers may be operated in the analog mode at high light levels, in which the output signal is the anode current in amperes. This signal is proportional to the irradiance on the photocathode. For low light level work they are normally used in a photoelectron counting mode, in which each burst of electrons arriving at the anode, associated with the production of one photoelectron at the photocathode, is simply counted with a pulse counter. It is the internal gain of the photomultiplier as described above that makes this possible. But incoming photons are not the only source of electrons in the device. Electrons in any of the dynodes (or photocathode), may randomly acquire sufficient thermal energy to escape from the material, and then be accelerated to produce an output pulse that is counted along with the photon pulses. Clearly, a smaller work function enhances this process, so in making near infrared measurements where the work functions are lower, the thermal contribution, called the *dark current* because it is the value measured in a dark environment, becomes more dominant. Since electrons released from an intermediate dynode are amplified less by the gain process, a pulse height discrimination process is used to select only those pulses with a height above some selected threshold. Modest cooling (−20°C or so) reduces the dark count of thermally released photoelectrons as well, so that dark count rates of about 1 count s^{-1}, are feasible, which is essentially negligible.

A photomultiplier is thus a device that counts single photons and has negligible noise contribution from the thermal background, but this does not mean that the signal has no noise. Rather it operates in the "photon noise limited region", where noise comes simply from the random arrival rate of the photons on the detector. Here random means that the arrival has a Poisson distribution, which can be characterized by a mean, and a standard deviation. If a "constant" irradiance is presented to the detector, and photons are counted for a 1 s duration and this is repeated many times, the individual values will not be the same, but a mean and standard deviation can be determined. It can be shown from statistical theory and demonstrated with measurements of this type that the standard deviation in the measurement of N photoelectrons is simply $\sqrt{N}$ (the time required to collect them is irrelevant). By accumulating sufficient counts one can always reach the desired accuracy; 10 000 counts for 1% error, for example. Because of this, it is the quantum efficiency of the photocathode that determines the signal-to-noise ratio and not the gain. This type of noise is called *photon noise*, or *shot noise*. As noted earlier, quantum efficiencies vary from $< 1\%$ to 20%, depending on the spectral region. The gain has no direct influence on the noise level but is merely a technical convenience. With modern integrated electronics capable of measuring small currents, or pulses of few electrons, this gain is not as important as it once was.

In the infrared, the situation is different; for two reasons. First of all, because of the low energies of the photons, the thermal noise can dominate over the photon noise. Since the thermal noise depends on the area of the detector (as well as the temperature), and not on the signal level, different considerations apply. Secondly, the thermal infrared radiation from optical elements within the instrument contributes to the measured signal and thus the dark current. For a significant dark current, the associated noise adds to the photon noise, increasing the overall noise level.Therefore infrared detectors may be photon noise limited, intrinsic-thermal noise limited or background-thermal noise limited. If these noise contributions can be reduced to some level, the detector could even be readout-noise limited. For a thorough review of detectors, the reader may consult Vincent [1990] or Bose [1995].

4.2 SCANNING SATELLITE IMAGER

4.2.1 Overview

A scanning imager is a photometer that obtains an image by scanning its field of view across the source in a *raster-type* scan, the same line-by-line scan as is used in a television display. One of the applications for which scanning imagers have been effectively used is from satellite platforms. Because all satellites that are not geostationary move with respect to the earth, the observing time must be taken into account. Suppose that the satellite is located at altitude z above the Earth, as in Figure 4.1. The solid angle of view is then $\Omega = p^2/z^2$ for a target of dimension p on the ground. For a satellite of rotation period T, where the photometer field of view is perpendicular to the spin axis, the observation time Δt is the time required for the field-of-view to be displaced by p, given by:

$$\Delta t = Tp/2\pi z \tag{4.1}$$

The responsivity of a photometer was earlier defined as the output divided by the input, where the output was defined as the number of photoelectrons per second. Here the output signal must correspond to the number of photoelectrons collected in one sample time, namely Δt; so that $N = RL\Delta t$, where, as before, R is the responsivity and L the radiance, or $N = LA\Omega t_s q \Delta t$. In what follows, only the variable terms are considered, just the product $\Omega\, \Delta t$, which is given by:

$$\Omega \Delta t = Tp^3/2\pi z^3 \tag{4.2}$$

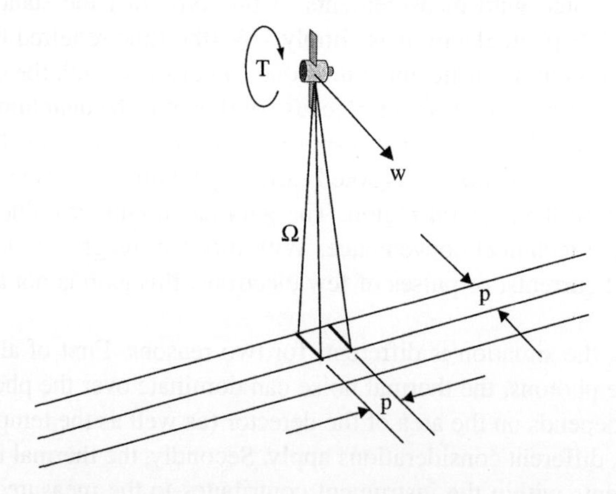

Figure 4.1. Scanning with a satellite photometer.

To achieve *cross-track scanning* the spacecraft spin axis is located in the plane of the orbit, and is parallel to the earth's surface underneath the satellite. After one rotation the satellite has moved forward in its orbit by an amount wT, where w is the spacecraft velocity, and it is about to scan another strip. If the requirement is made that one strip be exactly adjacent to the next, then it is necessary to have $T = p/w$. Substituting for T in the above the result is:

$$\Omega \Delta t = p^4/2\pi w z^3 \tag{4.3}$$

Thus the responsivity varies as the fourth power of the spatial resolution element p, introducing a severe penalty where high resolution is required. This type of imager also has the disadvantage that only one image is obtained per orbit. Nevertheless, the first imagers to take global images of the atmosphere were of this type.

4.2.2 The ISIS-II Satellite Imagers

The first satellite imagers to publish global views of the aurora were flown on the ISIS-II satellite, the last of the Canadian Alouette/ISIS (International Satellites for Ionospheric Studies) satellites, launched April 1, 1971. The spin scan concept, as applied here, originated with Dr. C.D. Anger who employed it in the *Auroral Scanning Photometer* (ASP) to acquire monochromatic images of the atomic oxygen 557.7 nm emission and the molecular nitrogen 391.4 nm emission [Anger *et al.*, 1973]. A companion instrument to measure the atomic oxygen 630 nm emission was developed by Shepherd *et al.* [1973]; it was called the *Red Line Photometer*, or RLP. The latter operated exactly as described in the previous section. The ISIS-II satellite had a circular orbit at a height of 1400 km, and the RLP had an angular field of view of 2.5°, corresponding to a dimension p of 60 km on the ground. For a spacecraft velocity of 7 km s^{-1}, $\Omega \Delta t$ is 1.07×10^{-4} sterad s. Using an aperture area of 12.9 cm^2, a system transmittance of 0.4, a quantum efficiency of 3% and assuming that the line is at the filter peak, a responsivity of $R_s = 1.66 \times 10^{-5}$ electron photon^{-1} cm^2 sr s is obtained. For a source of one rayleigh, the signal is 1.32 electrons.

The instrument was designed to detect sources as weak as 10 R, and auroras of a few hundred rayleighs were easy to detect. A schematic diagram is shown in Figure 4.2; the instrument had two oppositely directed inputs, one of 1 nm bandwidth for the signal, and one of 10 nm bandwidth for the background; the light from both inputs fell simultaneously on the detector. Since one field was directed to the dark sky while the other viewed the ground, the satellite rotation acted as a switch. In the same way, the ASP obtained alternate images of the $O(^1S)$557.7 nm and N_2^+ 391.4 nm emissions.

An example of results obtained for the RLP is shown in Figure 4.3. The data are shown as contours in geomagnetic coordinates with noon at the top. Here, at about 78° magnetic latitude, the solar wind low energy electrons penetrate to ionospheric levels through the magnetospheric *cusp*, preferentially exciting the $O(^1D)$ emission. During this magnetically quiet period, with little midnight auroral activity, the dayside red aurora dominates the global pattern. The hatched line is the spacecraft track.

The ASP had an additional feature; its photomultiplier was what is called an image disector, in which the sensitive area of the photocathode could be scanned up and down.

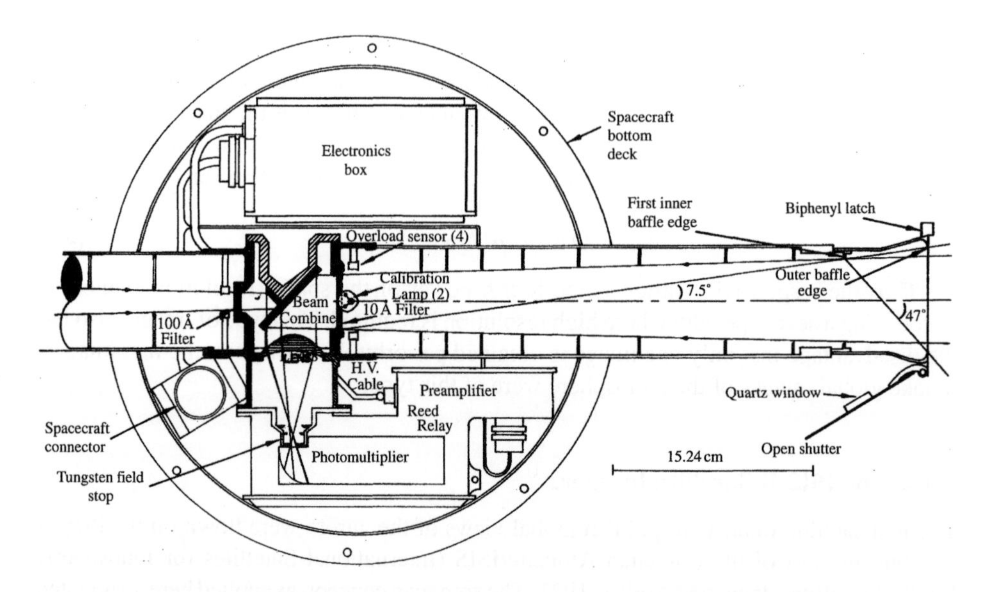

Figure 4.2. The ISIS-II Red Line Photometer. After Shepherd *et al.* [1973].

That allowed the basic scan strip to be further subdivided into vertical strips of 13 pixels; the target size was 13 km. Further details can be found in the paper by Anger *et al.* [1973]. An example of an image obtained is shown in Figure 4.4, and further examples of images along with a list of references can be found in the paper by Anger and Murphree [1976].

For downward viewing in the visible region from satellites, one observes not just the upgoing atmospheric emission, but the atmospheric emission reflected from the Earth as well. Over land cover and the oceans, the reflectance (called the albedo) is small, but over clouds or snow cover it can be large and must be accounted for in quantitative measurements; for aurora the auroral light scattered from the ground can smear the directly viewed auroral forms. The relevant aspects are described by Hays and Anger [1978]; they showed that under certain conditions the total radiance could be three times that of the emission itself.

4.2.3 Dynamics Explorer-1 Imager

The Dynamics Explorer-1 (DE-1) imager [Frank *et al.*, 1981] is also a scanning photometer but uses a cross-scan mirror to uncouple the satellite spin from the orbital motion. This allows the acquisition of multiple images per pass. The field of view is 0.32°, the scan step 0.25° and the scan width 30° so that dividing 0.25 into 30, 120 rotations are required for one image. With a period of rotation of 6 s the time to acquire one image is 12 minutes; this corresponds to a Δt of 5.3 ms. The value of $\Omega \Delta t$ is therefore 5.4×10^{-7} sr s. A schematic drawing is shown in Figure 4.5. Even though the collecting area of the optics is 20.3 cm^2, roughly twice that for the ISIS imagers, the responsivity is much lower; but it has produced a wealth of auroral images [Frank and Craven, 1988]. There were actually three separate imagers

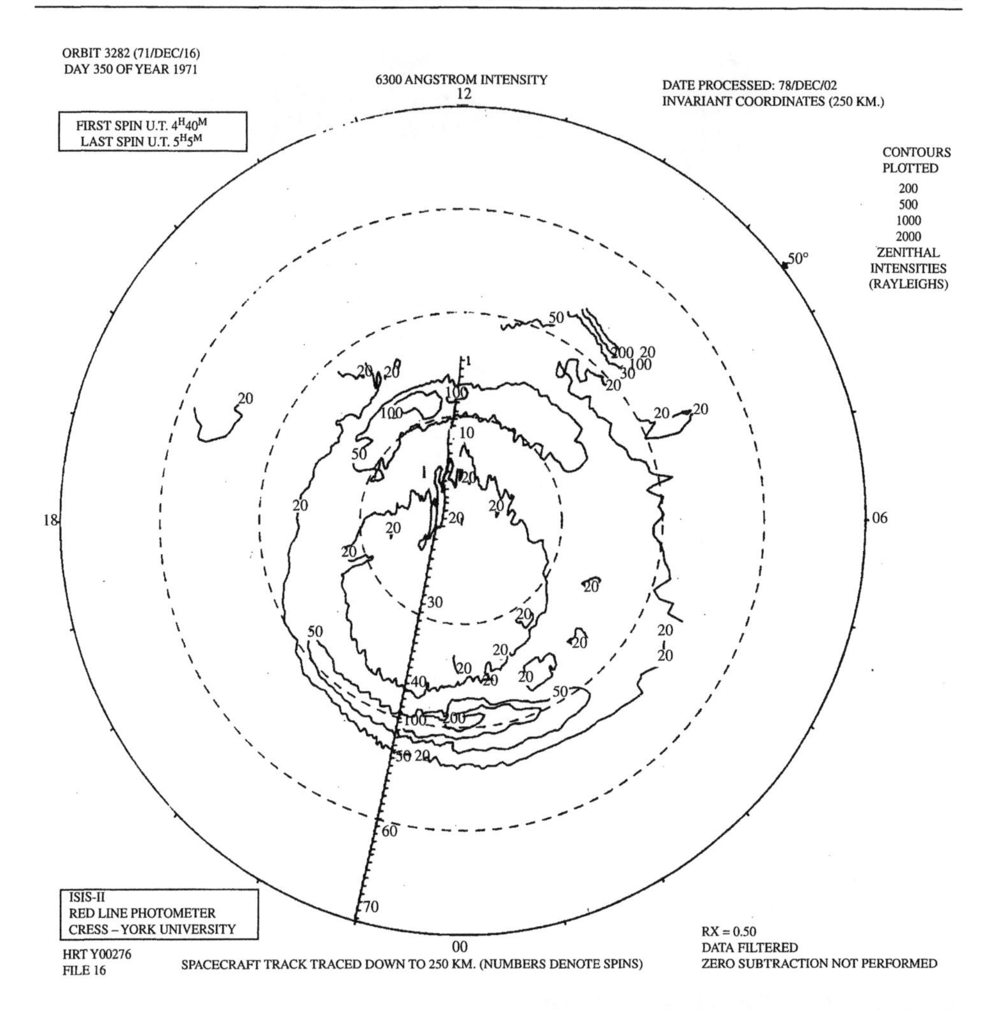

Figure 4.3. An image of the dayside aurora in the atomic oxygen 630 nm emission, obtained with the Red Line Photometer on the ISIS-II satellite. Magnetic coordinates are used with the sun at the top and midnight at the bottom. The outermost circle corresponds to 50° invariant magnetic latitude. The hatched line is the spacecraft track. Adapted from Shepherd [1979].

on board, two for the visible region and one for the ultraviolet. An example of an image sequence is shown in Figure 4.6. The most spectacular and controversial discovery made with this imager was that of *atmospheric holes* [Frank *et al.*, 1986]. These were called holes because they appeared as small dark regions against the airglow background, presumably due to absorption of light above the source of the airglow emission. The existence of these holes, proposed to be produced by the impact of small icy comets, was not confirmed by the co-investigators for the Viking ultraviolet imager, discussed later. This controversy is fascinating as it is based almost totally on the analysis of noise. Are these atmospheric

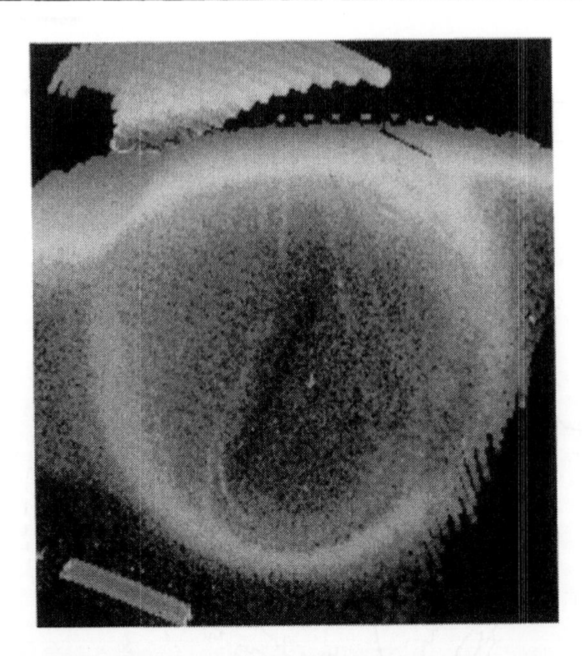

Figure 4.4. A global image of the aurora obtained with the Auroral Scanning Photometer on the ISIS-II spacecraft. Noon is at the top. The circular region defining the auroral oval is clearly evident, and the unusual looped feature inside is called a horse-collar aurora. Courtesy of L.L. Cogger (University of Calgary).

"holes" truly atmospheric features, or are they simply noise introduced by the instrument? The level of controversy shows that the analysis of noise, particularly in an imaging detector, is no simple matter.

4.3 WEATHER SATELLITE IMAGERS

The TIROS-1 (Television and InfraRed Observation Satellite) was launched on April 1, 1960, containing two vidicon television cameras, one narrow and one wide angle, and placed in a circular orbit at 644 km altitude, with an inclination of 48°. The inclination of the orbit is the angle between the orbit plane and the Earth's equatorial plane; Pease [1991] provides a discussion of satellite orbits. The cameras viewed parallel to the axis of the spin stabilized satellite, so that images were obtained only during the portion of the orbit when the cameras viewed towards the earth; it was a demonstration mission. This configuration was not changed until TIROS-9 in 1965, when the cameras were arranged to view perpendicular to the spin axis, with a 12 rpm spin rate. With the spin axis perpendicular to the orbit plane, in the so-called cartwheel configuration, images were possible all around the orbit. As the TIROS series developed and evolved into the NOAA (National Oceanic and Atmospheric Administration) series of satellites, the imaging capability was taken over by the AVHRR (Advanced Very High Resolution Radiometer). This device is a scanning

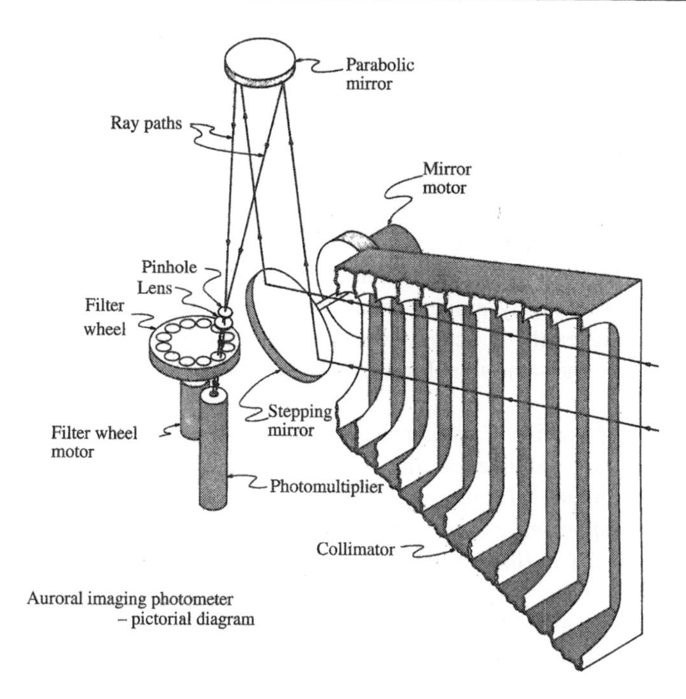

Figure 4.5. Schematic drawing of the Dynamics Explorer-1 Imager. After Frank *et al.* [1981].

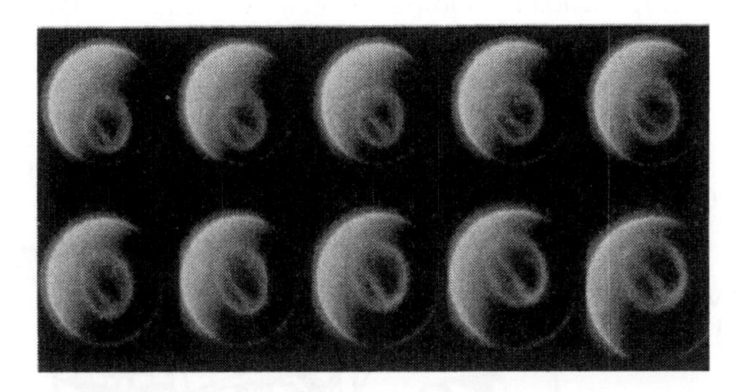

Figure 4.6. A sequence of auroral images taken with the spin-scan imager on the Dynamics Explorer-1 spacecraft. The sun is on the upper left, and the auroral oval is clearly seen to the right of the terminator. Within the auroral oval is a polar cap arc, seen drifting to the right. Courtesy of L. Frank (University of Iowa).

imager where the cross-scan (perpendicular to the direction of motion) is accomplished not with the satellite rotation, but with a scan mirror rotating at 360 rpm. The field of view is 1.3 mrad (milliradian), giving a resolution of 1.1 km at the subpoint, and secondary optics split the light into five spectral channels, using silicon detectors in the visible and near-IR. The 3.8 μm channel uses a InSb detector, while the 11 μm channels use HgCdTe detectors;

these IR detectors are cooled to 105 K. This instrument continues to be flown on the NOAA satellite series. More details are available in Rao [1990].

Another line of development, the NASA Applications Technology Satellite (ATS) program led to the NOAA Geostationary Operational Environmental Satellite (GOES) program. These satellites were based on the Spin Scan Cloudcover Camera (SSCC), making use of the rotation of the spin-stabilized geostationary satellite, combined with a scanning telescope and a photomultiplier detector. The telescope was displaced between each spin, so that an entire scene of the Earth disk was generated in 20 min, with a spatial resolution of 3.2 km. While ingenious, and relatively simple, when one remembers that the geostationary satellite is 41 000 km above the earth, the scanner views the earth for only 5% of the time, requiring a longer time to reach the same signal-to-noise ratio than would be the case for continuous viewing.

The GOES I to M series is an advance on this [Menzel and Purdom, 1994]. These satellites are three-axis stabilized so that one side continuously faces the earth, requiring two axes of scanning. There are five spectral bands: (a) 0.52–0.72 μm (visible); (b) 3.78–4.03 μm (short-wave infrared window); (c) 6.47–7.02 μm (upper-level water vapour); (d) 10.2–11.2 μm (long-wave infrared window), and (e) 11.5–12.5 μm (infrared window with slightly more water sensitivity). The resolution is 1 km for the visible channel, and 4 km for the IR channels. A schematic drawing of this imager is shown in Figure 4.7. GOES-J was launched on May 23, 1995 after which it was re-named GOES-9.

The Defense Meteorological Satellite Program (DMSP) was created and has continued in parallel with the programs described above, having objectives apparently related to lower light levels and higher resolution from low-earth orbit. A spectacular DMSP image of the global distribution of the city lights across Europe at night with the aurora to the north is

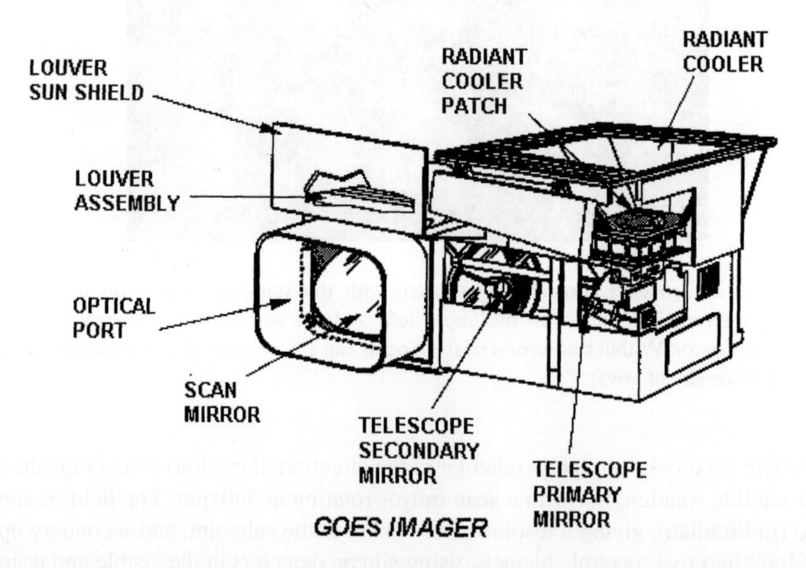

Figure 4.7. The GOES I-M Imager. Courtesy of NOAA's National Geophysical Data Center.

Figure 4.8. The city lights across Europe observed with the DMSP satellite. Image and data processing by NOAA's National Geophysical Data Center. DMSP data collected by US Air Force Weather Agency.

Figure 4.9. A Meteosat-7 image taken from above the equator at the Greenwich meridian on April 3, 2002 with the water vapour channel. Note the difference between the tropical and high latitude structures. Copyright 2002 EUMETSAT.

shown in Figure 4.8. The aurora is in segments because it is a mosaic from several orbits. Sicily is at the bottom of the image and Great Britain is near the middle left. Copenhagen and the rest of Scandinavia are near the top and the aurora is far to the north of Scandinavia; it is a relatively quiet aurora. Railway lines can be clearly seen in Russia. The image shows that the power manifested in the radiance of the aurora exceeds that of the city lights across Europe. The instrument is OLS (Operational Linescan System) and it uses a telescope and photomultiplier, with a side-to-side mirror scanner called the *whisk broom* or *pendulum* mode which, combined with the forward motion of the spacecraft, produces the required images. The orbit is circular at 830 km, and is sun synchronous. The resolution is 0.5 km.

Other countries have taken somewhat different approaches; India, Japan and Russia all have meteorological satellite programs. The European EUMETSAT organization relies on ESA (European Space Agency) to conduct its Meteosat Operational Program. These geostationary satellites are spin-stabilized, rotating at 100 rpm, and an array of detectors (250×250 silicon photodiodes) is used in the visible channel, using the satellite rotation for the East–West scan, and a stepping telescope for the North–South scan. For the IR, arrays of 70×70, HgCdTe detectors are used. The resolution is 2.5 km for the visible channel, and 5 km for the IR channels. A water vapour image taken above the equator at 0° longitude is shown in Figure 4.9; the equatorial region is well-defined by the water vapour.

4.4 INTRODUCTION TO ARRAY DETECTORS

An array detector consists of multiple photosensitive elements arranged in an orderly fashion. The elements are usually rectangular or square, and arrays are generally fabricated in two geometric shapes: linear and area. Linear arrays have a single row of elements which are laid out in a straight line or follow the circumference of a circle, as an annulus. Area arrays have elements laid out in x and y space as rows and columns, or could be sectors of the area of a circle. This is relevant to the circular interferometric fringe patterns to be discussed later, but since the application is so specialized, this format is not currently available on the market. Array detectors represent a means of obtaining vast amounts of information in the optical analog domain at rates that were never before attainable. This information can be quantified by the appropriate system electronics, and processed to provide specific results useful in many atmospheric spectral imaging investigations.

A large number of closely packed detectors presents a practical problem. How does one get the signal out of this array? The signal can only come out of a planar array in three ways: directly from the top or bottom, or along the plane at the edge(s). A jungle of individual wires and pre-amplifiers bonded to each detector element is impractical to deal with. The non-uniformity and power dissipation between the many amplifiers would cause difficulties. A solution to this problem is to use one or several amplifiers with a single common output line. Boyle and Smith [1970] of Bell Laboratories developed a clever process of moving the photoelectrons, in almost discrete packets, in a serial fashion to the output amplifier. This process is called charge coupling, performed by a charge coupled device (CCD). There were many early problems with this process such as lag or charge blurring during charge transfer. These have been largely overcome. At this point in time, charge transfer efficiencies (the

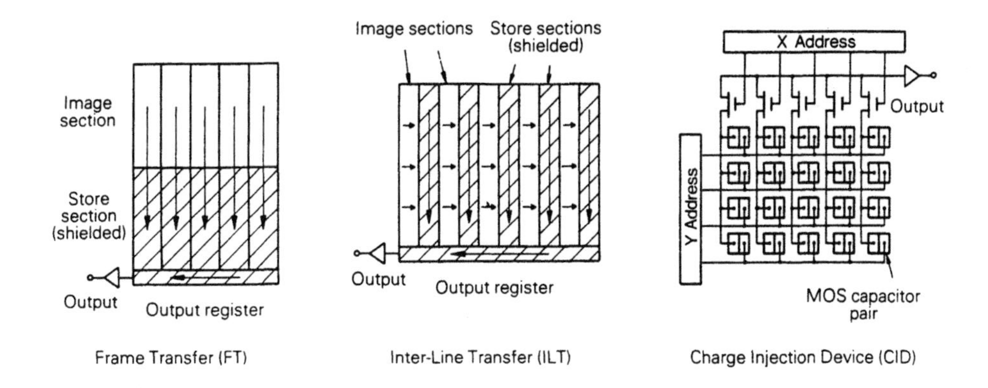

Figure 4.10. Illustrating the three main methods of array detector readout: (a) frame transfer; (b) interline transfer; and (c) x–y addressable (labelled charge injection device). Courtesy of Marconi Applied Technologies (formerly EEV), copyright.

fraction of electrons transferred to the total number existing in the well) are typically very high. Commercial devices by RCA, GEC and Fairchild appeared in 1980 with sizes roughly 400×500 pixels. By 1995 the sizes had increased to 2048×2048. In the mid 1980s infrared array detectors of 64×64 pixels appeared; Rockwell introduced their 2000×2000 device in the year 2000.

Charge coupling is not the only means of reading out the photo-generated charge. There are three main methods as illustrated in Figure 4.10. Frame transfer, as illustrated in the left panel, refers to the block movement of an array of charges from the image section to the storage section. This is a common method for the CCD detector and is described in detail in the next section. Interline transfer as shown in the middle panel involves the movement of the charges from a column of pixels into an adjacent column, from which the charges are shifted downward and read out. Frame transfer and inter-line transfer are both charge coupled techniques. The right panel illustrates the (x, y) addressable method, which is fundamentally different. This technique, using CMOS (Complementary Metal-Oxide Semiconductor) circuitry, is used both for charge injection devices and for infrared detectors. There is a grid of wires formed by one wire per column in the vertical direction and one wire per pixel row in the horizontal direction. By activating an (x, y) pair of wires, the charge from that device can be read. These alternative methods are discussed in later sections.

4.5 THE CHARGE COUPLED DEVICE (CCD) DETECTOR

4.5.1 Introduction to Semiconductors

The energy levels of atoms are changed when the atoms are bonded into a solid material. According to the band theory of solids, valence and conduction energy bands are formed

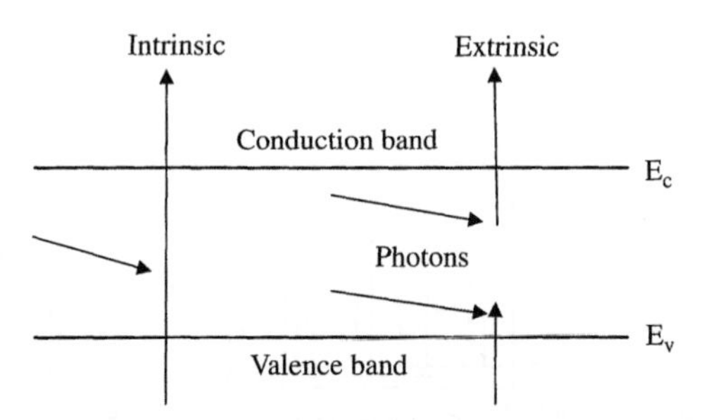

Figure 4.11. Illustrating the difference between intrinsic and extrinsic detectors. With intrinsic detectors, electrons must be raised from the valence band into the conduction band. With the extrinsic detector, electrons can be promoted from impurity sites.

with the conduction band, of higher energy, being separated from the valence band by a *band gap*. In an insulator the valence band is full of electrons and the band gap is around 10 eV, so electrons cannot move as they have no place to go. In a conductor the valence band is only half full and electrons move freely. In a semiconductor such as silicon or germanium, the band gap is about 1 eV, small enough for some electrons to be excited thermally across the gap, depending on the temperature; this is an intrinsic semiconductor. If a semiconductor material is *doped* with an impurity atom, e.g. arsenic, which has five valence electrons instead of four for silicon, then an electron is unbonded, so to speak, and the result is an n-type semiconductor. For an impurity with three valence electrons the result is electron vacancies, called *holes*; this is a p-type semiconductor. In either case it is an extrinsic semiconductor in which the doping atoms create energy levels very close to the conduction band, from which they can be easily excited, as illustrated in Figure 4.11.

4.5.2 Method of Operation

Fundamental to the charge storage and transfer operation of CCDs, is the unit cell of a MOS (Metal Oxide Semiconductor) capacitor with a silicon substrate. A positive voltage applied to the top metal electrode causes the underlying silicon to be in depletion and thus assume a positive potential which attracts negatively charged electrons. This effectively creates a deep potential well under the electrode, which is called the *channel*. The channel is laterally contained by appropriately doped inactive material forming *channel stops*, as illustrated in Figure 4.12. The number of electrons or charge that accumulate in the well during the exposure time constitute the signal. This is a method of almost discrete charge storage. It was discovered that if several of these MOS capacitors were placed close enough together, the almost discrete aggregations of charge could be transported in an orderly manner to an output gate and amplifier, if the appropriate clock potentials were applied as shown in Figure 4.13. The CCD unit cell may have a large number of metal electrodes or gates, but one to eight are common and three is typical. Each gate is driven by a separate clock and the clocks are phased such that the applied clock voltage influences the depth of the potential

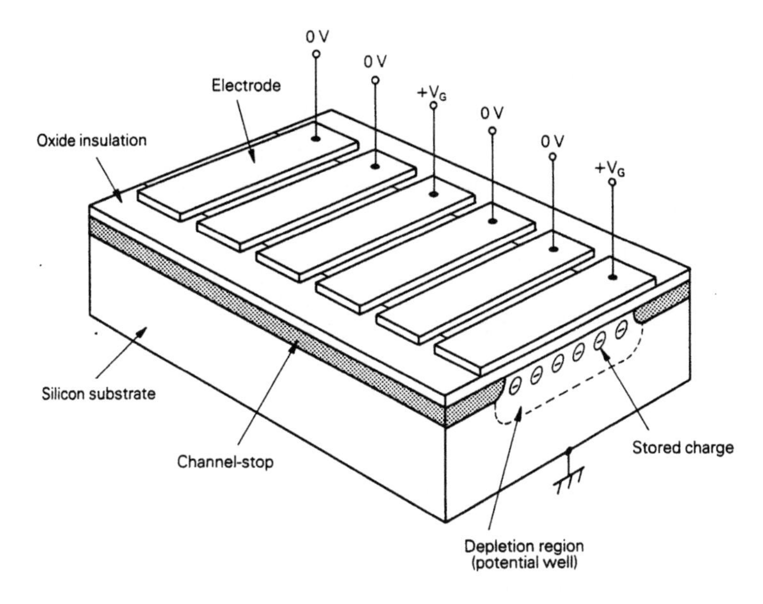

Figure 4.12. Basic structure of a CCD, showing two pixels, each of three electrodes. The depletion region and the stored charge are shown. Courtesy of Marconi Applied Technologies (formerly EEV), copyright.

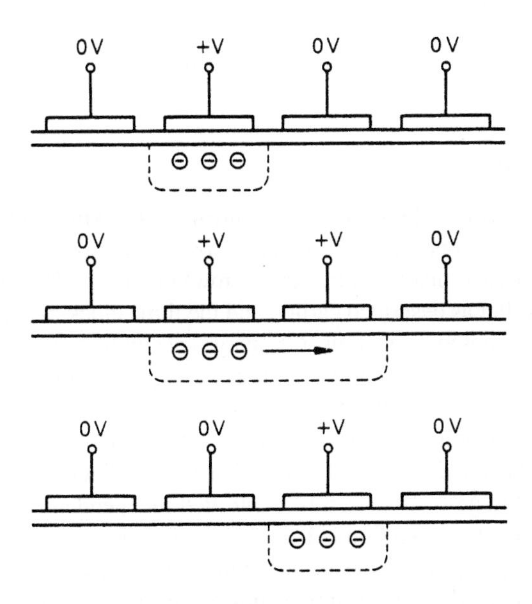

Figure 4.13. Showing how the electron charges can be shifted from one well to another, by changing the voltages on the electrodes. Courtesy of Marconi Applied Technologies (formerly EEV), copyright.

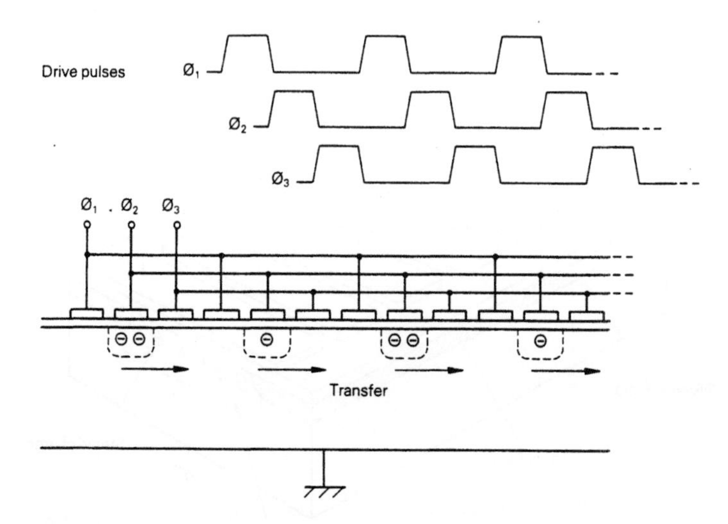

Figure 4.14. Showing how corresponding gates are connected in parallel and driven from the same voltage lines (phases). Courtesy of Marconi Applied Technologies (formerly EEV), copyright.

well under that gate. All of the gates in a functional block of unit cells are connected in parallel. Functional blocks can be storage registers, transfer buffers or readout registers. The charge packets shift through each unit cell in with respect to the cycling of each clock line. When connected in parallel, entire rows of pixels can be shifted at once, using multiple clock lines connected as shown in Figure 4.14 for a three-phase device.

4.5.3 CCD Readout

The schematic of an entire CCD is shown in Figure 4.15. Looking at the *image section* that constitutes the upper part one finds five pixels across (columns) and four vertical (rows), keeping in mind that each pixel requires three (row) elements. The elements in one row are connected in parallel as described earlier, and the three clock phases are shown. When the image section is exposed to photons, the charges accumulate in each pixel (well). The readout method shown here is that corresponding to the figure, namely *frame transfer*. Here, by rapid clocking, all of the charges in the image section are moved rapidly in parallel by rows into the storage section, which is covered so that it is not exposed to light. The frame transfer process is illustrated with the downward arrows in Figure 4.15.

The image and storage sections can be controlled independently, so a new image can be acquired while the storage section is being read out. This provides extremely efficient operation, in that a long exposure time may be required for weak signals, and a slow readout from the storage section reduces the sampling noise of the output values. The frame transfer process is very fast by comparison and probably negligible in comparison with the exposure and readout times. For reading out the storage area, the charges are shifted downwards so

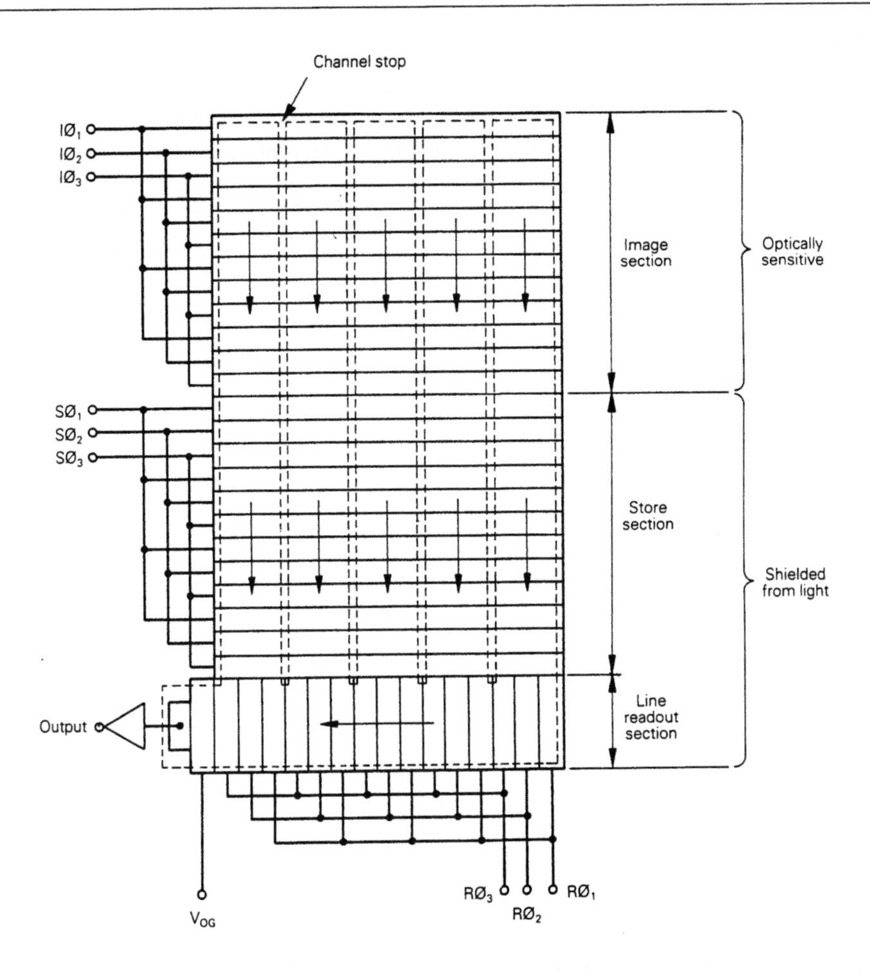

Figure 4.15. Schematic of an entire CCD of five horizontal and four vertical pixels, of three electrodes each. This is a frame transfer CCD with separate image and storage sections, along with a horizontal transfer register (labelled "line readout section" in the figure). Courtesy of Marconi Applied Technologies (formerly EEV), copyright.

that the bottom row of pixels is transferred into the *horizontal transfer register*, labelled "line readout section" in the figure, a set of five pixels, three electrodes each as before, except that the elements now are vertical rather than horizontal. In the same way, the charges are shifted horizontally out of this register, into an output register, from where the amount of charge is measured, digitized and stored.

The output charge is measured by transferring it to a capacitor and measuring the voltage across it, from $Q = CV$, where Q is the stored charge, C the capacitance and V the associated voltage. The accuracy of this process is normally enhanced by using a technique called correlated double sampling, in which the signal level is taken as the difference between the output voltage before and after charge transfer.

One extremely valuable characteristic of a CCD is the capability of binning. By shifting a row downwards without clearing the horizontal shift register, the contents of one pixel are added to the pixel below it. This can be done two, three or more times as desired; say three for example. Following this operation the horizontal shift register can be shifted three times without clearing the output register, thus accumulating the electrons of three side-by-side pixels. The result is the original image has been transformed into bins of 3×3 pixels, reducing the number of pixel elements by a factor of 9, and increasing the number of electrons in each bin. What is particularly valuable is that this "on-chip" binning is essentially noise free, because of the efficiency of charge transfer, yielding a result that is much better than would be obtained by reading out all the pixels and performing the binning in a computer. The reason is that in the latter case, the signal of each pixel has the readout noise superposed upon it, rather than on each bin.

4.5.4 CCD Characteristics

In this section some of the primary CCD characteristics are described. This is followed by an identification of the relevant noise sources, leading to a procedure for calculating the signal-to-noise ratio.

Spectral response and quantum efficiency

Quantum efficiency (q_λ) is the ratio of photogenerated carriers to the number of incident photons per pixel. To create an electron the photon must penetrate the electrode structure, so a photon transmittance must be taken into account. Photons of large wavelength penetrate more deeply into the silicon and a concentration gradient is built up within which diffusion can occur. Thus the silicon absorption coefficient, the electron diffusion length and the depth of the depletion layer must be considered. Values of 65% are readily achievable and values of 90% are attainable with anti-reflection coatings and thick depletion layers.

Charge transfer efficiency

Charge transfer efficiency (CTE) concerns the fraction of electrons moved from one pixel to the next during a transfer. For current devices the CTE is 0.99999; that is, the charge transfer inefficiency (CTI), is typically about 10^{-5}.

Dark current

Dark current is that produced by thermally generated carriers in the depletion regions (e.g., for silicon devices, between the substrate and the SiO_2 layer), which exists when the array detector is in the dark. The dark current I_D, in amperes, is represented by the diode equation given below.

$$I_D = C\, e^{-eV_{BG}/2kT} \tag{4.4}$$

where C is a constant, V_{BG} is the band gap of silicon (1.1 V), k is Boltzmann's constant, e is the electronic charge and T is absolute temperature. For silicon, cooling the device by 10°C reduces the dark current by at least a factor of two, so it is very sensitive to temperature. Dark current shot noise is equal to the square-root of the number of electrons in the dark signal, just as for photon noise, so that where η_D is the dark noise for an accumulated dark count N_D:

$$\eta_D = \sqrt{N_D} \tag{4.5}$$

Dynamic range

The full well capacity or saturation level of the pixel defines the electron charge holding capacity of the pixel potential well. The value depends on fabrication materials and size of the pixel. This value, combined with the dark current, determines the dynamic range of the measurement.

$$\text{dynamic range} = \frac{\text{well capacity}}{\text{dark current}} \tag{4.6}$$

Since the well capacity is measured in electrons and the dark current in electrons s^{-1}, the dynamic range is in seconds and thus is the exposure time over which the dark current alone will fill the well.

Photon noise

The noise η_{ph} in electrons associated with the accumulation of N_{ph} photoelectrons is as described earlier:

$$\eta_{ph} = \sqrt{N_{ph}} \tag{4.7}$$

Readout noise

When the pixels, possibly binned, are read out of the horizontal transfer register and the charge determined, and converted to a voltage, there is an associated noise that is signal independent. This is called the readout noise, η_R. The first commercial CCDs had readout noises in the range of 20–80 electrons rms, but current devices have readout noises of 4–6 electrons.

4.5.5 Signal-to-Noise Ratio

Based on the preceding information, an expression for signal-to-noise ratio (SNR) can be written, allowing for the binning of b pixels into each bin, as follows. The three noise components are combined as their quadrature sums.

$$SNR = \frac{F_e bt}{\sqrt{(F_e bt) + bN_D + \eta_R^2}} \tag{4.8}$$

where F_e is the pixel signal in electrons s^{-1}, t is the integration time, N_D is the dark current signal per pixel, which is the square of the noise, and η_R is the readout noise.

4.6 SPECTRAL RESPONSE AND MATERIALS

For atmospheric measurements, the following regions are important; the UV (ultraviolet), VIS (visible), SWIR, MWIR and LWIR (short, medium and long wave infrared respectively). Physically, the spectral response is limited by the energy gap of the semiconductor materials. Spectral response is determined by measuring the responsivity as a function of wavelength.

The desired spectral range of a particular application dictates the type of material used in an array. Detector materials are sensitive to wavelengths that can provide sufficient energy to excite a valence band carrier into the conduction band. For intrinsic detectors, this is defined by the band gap energy as shown in Figure 4.11, for extrinsic detectors by the doped impurity energy level, as illustrated in the same figure. Some of the materials currently used are discussed below.

The physics of silicon and its fabrication technology are well understood. Traditionally the spectral response of silicon devices is 0.4–1.1 μm. To avoid the transmittance loss of the electrode structure the light can be introduced from the back of the CCD. Such *back-side illuminated* silicon CCD array detectors, specially treated and thinned, have shown good sensitivity in the soft X-ray, and UV regions, but the silicon substrate must be thinned to make this effective. The expanded spectral range of silicon is now expected to be from 0.1 nm to 1.1 μm. In addition, CCDs are being used to directly detect electrons with energies of less than 500 eV. Between 0.9 and 1.1 μm, the absorption efficiency of silicon is low, resulting in low quantum efficiencies. Currently the best choice for this region appears to be InGaAs, which has good quantum efficiency to 1.7 μm and can be tuned to reach 2.5 μm.

At longer wavelengths, HgCdTe (also called MCT, Mercury Cadmium Telluride) is a versatile material since its cutoff wavelength is tunable over a spectral range of 1–30 μm. This is achieved by changing the molar concentration x of the cations according to the formulation Hg$(1-x)$Cd(x)Te. Since the cut-off wavelength is related to the energy band gap, this can be more precisely described by the mixture of CdTe, a semiconductor with a band gap of 1.16 eV and HgTe, a metal with a band gap of −0.302 eV. Here the conduction band and the valence band form an indirect gap.

As a detector array material, HgCdTe has some important qualities. It is intrinsic and thus operates at higher temperatures than extrinsic materials within the same spectral range. Thinner detectors can be made with HgCdTe since it has a larger photon capture cross-section or absorptivity than extrinsic materials, by about 50 : 1. Current devices of 2000 × 2000 pixels are possible, with a quantum efficiency of 50%, readout noise of 40 electrons rms and dark currents of 1 electron s^{-1} at 77 K.

4.7 CONSIDERATIONS SPECIFIC TO INFRARED ARRAY DETECTORS

4.7.1 Background Radiation

As discussed in Chapter 1, radiation is emitted from all surfaces having a finite temperature. This means that for an infrared instrument its walls are emitting radiation of the same wavelength as that of the atmospheric source. While radiation from the walls can be kept out of the beam, the optical elements such as lenses, filters etc. are all sources of radiation. This can be a dominant noise source unless the elements are cooled, or have low emissivity. The result is that radiation from this background creates a signal F_B at the detector. Equation (4.8) then must be modified as follows, where F_B is the thermal background electron rate per pixel, calculated from the summation of the radiance of all optical elements in the beam:

$$SNR = \frac{F_e bt}{\sqrt{(bt(F_B + F_e)) + bN_D + \eta_R^2}} \quad (4.9)$$

4.7.2 Infrared Detector Readout

Infrared devices operating in the 1–5 μm region are intrinsic photovoltaic devices operating in the capacitive discharge mode (CDM) as described by Finger [Philip *et al.*, 1995]. The electric field is that of a p–n junction. This is inherently non-linear, but one can operate in a linear regime if the photon generated current dominates the dark current and if only a fraction of the full well capacity is used.

Infrared detectors are not read out in the same way as for CCDs. Rather, the device is made in two parts, one used for photon detection and the other for readout as described above. The "upper" unit is composed of infrared-sensitive material which is divided into a grid structure by the construction of tiny junctions. The electrons are collected at the pixel location by applied electric fields. The depleted region then behaves as a capacitor which becomes progressively discharged as the charges accumulate. The lower unit is made of silicon, with a matching grid of unit cells in which the infrared detector is connected to the input gate of a source follower, with a "reset" FET (Field Effect Transistor) connected to the same node as described by McLean [Philip *et al.*, 1995]. The two units are connected with columns of indium, called *bumps*. The output from each unit cell is connected to an output bus and an output source follower by a MOS switch which is addressed by a system of shift registers, i.e. a multiplexer which can address each pixel in turn; that is, are (x, y) addressable. Thus the operation of the device is similar to that for a CCD, using clocked lines, level shifters, low-noise preamplifiers and analog-to-digital (A/D) converters. Because of the source follower, the readout is non-destructive and because of the MOS switches, the array is randomly addressable. Rockwell has developed NICMOS (Near Infrared Camera and Multi-Object Spectrometer) arrays for the Hubble Space Telescope; these are 2.5 μm cut-off MCT detectors optimized for low-readout noise. They have small full well capacities and slow readout rates as a result.

4.8 OTHER TYPES OF ARRAY DETECTORS

A variety of array detectors is available from a number of manufacturers. This section describes a few of the more common classes of array architectures.

4.8.1 Photodiode Arrays

Silicon photodiode arrays made with diffused p–n junction photodiodes, each with a capacitor and MOS multiplex switch, may use either of two readout schemes. Serial readout is facilitated with either a digital or an analog charge coupled device (CCD) shift register. Light integrated on the photodiodes is converted to charge, which is stored on the capacitor. The stored charge is proportional to the exposure (irradiance multiplied by the integration time) provided the accumulated signal remains below saturation. Readout consists of sequential biasing of the gates of individual MOS switches by the digital shift register, thereby placing the charge from each diode on a common video line, or block transfer of all charges from all diodes into the analog CCD shift register where they are clocked out sequentially. The analog output signal represents the irradiance distributed across the array during the integration period. The charge coupled photodiode arrays can provide effective data rates of 128 MHz. The silicon photodiodes have moderate dark currents and noise, which do not make them useful in low light level applications. Even with noise values of 500–1500 electrons these arrays have very large dynamic ranges on the order of 1000 : 1–16 000 : 1. The spectral range of these devices is about 200–1100 μm. Uniformity of response for these linear arrays is about ±7%.

4.8.2 Charge Injection Devices

The charge-injection device (CID), also a silicon based detector, uses different mechanisms from the CCD. The device is an x-y addressing type sensor with a pair of adjacent MOS capacitors at each pixel. Charge is accumulated in the pair of MOS capacitors at each sensing site and, during readout, is injected into the substrate. The resultant charge displacement is sensed by the CID. A schematic of a CID is shown in Figure 4.10. CID detectors are suitable for moderate to high light level applications. The CID has a low quantum efficiency (20–35%) and displays image lag. However, the dynamic range is large, the signal to noise ratio is high and the resolution capabilities are high.

The CID has non-destructive readout (NDRO) capability. The charge in a pixel can be reread many times because the CID readout process involves intracell charge transfer. The temporal read noise is greatly diminished when many successive NDROs are averaged. The CID architecture also allows random access to any pixel.

4.8.3 Intensifiers

Intensifiers are devices which create electron gain, as in the photomultiplier, but for an array detector. They may be used to minimize the effects of dark current and readout noise, but one then has to introduce a photocathode, which reduces the quantum efficiency. Photons generated in the photosensitive surface, which can be in any spectral range, are multiplied in the intensifier, normally a Micro Channel Plate (MCP) as illustrated in Figure 4.16, and allowed to fall on a visible region photosurface, within the spectral range of the CCD. This is particularly useful in the UV. At low light levels, with rapid readout, the intensified CCD can become a photon counting system. One can show that there is a threshold value of accumulated photons, below which the intensified CCD is the better choice. Above this level, the bare CCD is preferable.

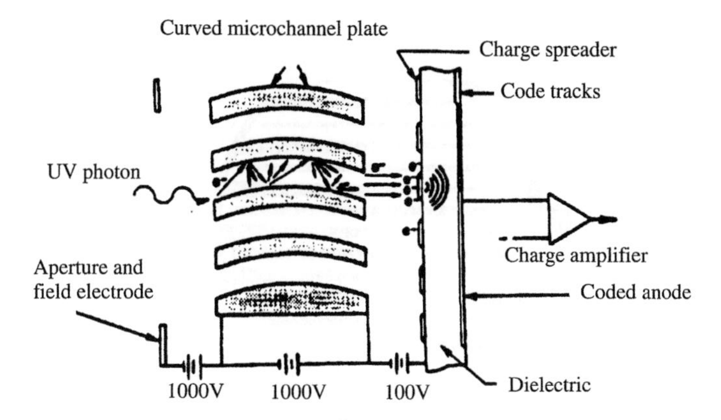

Figure 4.16. A schematic drawing of a microchannel plate with three curved channels. The electron multiplication takes place on the walls of the channels because of the potential difference along the channel. Channels may be straight, V-shaped (chevron) or curved as shown here. After McClintock *et al.* [1982].

4.8.4 Position Sensitive Arrays

There is a class of array detector that may be called *position sensitive*. These are devices which detect one electron at a time, and which determine the location on the area detector on which the electron landed. By recording these locations for a period of time, an image is built up. These devices are normally used with intensifiers, having gains of about 10^6. The first type is the *resistive anode* [McWhirter *et al.*, 1982], which is most readily explained for a one-dimensional array. Suppose that the array consists of a conductive (i.e. resistive) strip with readout electronics at each end, and the charge cloud of 10^6 electrons lands in the middle. An equal amount of charge diffuses in both directions and the signals recorded at each are equal. If the charge lands one-quarter of the way from one end, then the signal at that end will be larger than at the other; the ratio of the two signals yields the position on the strip on which the electrons landed. For a two-dimensional detector the signals are measured at the four corners. The wedge and strip detector [Siegmund *et al.*, 1984, 1986] operates on a similar principle except that the anode pattern is in the form of alternating wedges and strips. All of the wedges are connected into one anode, the strips into another, and the area behind constitutes a third anode, as shown in Figure 4.17. In the downward direction on the figure, the widths of the strips increase linearly with distance. The widths of the wedges increase linearly from left to right. The anode is placed about 15 mm behind the MCP, allowing the multiplied cascade of electrons to spread into a cloud that is large enough to register on all three electrodes, regardless of where it impacts. From the ratios of the charge collected on the three anodes, the impact location can be determined. As for all such devices, the charge must dissipate before the next electron arrives.

The CODACON (Coded Anode Converter) device is described by McClintock *et al.* [1982] where an electron landing on the microchannel plate produces an electron cloud which impacts on the rectangular grid of wires, called charge spreaders, in each of the

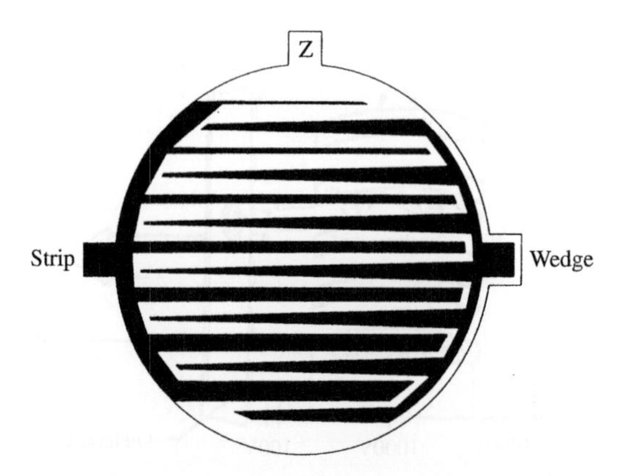

Figure 4.17. Schematic drawing of a three-element wedge and strip anode. The widths of the wedges and the strips vary linearly with distance in the x and y directions. After Siegmund *et al.* [1986].

x and y directions. Above the array are N conducting strips called code tracks. Each strip has two parts, one wide and one narrow, which are codes for the binary bits 0 and 1. The N code strips have 2^N codes so that the wire on which the electrons land can be identified by its N-bit code. Similarly for the horizontal direction, so that the x-y position is uniquely defined by the corresponding codes. Each wire requires a separate amplifier, but this is not excessive as 10 amplifiers are adequate for a 1024 pixel array. These devices have been used mostly in the ultraviolet for low-light-level applications, where the photon energies are large, and the dark count can be low, 2–3 electrons s^{-1}. Count rates up to 10^5 are feasible.

4.9 EARLY ARRAY DETECTOR IMAGERS

4.9.1 Elementary Imagers

Imagers are as old as the first photographic cameras which took, among other things, very beautiful pictures of tropospheric clouds. Film cameras were first employed in a systematic way to globally image the visible region non-thermal atmosphere when, during the International Geophysical Year of 1957/58, it was realized that a step towards global imaging of the aurora could be achieved by a widespread network of cheap, simple, reliable all-sky cameras. These involved a convex mirror with a vertical axis and a movie camera looking down onto it from above to take pictures of the entire sky. The movie camera was slowed to rates of about one image per minute. They were distributed in fairly dense networks over North America, Scandinavia, the U.S.S.R, and Antarctica. The results led to the concept of an auroral oval fixed in the framework of the sun, under which the Earth rotated [Feldstein, 1964]. Akasofu [1963] recognized that this pattern was subject to temporal variations, specifically the global dynamics of the *auroral substorm* which produces the spectacular displays which fortunate readers will have witnessed. Photographic film was

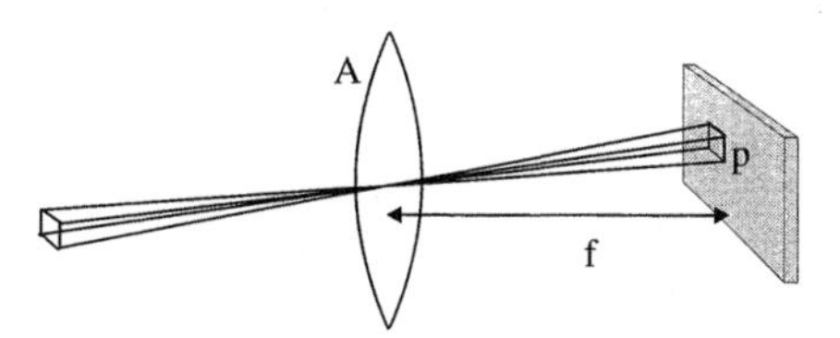

Figure 4.18. A simple imager of pixel dimensions p, lens area A and diameter D.

first used to image gravity waves in the hydroxyl airglow emission by Petersen and Kieffaber [1973] and later by Clairemidi *et al.* [1985]. Photographic film was first replaced by video tape for aurora by Davis [1969], which was more sensitive and convenient, but still lacked the ability to make accurate quantitative measurements. Since then, video systems have been replaced by array detectors as just described, which provide truly quantitative imaging. For wide-angle systems the distortion of the simple convex mirror has been eliminated by modern fish-eye lenses. An example of an airglow image recorded with such a system was shown in Chapter 1.

An imaging system involves considerations which are different from the photometer described in Chapter 3; it consists of a lens which can be very simple or very complex, followed by an array detector, illustrated with a simple lens in Figure 4.18. Each pixel of side p has an area of p^2, so the solid angle subtended is $\Omega = p^2/f^2$, where f is the focal length of the collecting lens having an area A. The étendue of the pixel is:

$$A\Omega = \frac{Ap^2}{f^2} = \frac{\pi D^2 p^2}{4f^2} = \pi \frac{p^2}{4}\left(\frac{D}{f}\right)^2 \tag{4.10}$$

where D is the diameter of the aperture in front of the lens. But f/D is what photographers call the f/number, in the sense that an $f/5.6$ aperture corresponds to $f/D = 5.6$, so the étendue of the pixel is determined solely by the f/number of the lens once the CCD has been selected. This suggests that a search for a responsive imager involves a fast lens and large pixels. Fortunately, one of the advantages of the array detector is that its binning can be adjusted during operation to change the effective pixel (bin) size, and so dynamically adjust the response, though at a loss of spatial resolution.

4.9.2 Spectral Imagers

Monochromatic imagers involve placing an interference filter in the light path at some convenient point. This must be done with care, because the transmittance of an interference filter (a type of Fabry–Perot etalon) is a function of off-axis angle as was discussed in Chapter 2. The goal of minimizing the angular spread through the filter leads to the telecentric system [Mende *et al.*, 1977]. An example of a modern ground-based all-sky camera system is shown in Figure 4.19. The spectral selection is through the use of a *filter-wheel*, a rotatable wheel containing a number of interference filters around its periphery.

Ground-based auroral imaging has gone in several directions. One of these is to achieve the highest possible spatial resolution; Trondsen and Cogger [1998] have recorded auroral features smaller than 100 m. They also have studied what is called *black aurora*, voids in

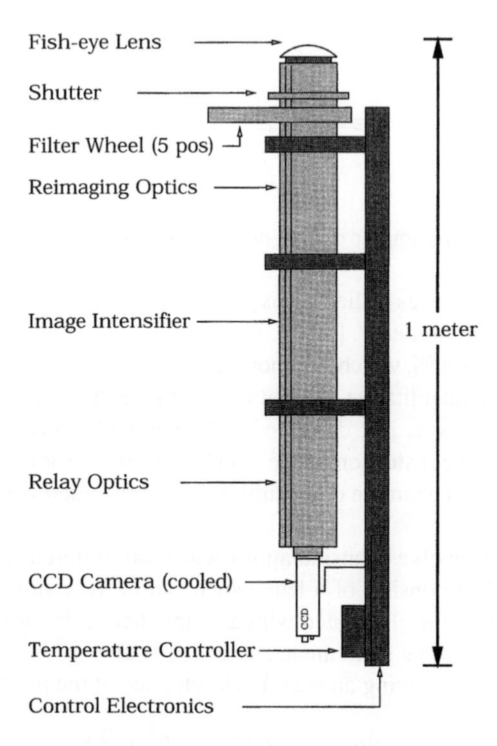

Figure 4.19. Schematic drawing of a modern all-sky camera. Courtesy of T.S. Trondsen (University of Calgary).

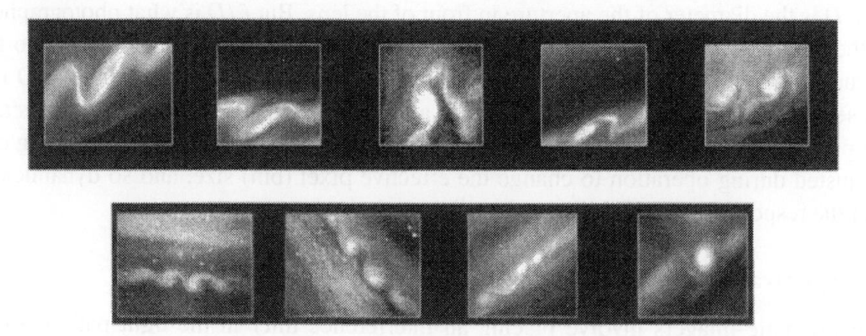

Figure 4.20. Narrow field auroral image taken with a CCD imager and an exposure time of 17 ms, showing auroral structures smaller than 100 m. Auroral arcs (above) and auroral curls (below). Courtesy of T.S. Trondsen and L.L. Cogger (University of Calgary).

the aurora [Trondsen and Cogger, 1997]. Examples of narrow-field images are shown in Figure 4.20; these images were taken with an exposure time of 17 ms. Another direction is to use lines of multiple cameras in communication with a central-site computer so as to achieve real-time tomography. Such a system called ALIS (Auroral Large Imaging System) was pioneered by Åke Steen in Sweden [Aso *et al.*, 2000].

4.9.3 The KYOKKO Auroral Imager

The first true satellite auroral imager was an ultraviolet imager flown by the Institute for Space and Aeronautical Science (ISAS) of Japan on the KYOKKO satellite [Kaneda *et al.*, 1981], and was launched in 1978. It used a television-type vidicon similar to that of the early TIROS satellites, in a specially designed camera with a UV sensitive photocathode. The imaging camera used a MgF_2 window and a KBr photocathode to define the spectral passband; it imaged the auroral oval from a satellite stabilized with a bar magnet to point along the Earth's magnetic field. One image was obtained every two minutes with a field of view of 60°. Few results were published from this mission, but more results were obtained using the same technique with the subsequent Akebono mission [Oguti *et al.*, 1990; Kaneda and Yamamoto, 1991].

4.10 CCD SATELLITE IMAGERS

4.10.1 The Viking Ultra Violet Imager (UVI)

The Swedish Viking satellite was launched on February 22, 1986, into a polar orbit with an apogee (the largest radial distance from the Earth) of 15 000 km above the northern auroral region and a perigee (closest radial distance) of 800 km. The Canadian Ultra Violet Imager (UVI) [Anger *et al.*, 1987] on board, was designed for ultra-violet emissions to allow the imaging of auroral forms under sunlit conditions. This is possible first of all because the Rayleigh scattered sunlight is weak since the solar irradiance is low in the FUV region. Second, the FUV radiation is absorbed by the atmosphere below, so that there is no Earth reflection of the auroral light. Finally, there are strong auroral emissions in the FUV region, namely, N_2 emissions in the Lyman–Birge–Hopfield (LBH) system (132.5–150 nm) and the atomic oxygen OI (triplet 130.4 nm and doublet 135.6 nm) that may be used as source emissions.

The instrument consisted of two cameras, one to exploit the LBH system and the other the OI lines as noted above. The other important difference from the ISIS imagers was that the Viking imager obtained one image per satellite rotation, compared to one image per orbit with ISIS. This was possible because of the array detector used, as described below.

A schematic of the camera, employing an inverse Cassegrain optical system is shown in Figure 4.21. It is "inverse" because the light strikes the smaller primary mirror first, before being reflected from the larger secondary mirror to the spherical microchannel plate (MCP) surface. With a field of view of 25° and reflecting surfaces throughout, this compact $f/1$ system yields good responsivity in the FUV. The spherical surfaces, including the spherical MCP in the focal plane all have a common centre of curvature; this arrangement also gives excellent correction for spherical aberration while off-axis aberrations are minimized by the concentric design. The conceptual design was done by Alister Vallance Jones, and the detailed design by Harvey Richardson.

The two separate cameras have passbands from 123 to 160 nm (OI camera) and from 134 to 180 nm (LBH Camera). Their passbands were achieved by a suitable combination of filters and photocathodes: for the LBH camera; a BaF_2 filter and CsI photocathode; and for the 130.4 nm camera a CaF_2 filter, KBr photocathode and reflective coatings on the secondary

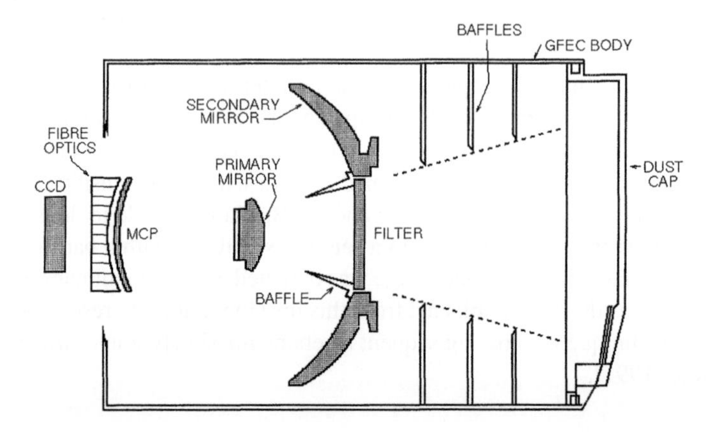

Figure 4.21. Schematic view of the Viking ultraviolet inverse Cassegrain camera. After Anger *et al.* [1987].

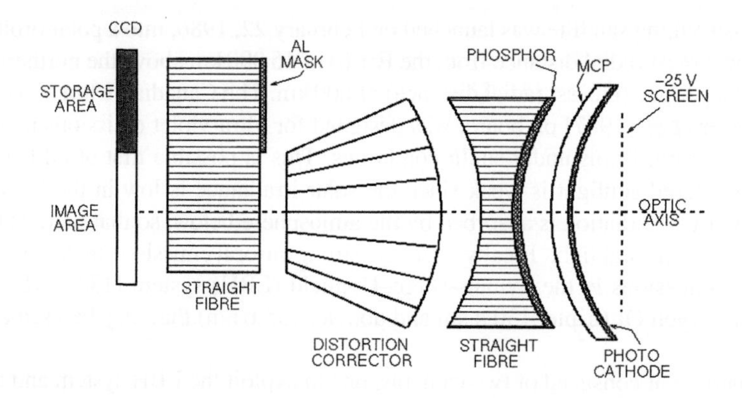

Figure 4.22. A schematic cross section of the intensifier assembly detailing the electron screen, the curved mcp and photocathode, the phosphor screen, sections of the fibre optic distortion corrector, and the CCD. After Anger *et al.* [1987].

mirror. The filters have a short wavelength cutoff with uniform transmittance above the cutoff wavelength, while the photocathodes have a relatively uniform response up to the long wavelength cutoff in quantum efficiency and the responsivity depends on the product of the two. By choosing the cutoff wavelength appropriately, the responsivity is limited to a narrow spectral region with the same effect as if a band-limiting filter had been used.

Concerning the detector system, a schematic of the intensifier assembly is shown in Figure 4.22. The photocathode material is deposited directly on the front surface of the MCP. Photoelectrons from the photocathode initiate an electron cascade in a channel of the MCP with the resulting electron image collected on a phosphor screen deposited on a concave depression in the fiber-optic faceplate. In front of the MCP is a screen held at −25 V. In combination with a +25 V potential on the front surface of the MCP, this screen

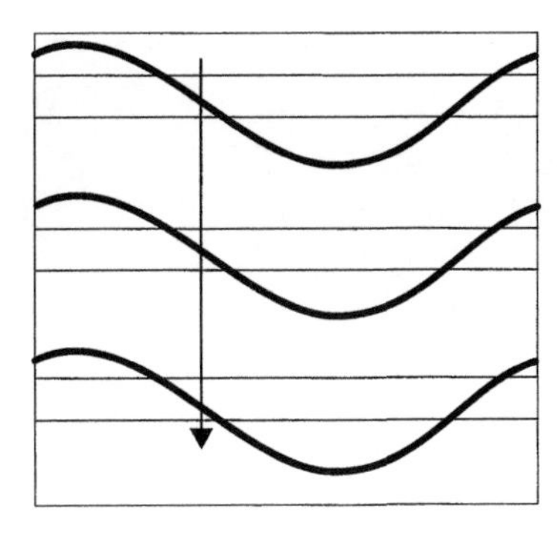

Figure 4.23. Illustrating the "Time Delay and Integrate" mode of scanning the Viking UVI.

provides a modest barrier against low-energy charged particles. The potential also pushes photoelectrons back to the MCP. Visible light from the image undergoes multiple scattering in the channels of the MCP, reducing its radiance on the detector by a factor of about 5000. Electrons from the MCP are accelerated to the phosphor by a potential difference of 3500 V and the light produced there is conducted to the CCD by the fiber-optic output faceplate. A tapered optical fiber assembly serves as a corrector to remove distortions as a result of mapping a "spherical image" onto a 2-D flat surface. When projected onto the MCP, the CCD pixels have an apparent size of 30 μm square (total of 288 × 385 pixels). An opaque layer was used to mask the storage area of the CCD. Perhaps the most innovative idea in the concept for this camera was compensation for the spin of the satellite. There was a fundamental problem involved because the satellite was spinning, whereas one would like to have the camera fixed on the image during the exposure time. A "snapshot" from a rotating satellite could have an exposure time of only a few ms. To circumvent this problem the image was effectively frozen for about 1 s electronically by stepping the charges in the CCD from row to row in synchronization with the motion of the image across the array as illustrated in Figure 4.23; this is called the *Time Delay and Integrate* (TDI) mode of operation. This provides a much greater exposure time than would be obtained with a simple spinning camera.

The camera has $f/1$ optics so that $f = 22.5\,\text{mm}$ (focal length) corresponds to $d = 22.5\,\text{mm}$ diameter. The pixel dimensions are $30 \times 30\,\mu\text{m}$, yielding an Ω of 1.8×10^{-6} sr, and with a lens area of $3.9\,\text{cm}^2$, an étendue of $7.1 \times 10^{-6}\,\text{cm}^2$ sr. The value of $\Omega\,\Delta t$ is 2.0×10^{-6} sterad s, compared to 5.4×10^{-7} for the DE-1 auroral imager and 1.07×10^{-4} for the ISIS-II RLP. Including the transmittance of the mirror pairs as $(0.8)^2$, a 30% obstruction loss and $q = 30\%$ results in 27.7 counts kR^{-1}. An auroral image obtained with the Viking UVI is shown in Figure 4.24. The auroral *oval* appears to be a circle when viewed from space; the oval designation from ground-based measurements comes from the fact that the centre of the circle is offset from the geomagnetic pole, towards the Earth's night-side. Noon is at the top, the cusp that was evident in Figure 4.3 corresponds to a gap here, because these ultraviolet emissions are preferentially excited by more energetic electrons whose effect is otherwise seen over the complete range of local times. The smooth ring is formed in part by

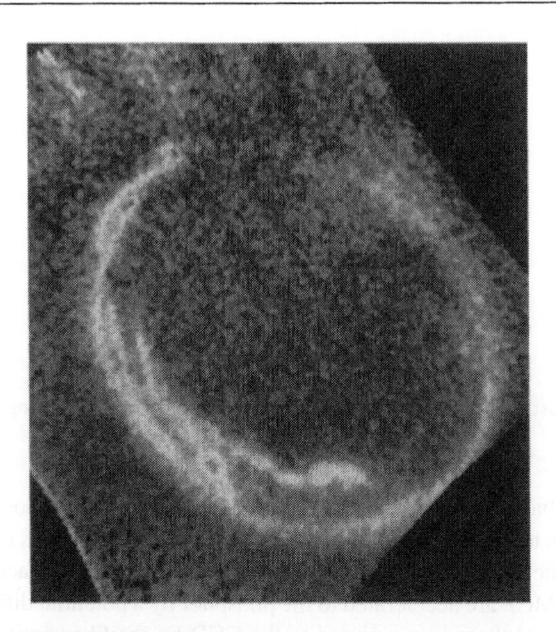

Figure 4.24. An image of the auroral oval obtained with the Viking Ultra Violet Imager. Courtesy of L.L. Cogger (University of Calgary).

quasi-trapped electrons drifting in the Earth's magnetic field, while the aurora inside this ring is caused by more impulsive electron precipitation from the magnetotail.

4.10.2 The Polar VIS Imager and the IMAGE Satellite

VIS is a pair of auroral imagers mounted on the Polar spacecraft, which is one of the ISTP/GGS International Solar Terrestrial Program/Global Geosphere System. It is a hybrid instrument that uses an area detector, a CCD, but uses a pointing mirror as well to enlarge the field; here only the medium resolution camera is described, which has a pixel size of $0.011° \times 0.013°$. This yields a field of view $2.8° \times 3.3°$ which, with the mirror pointer, can be positioned anywhere within the $20° \times 20°$ field of the collimator. A diagram of the system is shown in Figure 4.25. The instrument is mounted on a de-spin platform on the spacecraft that allows it to point to the Earth continuously. The sensors used are CCDs with image intensifiers in front that provide a gain of about 10^5. This allows single photon events to be detected by the CCDs. Details are provided by Frank *et al.* [1995].

The IMAGE satellite, launched on March 25, 2000, images not only the aurora, but the magnetosphere, the Earth's magnetic field environment, as well. This is done in two different ways. The first is to record images from the He^+ line at 30.4 nm; since this ion is a constituent of the magnetosphere, the technique produces images of the magnetosphere. Previously most of the information about the magnetosphere came from *in situ* measurements of energetic particles, but building up an image of the magnetosphere on

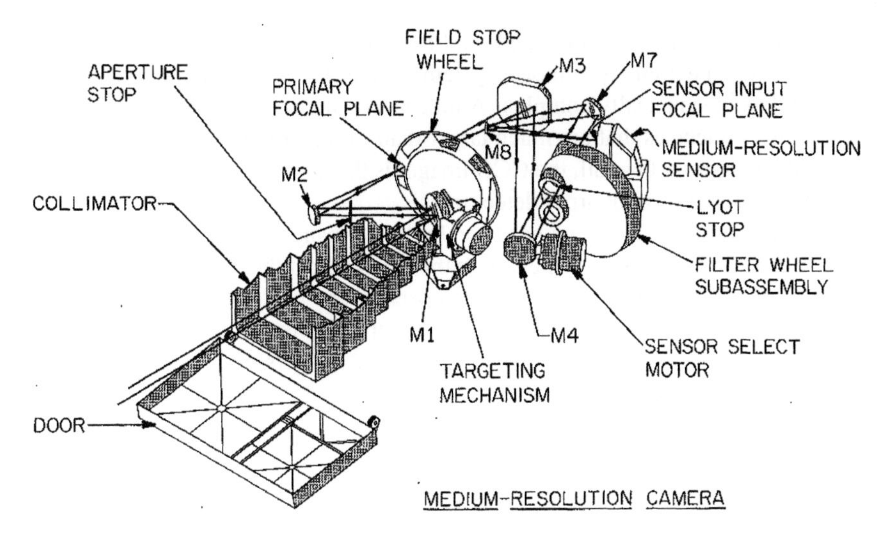

Figure 4.25. Schematic diagram of the Polar VIS medium-resolution camera. After Frank *et al.* [1995].

a point-by-point basis is rather slow so remote sensing is advantageous. The other way is not optical, but constructs images from particle detectors that capture fast neutral H atoms. These are formed through charge exchange in the magnetosphere where, in collisions between energetic protons and the atmosphere, charges are exchanged between species, leading to fast H atoms. Once losing their charge, the H atoms are no longer bound by the magnetic field and move in straight lines from the source to the detector, making imaging possible.

4.11 SUMMARY

In this chapter two types of imagers were discussed, scanning imagers and array detector imagers. For the scanning imager an image is obtained by mechanically scanning a photometer in a raster-type scan, either by moving the photometer itself, or by using scanning mirrors. This means that the pixels are not observed simultaneously, which may be a problem in some applications, and at first sight it seems inefficient. But if the array detector and the photomultiplier were to have the same dimensions, then the light that would fall on all M pixels of the array detector is the same as that falling onto the photomultiplier, which provides M times the signal and exactly compensates for the fact that each sky point is viewed for only $(1/M)$th of the time. There are still some important differences. Since the photometer "pixel" is larger in linear dimensions by $\sqrt{M}$, the focal length and lens area must be increased in the same ratio for the same angular resolution, that means a bulkier system. On the other hand, the angular spread through an interference filter placed in the system is that associated with one pixel, not the whole image, so smaller spectral bandwidths are achievable, though with a physically larger filter. The advantages of the array imager are that mechanical motions are eliminated, and that the small f/number requirement forces the designer into an extremely compact system, whereas scanning photometers tend to be bulky as well as mechanically complex. Another very significant factor is that CCDs can have quantum efficiencies that are much larger than for photomultipliers, especially in the red end of the spectrum. Overall, the CCD imager is the most effective choice, or for other spectral regions some form of array detector.

An amazing variety of imagers has been built and many of these have flown in space. Only a few of these have been touched on in this chapter, but the intent was to describe the fundamental principles and the scope of the observations made to date. These principles span all spectral regions. For the most part, the spectral discrimination of the imagers described here has been limited, for obvious reasons. Imagers of greater spectral content and discrimination are described in later chapters.

4.12 PROBLEMS

4.1. A digital camera is constructed by placing an f/2.8 lens with a 25 mm focal length and a transmittance of 0.9 in front of a CCD array having $10\,\mu m \times 10\,\mu m$ pixels and a quantum efficiency of 35%. Take the solar irradiance at the Earth's surface as 3.4×10^{21} (green) photons $s^{-1}\,m^{-2}$ from Chapter 1 and assume a typical photogenic object has a reflectance of 0.1.

(a) Calculate the exposure time to half-fill the pixels which have a full well capacity of 10^5 electrons.

(b) If the CCD has 1024×1024 pixels, calculate the angular dimensions of the field of view.

4.2. A satellite scanning imager, whose scan strips are precisely adjacent to one another, has an objective lens 20 cm in diameter, a system transmittance of 0.7 and a quantum efficiency of 5%. For a fixed pixel size of $1\,km \times 1\,km$ on the ground, plot the responsivity as a function of altitude from 500 to 5000 km, bearing in mind that the spacecraft velocity w is a function of satellite altitude. The responsivity is as defined in Chapter 3, with units of electron m^2 sr.

4.3. For a ground illumination of 3.4×10^{21} (green) photons $s^{-1}\,m^{-2}$ from Chapter 1, and the photometer characteristics of Problem 4.2.

(a) Calculate the observation time to reach a photon count of 10^4 electrons as a function of altitude from 500 to 5000 km.

(b) Plot the actual observing time according to Eq. (4.1) for a period of 20 s.

(c) From these superposed plots, determine the altitude range over which the desired count rate may be achieved.

4.4. The TIROS satellites were at 644 km altitude and the AVHRR scanned cross track with a 360 rpm rotating mirror, with adjacent scan tracks as described in the text. GOES is in geostationary orbit and images the whole Earth by scanning a square region just containing the Earth, in 20 minutes. Calculate the ratio of the TIROS responsivity to the GOES responsivity, assuming that both achieve the same integrated signal level (in electrons). The spatial resolutions on the Earth are given in the text.

4.5. The Viking satellite made one rotation in 6 s and the UVI had a field of view of $25^\circ \times 25^\circ$.
(a) Calculate the "exposure time" in the TDI mode, in which electron integration for a given picture point continues as long as an auroral form is in the field of view.
(b) Compare this with the "exposure time" if the same camera were spinning at the same rate but without the TDI mode. The CCD has 288 pixels in the relevant direction.

4.6. Repeat the calculations of Problem 4.4, replacing the CCD with an infrared detector, and also changing the wavelength. Assume that the instrument responds to a band of width 1 μm centred at 8 μm. All other parameters may be left the same, but a thermal background is introduced. It is difficult to calculate the thermal background radiation reaching the detector, but for simplicity assume that the detector "sees" thermal emission filling the field of view, and corresponding to an emissivity of 0.1. This would be the case if any lenses, windows or mirrors have a cumulative emissivity of this amount, and no radiation reaches the detector from outside the field of view.
(a) With the instrument at 300 K, calculate the signal-to-noise ratio. You may use the results available in Figure 1.2. Comment on the possibility of making measurements at this temperature.
(b) Determine the instrument temperature required to achieve an SNR of unity. To do this it is necessary to derive the inverse Planck equation, giving temperature as a function of radiance. The Planck equation may be found in Problem 1.1.

5

THE FABRY–PEROT SPECTROMETER

5.1 INTRODUCTION

In Chapter 2 the diffraction grating and Fabry–Perot spectrometers were introduced and in Chapter 3 their superiorities were derived and compared. In Chapter 4 it was shown how the signal-to-noise ratio is calculated. With these fundamental concepts established, it is now appropriate to examine the Fabry–Perot spectrometer in more detail. In this chapter the basis for the design of real instruments is outlined, and a number of applications to atmospheric measurement is described.

Fabry and Perot [1896, 1897] carried out a thorough investigation of the properties and use of a device that was later called the Fabry–Perot interferometer, and followed these publications with many others describing its application to spectroscopic measurement. For complete historical details the reader is referred to the work by Hernandez [1986b]. This new interferometer revolutionized spectroscopy by making high resolution measurements readily feasible. Just after World War II, the device was revolutionized again in France, beginning with Jacquinot and Dufour [1948], who replaced the film method of recording data by one using a photoelectric detector. The characteristics of this new instrument, which is best called a Fabry–Perot *spectrometer* (FPS) because it produces a spectrum, not an interferogram, were explored in detail by Chabbal [1953]. This pioneering work done in Pierre Jacquinot's laboratory, stimulated an intense period of discovery of new systems for interferometric spectroscopy world-wide, yielding a wealth of ideas on how to build better spectrometers and interferometers. For a detailed account of the origins of this work, the CNRS [1958, 1967] symposium proceedings may be consulted.

One of the early atmospheric photographic applications was by Babcock [1923], who used an FPS to measure the wavelength of the atmospheric "auroral" green line emitted also in the airglow. McLennan and McLeod [1927] produced what appeared to be the same $O(^1S)$ emission line in the laboratory and used an FPS interferometer to measure its wavelength, demonstrating that it had, to high accuracy, the same wavelength as the atmospheric line. This important result proved the existence of atomic oxygen in the upper atmosphere. In the last decade the FPS has become a popular instrument for measuring atmospheric winds through the measurement of Doppler shifts of emission line wavelength, but it is also used at low resolution, as a kind of tunable filter.

For his MSc thesis the author used a diffraction grating spectrometer [Hunten, 1953] with a photomultiplier detector, one of the first used for acquiring spectra of the aurora and airglow. For his PhD thesis work on Raman spectroscopy he used a prism spectrometer, a retrograde step as he was to learn later. Then, as a new professor, Donald Hunten gave him a pair of 2.5 cm diameter etalon plates which he made into a FPS. This tiny instrument yielded better spectra of the sodium airglow than had the much larger grating spectrometer, demonstrating the power of the FPS, and inspiring him to continue exploring the wonders of interferometric spectroscopy.

Today, the FPS is still being improved, and taking on new forms. Hernandez [1986b] follows the developments closely into the 1980s, but his description stops just as the CCD and other array detectors began to come into common use, with a further revolutionary impact. In this chapter the basic principles of the FPS are reviewed, and the evolution of this device is further followed towards its ultimate configuration.

5.2 THE IDEALIZED ETALON

The Optical Path Difference (OPD) between adjacent rays, here denoted d, arises from the difference between the optical paths $\underline{BDE}$ and $\underline{BC}$ as shown in Figure 5.1. Light emerging at B passes through an OPD of $n'\underline{BC}$, where n' is the refractive index of the medium outside the etalon. Ray BDE travels an OPD of $n(\underline{BD}+\underline{DE})$, where n is the refractive index inside the etalon, yielding,

$$d = n(\underline{BD} + \underline{DE}) - n'\underline{BC} = 2n\underline{BD} - n'\underline{BE}\sin i \tag{5.1}$$

where i is the incident angle and θ the refracted angle inside the cavity. Noting that $\underline{BD} = t/\cos\theta$, $\text{BE} = 2\underline{BD}\sin\theta$ and applying Snell's law in the form $n\sin\theta = n'\sin\theta'$ the result is:

$$d = \frac{2nt}{\cos\theta} - \frac{2nt\sin^2\theta}{\cos\theta} = 2nt\frac{(1-\sin^2\theta)}{\cos\theta} = 2nt\cos\theta \tag{5.2}$$

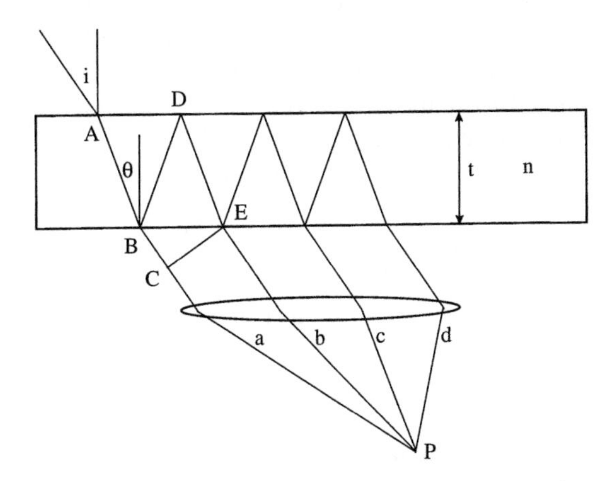

Figure 5.1. Transmission and reflection in a planar Fabry–Perot etalon.

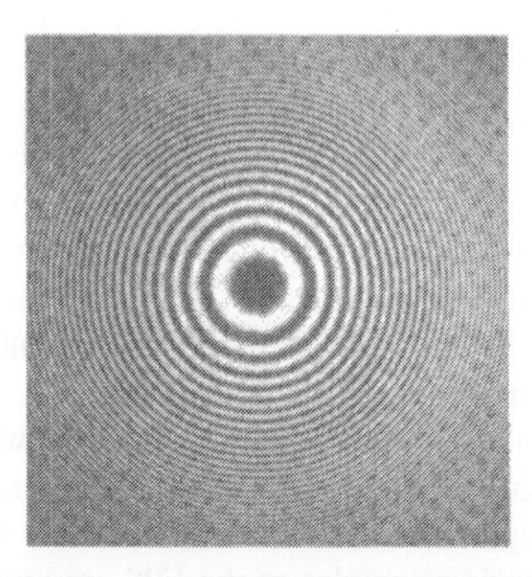

Figure 5.2. Photograph of Fabry–Perot fringes.

The optical path difference between adjacent rays has its maximum value of $2nt$ for normal incidence, and decreases with increasing incident angle.

Now if the etalon is illuminated with monochromatic light of wavelength λ and if $d = m\lambda$ for some angle θ, where m is an integer, $m = 1, 2, 3, \ldots$, then the light has the same phase at C as at E, and rays a, b and c arrive at the image plane in phase, giving a maximum radiance at point P. If, on the other hand, $d = (m + 1/2)\lambda$, rays a and b arrive in the focal plane with a phase difference of π, and so will c and d, causing each pair to interfere destructively so that point P is dark. Since the condition $d = m\lambda$ holds for a fixed θ then all azimuths at this angle θ will have maximum radiance and a bright ring results, as shown in Figure 5.2. This ring is enclosed by a dark one of larger diameter, as θ increases to make d equal to $(m - 1/2)\lambda$. Moving still farther out there is another value of θ such that $d = (m - 1)\lambda$ so that another bright ring results and so on. Thus the equation describing the bright rings is:

$$m\lambda = 2nt \cos\theta \tag{5.3}$$

In the above analysis the light is monochromatic so that λ is fixed and m is allowed to vary (denoted here as case 1). However, there is another way of applying this equation. If m is fixed, then λ can be considered as the variable so that as θ is decreased, the wavelength of light having maximum transmittance is larger (case 2). If the OPD is changed by one order, then in case 1 the change is from $m\lambda$ to $(m + 1)\lambda$, and in case 2 the change is from $m\lambda$ to $m(\lambda + \Delta\lambda)$. Since these must be equal it follows that $(m + 1)\lambda = m(\lambda + \Delta\lambda)$, or

$$\Delta\lambda = \frac{\lambda}{m} = \frac{\lambda^2}{2nt\cos\theta} = \frac{\lambda^2}{d} \tag{5.4}$$

In wavenumbers, (5.4) takes an even simpler form, because $\sigma = 1/\lambda$ leads to $d\sigma = -d\lambda/\lambda^2$ or:

$$\Delta\sigma = \frac{|\Delta\lambda|}{\lambda^2} = \frac{1}{d} \tag{5.5}$$

$\Delta\sigma$, or $\Delta\lambda$, is called the free spectral range (FSR); it is determined only by the separation of the plates and the refractive index between them. Within one free spectral range the wavelengths from an extended polychromatic source are distributed in spectral space and so the system is quite properly called a spectrometer, the Fabry–Perot spectrometer (FPS). However, the spectrum can be complicated to interpret since different spectral lines in this range may have different orders. The adjacent FSR provides no new information since the pattern simply repeats. Thus the FPS is normally used to observe spectra of limited range that fall within one FSR. In other words the gain in responsivity identified in Chapter 2 comes at the cost of a limited spectral range for a single etalon.

In Chapter 2, the passband response function was derived; the result is reproduced here except that it is denoted $A(\sigma)$ rather than $W(\sigma)$ to emphasize that it is a function depending solely on the etalon, which is assumed to be perfect.

$$A(\sigma) = \frac{[t_a/(1-r_a)]^2}{1+\left[4r_a/(1-r_a)^2\right]\sin^2 \pi\sigma d} \tag{5.6}$$

For $\sin^2(\pi\sigma d) = 0$, i.e. $\pi\sigma d = 0, \pi, 2\pi, \ldots = m\pi$; it follows that $\sigma d = m$, where m is an integer so that $A(\sigma)$ has a maximum value. The passband peaks occur at wavenumbers of $\sigma = m/d$ and so are spaced by $(1/d)$ which is just the free spectral range (FSR), in agreement with the preceding analysis. Where $\sin^2(\pi\sigma d) = 1$ corresponding to $\sigma d = (m + 1/2)$, $W(\sigma)$ has a minimum value, but the minima are broad rather than sharp. The sequence of transmittance peaks is clearly evident in Figure 5.3 which has been computed from Eq. (5.6) for $r_a = 0.7$, as well as the broader minima.

The spectral resolution of the etalon is determined by the spectral width of the peaks, which are best described in terms of the *full width at half maximum* (FWHM) or *half-width* for short, though the latter term is ambiguous since it can be confused with the half-width at half maximum.

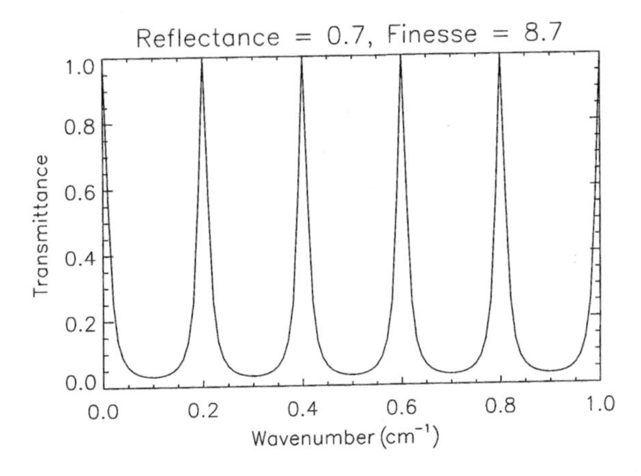

Figure 5.3. Etalon passband function for $r_a = 0.7$, a finesse of 8.7.

The Airy function reaches half its maximum value when $\sigma = \sigma_h$ with respect to σ_o for the peak such that:

$$\frac{4r_a}{(1-r_a)^2}\sin^2 \pi d(\sigma_h - \sigma_o) = 1 \tag{5.7}$$

For small arguments this becomes:

$$4r_a\pi^2 d^2(\sigma_h - \sigma_o)^2 = (1-r_a)^2 \tag{5.8}$$

$$\sigma_h - \sigma_o = \frac{1-r_a}{2\sqrt{r_a}\pi d} \tag{5.9}$$

The full width at half maximum, FWHM, is just twice $(\sigma_h - \sigma_o)$, or

$$\text{FWHM} = \frac{1-r_a}{\pi d\sqrt{r_a}} \tag{5.10}$$

An important characteristic of the FPS results from dividing the free spectral range (FSR $= 1/d$), by the FWHM; the resulting ratio is a function only of the etalon reflectance and is called the reflective, or Airy, finesse, $\mathfrak{F}_A$. It is a measure of the sharpness of the passband peaks.

$$\mathfrak{F}_A = \frac{\text{FSR}}{\text{FWHM}} = \frac{\pi\sqrt{r_a}}{1-r_a} \tag{5.11}$$

As the free spectral range is expanded or contracted by changing the etalon spacing, the passband width is increased or decreased in the same ratio. Thus extremely narrow passbands can be obtained with an FPS provided a small FSR can be accepted. This is certainly the case where the spectrum consists of a single spectral line, and will often be true where the spectrum contains just a few closely spaced lines. The finesse for a functional etalon is usually in the range ten to thirty. In Figure 5.3, the FPS transmittance (Airy function) is shown for a finesse of 8.7 and in Figure 5.4 it is shown for a finesse of 29.8. Note that for

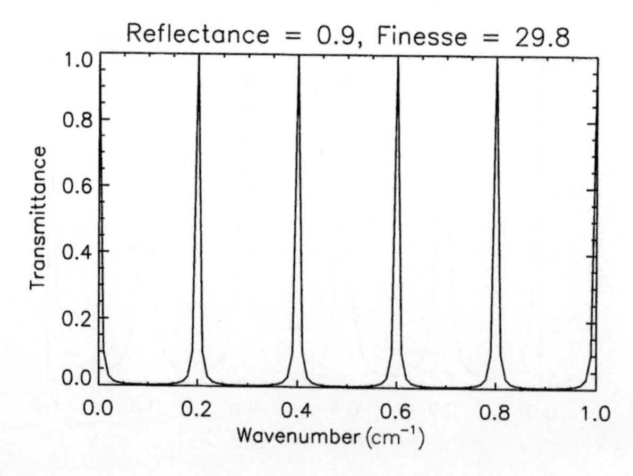

Figure 5.4. Etalon passband function for $r_a = 0.9$, a finesse of 29.8.

a finesse of 8.7 the etalon transmittance does not closely approach zero between orders as it does for the etalon of higher finesse.

5.3 THE REAL ETALON

The discussion of the preceding sections has presumed a perfect etalon, and in this section the characteristics of a real etalon are introduced. To begin with, real etalon surfaces are not perfectly flat. This may be illustrated in a simple way by assuming that all of the error is in one surface and that this surface has a spherical shape with a radius of curvature r_s. This means that the entire etalon cannot be characterized by a single value t_o of etalon spacing, but that t varies about t_o as one moves across the etalon. What this means is that, for a spacing interval between t and $t + dt$, there is a corresponding area of the etalon which is only a fraction of its entire area. This portion of the etalon has peak transmittance for wavenumbers between σ and $\sigma + d\sigma$. Chabbal [1958] introduced an imperfection function D defined by $D(\sigma) = dA'/d\sigma$ where dA' is the fraction of the area of the etalon that transmits radiation lying between σ and $\sigma + d\sigma$. If the total area of the etalon is $A = \pi R_m^2$ then the fractional area dA' corresponding to a radial increment dR is as follows.

$$dA' = \frac{1}{\pi R_m^2} 2\pi R\, dR = \frac{2R\, dR}{R_m^2} \tag{5.12}$$

From the geometry of the situation and similar triangles, $dt/dR = R/r_s$, so that:

$$D_S(\sigma) = \frac{dA'}{d\sigma} = \frac{2R}{R_m^2}\frac{r_s}{R}\frac{dt}{d\sigma} = \frac{2r_s}{R_m^2}\frac{t}{\sigma} \tag{5.13}$$

where Eq. (5.3) has been rewritten as $m = 2nt\sigma\cos\theta$, so that the following relation has been used:

$$\frac{dt}{t} = \frac{d\sigma}{\sigma} \tag{5.14}$$

From Eq. (5.13) $D_S(\sigma)$ is approximately constant, so that equal fractional areas correspond to equal spectral intervals. This has two important consequences. The first is that light corresponding to a narrow spectral interval is transmitted through only a fraction of the area of the etalon; the effective transmittance is reduced accordingly. The second consequence is that the spectral response function associated with the spherical error is flat; it is a rectangular function with a width as determined below.

Let the maximum dt corresponding to the difference in t between the centre of the etalon and its edge be written as λ/M, that is, as a fraction of a wavelength. This is a conventional way of specifying etalon flatness; $\lambda/200$ is a good etalon, for example. Using Eq. (5.14) the corresponding $d\sigma$ is given by:

$$d\sigma = \frac{\sigma}{t}\frac{\lambda}{M} = \frac{1}{tM} = \frac{2\Delta\sigma}{M} \tag{5.15}$$

Thus as for the reflective finesse it turns out that $d\sigma/\Delta\sigma$ is constant for a given etalon so that a spherical surface finesse, $\mathcal{F}_S$, may be defined as:

$$\mathcal{F}_S = \frac{\Delta\sigma}{d\sigma} = \frac{M}{2} \tag{5.16}$$

That is, a spherical error of $\lambda/100$ gives a surface finesse of 50.

The surfaces of the etalon plates are not perfectly smooth, but have minute roughness. Normally the surface roughness is assumed to be represented by a Gaussian distribution, and the corresponding imperfection function is denoted $D_G(\sigma)$. The associated finesse is denoted $\mathcal{F}_G$.

Another characteristic of the real etalon is its finite size. Especially for large off-axis angles, high resolutions and large reflectances, the multiply-reflected beams "walk off" the edges of the etalon plates, reducing the responsivity in a way that must be taken into account.

5.4 ELEMENTARY FABRY–PEROT SPECTROMETER CONFIGURATION

To make an elementary Fabry–Perot spectrometer, the etalon is placed in front of the objective lens of the photometer described in Chapter 3. The field stop at the focal point of the objective lens introduces another spectral passband contribution that must be considered. In Section 3.7.4 it was shown that for a wavenumber interval defined by σ_1 and σ_2:

$$\sigma_1 - \sigma_2 = \sigma_o \frac{i_1^2}{2n^2} - \sigma_o \frac{i_2^2}{2n^2} = \sigma_o \frac{i_1 + i_2}{2n^2} \times (i_1 - i_2) = \sigma_o \frac{\Omega}{2\pi n^2} \tag{5.17}$$

where Ω is the solid angle of the aperture defined by off-axis angles i_1 and i_2. This equation shows that for a given wavenumber element $d\sigma = \sigma_1 - \sigma_2$, the corresponding solid angle is independent of i. This means that the instrument responsivity is the same for all values of σ, which is an extremely important result. As an interference ring (spectral element) in the focal plane gets larger, it gets thinner in such a way that its area remains constant as seen in Figure 5.2.

This fact allows a simple definition of the aperture function. Just as was done for the imperfection function D, an aperture function $F(\sigma)$ may be created, corresponding to the fraction of the aperture that transmits wavenumbers between σ and $\sigma + d\sigma$. Since, according to Eq. (5.17), this quantity is independent of angle, $F(\sigma)$ is a rectangular function with a spectral width $\sigma_1 - \sigma_2$ corresponding to the solid angle Ω for the aperture, or normally to $\sigma_1 - \sigma_o$ for a circular aperture.

The overall instrument passband function of an FPS is thus given by the cross-correlation of all the contributing spectral passband functions, where the wavenumber is referenced against σ_o which is determined by the wavenumber centre of the passband selected by the etalon,

$$W(\sigma) = A(\sigma - \sigma_o) \star D_G(\sigma) \star D_S(\sigma) \star F(\sigma) \tag{5.18}$$

and the $\star$ denotes cross-correlation. The complexity of using an FPS, unlike a grating spectrometer, is that optimum values of all of these functions must be chosen. To a first approximation, they may all be made approximately equal in width, as a combination of a narrow and wide function gives the resolution corresponding to the wide function and the responsivity corresponding to the narrow one, which is the worst possible situation. In particular, if very high reflectance coatings are used to obtain a narrow Airy function, but the etalon surfaces are not of good quality, then only a small fraction of the etalon area transmits that narrow Airy function and the effective transmittance is very low.

The observed spectrum involves one further cross-correlation, $Y(\sigma) = F(\sigma) \star W(\sigma)$. To recover the true spectrum $F(\sigma)$ from the measured $Y(\sigma)$ it is necessary to know $W(\sigma)$. While, in principle, this can be calculated from the cross-correlated components above, it is normally determined through the observation of a laser spectral line which is equivalent to a Dirac delta function so that $Y(\sigma) = W(\sigma)$ in this case. Then for an unknown spectrum, $F(\sigma)$ is determined through a deconvolution procedure (the term that is normally used even if it is an inverse cross-correlation). In principle this can be done by taking the relevant Fourier Transforms, dividing them and doing an inverse transform. But there are other numerical procedures that may give better results.

Thus one normally begins by buying the flattest and smoothest etalon plates that one can afford. The best obtainable is a surface flatness of about $\lambda/200$, corresponding to an imperfection finesse of 100. Then the surface reflectance must be chosen; the reflectance finesse will normally be lower than the imperfection finesse for the reason given above. The coatings are created with dielectric multilayers and because they comprise a specific odd number of layers, only a few discrete reflectance values are available; five layers gives a reflectance of about 0.8, while nine layers gives a reflectance of about 0.95. This is made more complicated by the fact that the reflectance of the multilayer is a function of wavelength. However, custom designed coatings can produce the desired reflectance over a wide spectral range.

Although the etalon has now been defined (and perhaps purchased), the passband width has not yet been chosen; that is an important advantage of the FPS; there is a lot of flexibility in its use. If the objective is to measure the width of an atomic line, then the passband width must be some fraction of the line width; one-third is a reasonable choice so if the linewidth is $0.045\ \text{cm}^{-1}$ an instrumental passband width of $0.015\ \text{cm}^{-1}$ is required. Thus if the etalon has an overall finesse of 20, a free spectral range of $0.3\ \text{cm}^{-1}$ is needed. From Section 5.2, this determines the etalon separation as $1/(2 \times 0.3) = 1.67\ \text{cm}$. Finally the aperture size, to be compatible with the above, would also be chosen to be $0.015\ \text{cm}^{-1}$. For $\sigma = 15\,000$, $d\sigma/\sigma = 10^{-6}$ and the resolving power $\mathfrak{R}$ is 10^6. Using Eq. (5.17) the value of $0.015\ \text{cm}^{-1}$ is equal to $i^2/2$ for $n = 1$ (an air spaced etalon), so that $i = 1.4 \times 10^{-3}$ rad. While this approach will produce a workable instrument, a more detailed numerical simulation analysis is recommended to configure an optimum instrument for a given situation. The procedures for this are described and illustrated by Hernandez [1979, 1986b].

5.5 THE SPHERICAL FABRY–PEROT SPECTROMETER

Connes [1956, 1958a] recognized that a Fabry–Perot spectrometer need not be restricted to planar surfaces; he demonstrated that a spherical cavity would work as well. In Figure 5.5

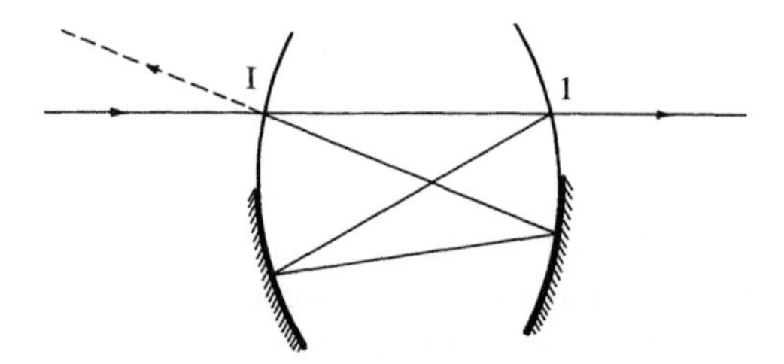

Figure 5.5. Connes' masked spherical etalon, after Hernandez [1986b].

a confocal etalon is shown, one in which the centre of curvature of one mirror lies on the surface of the other. Light entering at I is partially transmitted at 1, with partial reflections emerging at 1 having made four trips across the cavity. The optical path is thus $4nt$, where t is the separation between the partially reflecting spherical mirrors. This "four-trip" etalon also differs from the planar "two-trip" etalon in that there are four reflections rather than two; this means that in the transfer function the reflectance term is r_a^2 rather than r_a. Hernandez (1986b) has shown how to recast the equations into a "device-independent" form, applying equally to planar or spherical configurations for a single beam. The configuration described by Connes is a *masked system* because of the fully reflecting lower halves of the etalons. An unmasked system transmits two beams, as described by Hernandez [1986b]. If the two beams are allowed to superpose, the effects of a two-beam interferometer are superimposed on those of the Fabry–Perot etalon.

For a real spherical etalon there is a more profound difference to consider. As described by Hercher [1968], the paths for rays more distant from the axis do not return to the same point, because of spherical aberration. Thus the spherical etalon has properties which are inverse to those of the planar instrument. The latter is restricted in angular extent because of the optical path difference dependence on off-axis angle, but has no restriction on the etalon dimensions, while the spherical etalon can accept a full hemisphere in angular extent for rays passing through its centre, but is limited in its physical dimensions because of spherical aberration. Thus, in terms of the étendue $A\Omega$, the planar etalon is "omega-limited", while the spherical etalon is "area-limited".

Thus analogous to the planar instrument where the characteristics were determined in terms of the difference in optical path difference between a normal ray and one having a limiting angle (formed by the lens/aperture combination, for example), for the spherical etalon one is concerned with the difference in optical path difference for normal rays passing through the centre of the mirrors, and for a parallel ray at a distance r from the optical axis. From Hernandez [1986b]:

$$\Delta - \Delta_o = \frac{2\pi\sigma r^4}{t^3} \tag{5.19}$$

where t is the etalon spacing and r is the distance of the ray from the axis. This describes in quantitative terms how rapidly the path difference changes as rays parallel to the axis are moved away from the axis; the OPD change is rapid, increasing as r^4. The luminosity resolving power product has been derived by Connes [1958a], and Hercher [1968], and with further modification by Hernandez [1986b] is expressed as:

$$\mathrm{LRP} = \frac{\pi^2 t^2}{2} \frac{f t_s}{d\sigma} \tag{5.20}$$

where f is the spectral width of the aperture and $d\sigma$ is the total instrumental width. To be consistent with the earlier discussion of the planar FPS, f is set equal to $d\sigma$, with the physical meaning that the total spectral width is determined solely by the aperture in the focal plane of the lens. LRP is Jacquinot's luminosity multiplied by the resolving power $\mathcal{R}$, where luminosity $= A\Omega t_s$. Superiority is simply Ω multiplied by $\mathcal{R}$, and so superiority is obtained through dividing LRP by At_s. The result is the superiority of the spherical FPS, denoted S_{SFPS}, where D is the diameter of the etalon:

$$S_{SFPS} = 2\pi t^2/D^2 \tag{5.21}$$

Since the superiority of the planar etalon, S_{PFPS}, is just 2π where the refractive index in the cavity is equal to unity, the superiority gain of the spherical FPS over the planar FPS is given by:

$$\frac{S_{SFPS}}{S_{PFPS}} = \frac{t^2}{D^2} \tag{5.22}$$

From this equation an increase in D for fixed t increases the superiority of the planar instrument with respect to the spherical, but for fixed D and increasing etalon separation, t, the superiority of the spherical etalon increases relative to the planar etalon. For $D = t$, which corresponds to a complete sphere for the spherical etalon, the two have equal superiority. An etalon with $t = 1$ cm has moderately high resolution but with a diameter D of 1 cm the responsivity is very low for either etalon. To have a superior spherical etalon it is necessary to increase t to 5 cm, for example, without increasing D. What this means is that for etalons of the same size, the superiority of the spherical etalon does not overtake that of the planar etalon until the resolution exceeds that required for atmospheric applications. Hernandez [1985] compared planar and spherical etalons of the same wind error and noted that the performance of the planar etalon can be matched by a spherical etalon of much smaller dimensions, leading to a very compact instrument. This was taken advantage of by Blamont and Luton [1972] in a satellite instrument that measured thermospheric temperature from the line width of the atomic oxygen $O(^1D)$ line at 630 nm. It was flown on NASA's Ogo-6 (Orbiting Geophysical Observatory) satellite, was very successful, and the results agreed with later models [Hedin and Thuillier, 1988]. In this application the spherical FPS had two distinct advantages. The first was the small size for a given responsivity, as already noted. The second was that, as the piezoelectrically scanned mirror moved, small tilts of the mirror would only move the centre of curvature of the moving mirror along the surface of the fixed mirror, causing no maladjustment. No other atmospheric application has been implemented since then, to the author's knowledge. For lasers, the spherical cavity has obvious high resolution advantages, which is where it is mostly employed.

5.6 SCANNING METHODS FOR FABRY–PEROT SPECTROMETERS

From Section 5.2, the position σ of a passband peak for a Fabry–Perot spectrometer is given by the basic FPS equation: $m = 2nt\sigma \cos\theta$. Scanning consists of varying σ, which can be achieved by varying one of the other variables in the equation. Since scanning is meaningful only within one free spectral range, corresponding to fixed m, this means that n, t, or θ must be varied. The first is called refractive index scanning; it can be accomplished by putting the etalon into a chamber within which the gas pressure is varied. By differentiating it is found that:

$$\frac{d\sigma}{dn} = \frac{\sigma}{n} \tag{5.23}$$

Since the refractive index of a gas is given by $n = 1 + CP$, where P is the pressure and C a constant,

$$\frac{d\sigma}{dP} = \frac{C\sigma}{n} \tag{5.24}$$

That is, for a fixed pressure change which would characterize a given cell, the scanning range $d\sigma$ is independent of the etalon spacing t. Thus for any free spectral range, the scanning range is fixed, which for pressure changes of an atmosphere or so, corresponds to a fraction of a cm^{-1}. In other words, refractive index scanning is useful only for high resolution applications, such as measuring line widths, where the scanning range is small in an absolute sense. An example of a Fabry–Perot spectrometer designed for this purpose has been given by Nilson and Shepherd [1961].

For the second scanning method, t must be varied and this can be done in a number of ways, some of them strictly mechanical, which is difficult. A pneumatic method was described by Shepherd [1960], and is illustrated in Figure 5.6. This was the first interferometer built by the author and the design was based on the advice of Robert Chabbal who made a visit to Saskatoon at the time. Chabbal came there from Madison, Wisconsin, where he had been working on PEPSIOS but a problem arose in North Dakota when he found there were no buses crossing into Canada. Resourceful as always, he hitchhiked across the border.

Others have used magnetostriction but piezoelectric systems are currently the most popular; these are described later. Again differentiating it is found that:

$$\frac{d\sigma}{dt} = \frac{\sigma}{t} = 2n\sigma \text{FSR} \tag{5.25}$$

where FSR is the free spectral range. If the spectral scan is restricted to one free spectral range, the maximum dt required is $\lambda/2$, so that the range of the spectral scan, $d\sigma$, depends only on the FSR (and n). For a high resolution application the FSR is small, so the scanning range is small, but for large FSR, the scanning range can be large. Thus mechanical scanning can be particularly advantageous for low resolution Fabry–Perot spectrometers.

The last method, in which θ varies, has been called spatial scanning by Shepherd *et al.* [1965]. The basic equation has already been given as:

$$\sigma - \sigma_o = \frac{\sigma i^2}{2n^2} \tag{5.26}$$

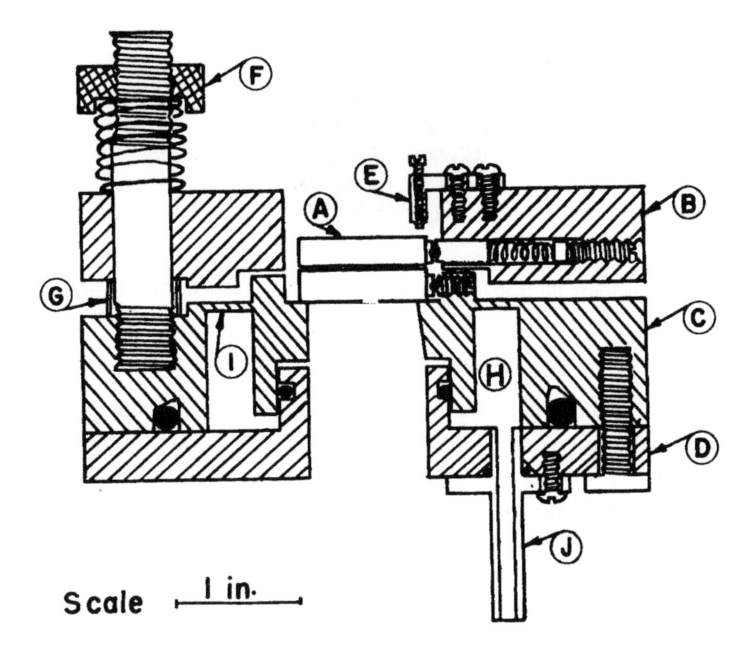

Figure 5.6. A pneumatically scanned Fabry–Perot spectrometer, after Shepherd [1960].

Spatial scanning has some important advantages. The wavelength is independent of t, which means that if etalons are put in tandem, they will scan synchronously. If they don't have the same n, as is the case if an interference filter is coupled to an air-spaced etalon, an angle transformer (that is, a telescope having a magnification n) can be placed between the two etalons to allow synchronous operation.

Interference filters are Fabry–Perot etalons in which both the spacer and the reflecting surfaces are made of dielectric materials; they are solid so that only spatial scanning can be used. Lissberger [1968] has provided a useful formula giving the transmittance of an interference filter:

$$t_f(\lambda, i) = \frac{t_{fMax}}{1 + \left[2(\lambda - \lambda_o)/(\Delta\lambda)_o + (\lambda/(\Delta\lambda)_o)(i^2/n_e^2)\right]^2} \tag{5.27}$$

where i is the off-axis angle of the entering ray, t_{fMax} is the filter transmittance at $i = 0$, $(\Delta\lambda)_o$ is the full-width at half-maximum of the filter passband and n_e is the effective refractive index of the filter.

Since interference filters are simply low resolution high finesse etalons, these ideas can be applied to make multi-channel photometers. One of the first of these (if not the first) was developed by J.R. Miller for rocket flight. It had two channels, one formed by a plastic cone, the other by a rod that passed through the side of the cone [Miller and Shepherd, 1968]. A more elaborate rocket photometer was made by Deans and Shepherd [1978], using a stepping mask, illustrated in Figure 5.7. A photomultiplier centred on any of the above

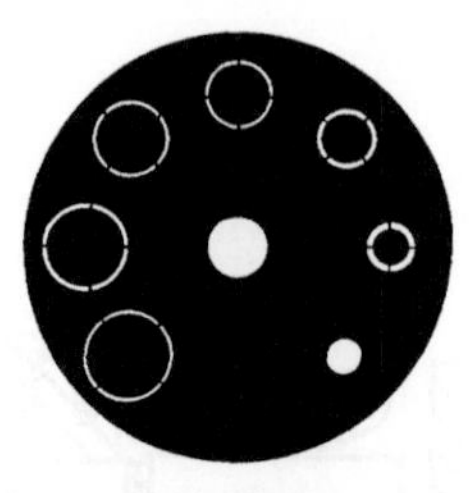

Figure 5.7. A stepping mask for a spatial scanning Fabry–Perot spectrometer, after Shepherd *et al.* [1978].

apertures sequentially received the signal passing through each aperture in turn. While this appears to be a bulky design for a rocket instrument, it took advantage of the fact that the same mask could serve four separate photometers with different interference filters. These were separated by 90°, i.e. top, bottom, left and right in Figure 5.7, and the instrument thus provided seven spectral channels and one dark channel for each of the four interference filters.

5.7 THE APPLICATION OF FABRY–PEROT SPECTROMETERS

5.7.1 Multiple Ring Aperture Instruments

At the oral examination of the author's first M.Sc. student to build an FPS and measure auroral Doppler temperatures with it, an examiner said, "You have been telling us what a large sensitivity this new instrument has, but how can you reconcile that with the fact that all the light is collected through such a small aperture". The student quite correctly said that even though the aperture was small, a few mm, it was circular and the associated solid angle was much larger than for a grating spectrometer having a long, but very narrow slit, operating at the same resolving power. True, but the questioner had a good point. What about all the light that was not going through the aperture and was being wasted? After the first elementary FPS instruments had been put into use, essentially the simple photometer of Chapter 3, but with an etalon in place of the filter, the users began to think of ways to waste fewer photons.

The most obvious way to begin this process is to acquire photons from different orders onto the same photomultiplier. Since for a high resolution instrument there are many orders available in the focal plane, and each order carries the same information, adding them together onto one photomultiplier increases the signal by a factor equal to the number of orders added. Effectively the mask is fabricated to replicate the ring pattern that would be reproduced with the same system, for monochromatic light of the same wavelength. But this is not all that simple, since the outer rings are very narrow, and distortion in the lens needs to be accounted for. In fact, the most accurate masks are made using a photographic method and imaging the rings with the same etalon and lens combination. While mechanically fabricated masks are limited to responsivity gains of about 10, the photographic ones can reach a gain near 50. Such systems have been used by Hirschberg and Platz [1965], Hirschberg *et al.* [1971], Meaburn [1976], Dupoisot and Prat [1979], Okano *et al.* [1980], and Biondi *et al.* [1985]. Some of these methods take remarkable forms. Hirschberg used a set of nested tubes, each having a cross-section corresponding to a Fabry–Perot ring. The end of the set of tubes was cut at a small angle from the perpendicular, and each tube then rotated by an angle of $360°/N$, where N is the number of tubes. The reflecting ends of the tubes each reflected light in a different direction, to a different photomultiplier, for each spectral element.

A different approach is to sample the entire focal plane (as is now done with array detectors) using spatial scanning. The author's students began building such a Fabry–Perot spectrometer in Saskatoon and continued the development in Toronto. The basis of scanning was a fibre optic bundle, in which the fibres were divided into ten sub-bundles, each corresponding to an equal-area Fabry–Perot ring as described by Eq. (5.17); the set of ten thus corresponding to a Fabry–Perot pattern with ten adjacent rings. In this sense it makes no difference as to whether a Fabry-Perot "ring" is considered a spectral element, or an order; the pattern is the same in any case. Here the ten rings were arranged to divide one FSR into ten spectral channels. The other ends of the sub-bundles led off to ten photomultiplier detectors. In this way, the device became a multiplex ten-channel Fabry–Perot spectrometer, in which the light in all ten channels was measured simultaneously. A full description of this instrument has never been published, but the individuals who successively made it work were L.L. Cogger, S. Peteherych and R.N. Peterson. Results obtained with this instrument were published by Cogger and Shepherd [1969], and Peteherych *et al.* [1985].

A much more compact approach to the same problem was realized at the University of Michigan in fabricating the first FPS to measure upper atmospheric winds from space. The instrument, called the Fabry–Perot Interferometer (FPI) [Hays *et al.*, 1981], used a multi-ring anode detector that was custom fabricated by the supplier [Killeen *et al.*, 1983]. Twelve annular anodes were constructed inside one detector system, each anode corresponding to one spectral element. This was the same principle as the fibre-optics bundle described above, but requiring only one detector unit, not ten. It is characteristic of space instrument design that solutions can be found that would be too expensive for ground-based systems, for which budgets are more limited. This is because incorporating a new technology can have a major impact on the instrument's mass and power requirements, thus radically reducing the demand on spacecraft resources. There was a three-fold purpose in introducing this system. The first was to increase the responsivity, as already described. The second was to measure all spectral elements simultaneously, so that the spectrum could not be influenced by temporal variations occurring during the exposure time on a rapidly moving spacecraft. The third was to allow the use of an etalon of fixed spacing, as a highly stable and rugged etalon, as described by Rees *et al.* [1982], was required. With this detector system it was possible to use an etalon with an effective diameter of only 3.3 cm and produce excellent results. There was a limitation in that, owing to restricted spacecraft resources, it was possible to have only one field of view, so that only one wind component could be measured. Fortunately, a neutral mass spectrometer did measure the other component of the wind so that vector winds were obtainable. A schematic diagram of this very compact instrument is shown in Figure 5.8.

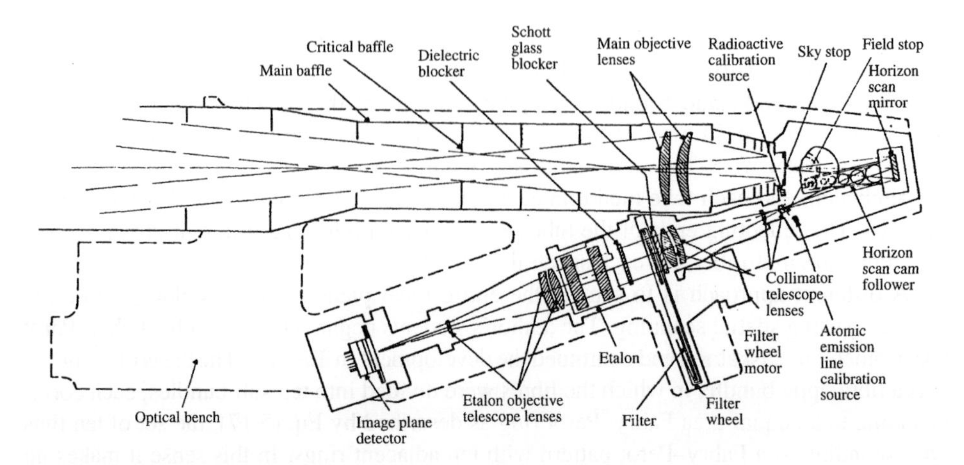

Figure 5.8. The Dynamics Explorer Fabry–Perot Interferometer, after Hays *et al.* [1981].

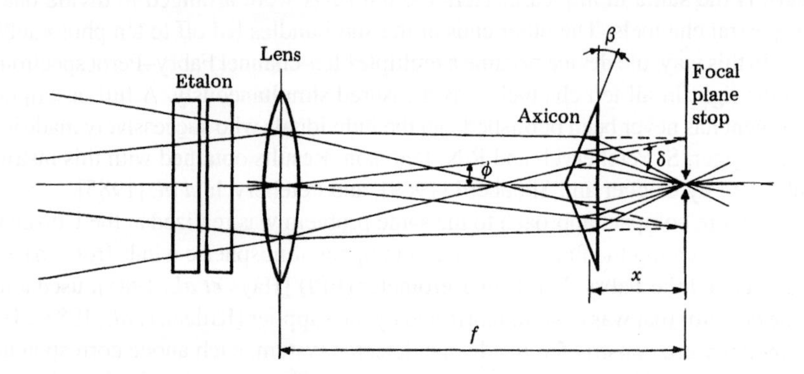

Figure 5.9. FPS scanning with Katzenstein's axicon, after Katzenstein [1965].

5.7.2 Axicon System

One of the more ingenious Fabry–Perot configurations is the axicon introduced by Katzenstein [1965], shown in Figure 5.9. For one position of the conical prism, light of one off-axis angle is brought to the aperture. As the axicon is moved along the optical axis, light of different angles is brought to the aperture and detector, generating a spectrum. Essentially, the axicon forms images at different points along its axis. The fundamental properties of the axicon were described earlier, by McLeod [1954].

5.7.3 Low Light Level Applications

FPS instruments are capable of reaching low light levels, allowing some unusual atmospheric emissions to be studied. One of these is the helium 1083 nm triplet, which is described by Kerr *et al.* [1996]. This is not only relatively weak, at about 200 R, but is in fact difficult to observe because photomultipliers are not effective in this spectral region; an intrinsic Germanium detector was used instead. Comparatively high resolution must be used because at low resolution the helium emission is overwhelmed by the OH (hydroxyl) airglow emission. Kerr *et al.* [1996] present an airglow spectrum showing those superposed lines along with a longer wavelength feature that is mostly hydroxyl, blended with one of the helium triplet components. The hydroxyl line is at 10831.34 Å which is not where it appears in the spectrum; it is in an adjacent order and appears because the width of the interference filter used (3.9 Å FWHM) is wider than the free spectral range of 2.4 Å. However, a useful measurement of the helium line which is produced fundamentally through fluorescence of sunlight in late twilight can be made from the shorter wavelength component.

A somewhat different example is the measurement of the Λ-doubling of two lines in the hydroxyl airglow Meinel (6-2) band near 840 nm as reported by Greet *et al.* [1994]; the instrument was described by Jacka [1984]. They obtained spectra of the $P_1(3)$ and $Q_1(2)$ lines using an etalon separation of 5.57 mm corresponding to an order of 13235 for the $P_1(3)$ line, which had to be determined accurately using several different spectral lines in order to determine the hydroxyl line separations. These were determined to be 5.50 pm (picometers) for the $Q_1(2)$ line and 19.17 pm for the $P_1(3)$ line. The Doppler temperatures were also determined from the linewidths, at 124 K corresponding to 87 km. However, as shown later, the rotational temperature provides more accurate values of temperature for the hydroxyl airglow. The Λ-doubling in the OH lines had also been measured by Hernandez and Smith [1984], some years before.

5.7.4 Tandem Fabry–Perot Spectrometers

The tandem Fabry–Perot spectrometer was used by Nagaoka [1917], and by Houston [1927]. The application in the contemporary form was suggested by Bradley and Khun [1948], investigated by Khun [1950] and studied in detail by Dufour [1951]. Practical instruments were implemented by Robert Chabbal in Pierre Jacquinot's laboratory. (Jacquinot later became the director of the French Centre National de la Recherche Scientifique and still later, Chabbal held the same position.) Like his predecessors, Chabbal recognized that the advantages of the Fabry–Perot were, for many applications, offset by the fact that for a single etalon the scanning range was limited to one free spectral range. This problem could be overcome by putting several etalons in tandem, in such a way that a single effective scanning passband would result. He called this the Integral Fabry–Perot Spectrometer [Chabbal, 1958]. It thus became a goal to generate a unique passband spectrometer by putting several etalons in tandem using refractive index scanning.

It was very difficult to keep them synchronized using methods available at that time. The idea developed into a collaboration with the University of Wisconsin; a pilot model was developed there by Mack *et al.* [1963] and implemented by Roesler and Mack [1967]. The name PEPSIOS became applied to this scheme (Purely Interferometric High Resolution Scanning Interferometer, in English). It was first applied to the measurement of atmospheric sodium in the daytime. Clyde Burnett applied it to the measurement of total integrated concentration of atmospheric OH, using daytime absorption measurements [Burnett and Burnett, 1983]; the name PEPSIOS survives with this instrument. There are two ways of putting etalons in tandem in order to achieve a unique passband. This first is to stack etalons of sequentially increasing spectral width. Thus three etalons, one of high resolution, one medium and one low can achieve a unique passband when used in conjunction with an interference filter. The other approach is called the vernier method, in which all (three, say) etalons have roughly the same passband width, but have slightly different values of the FSR. Thus the passbands "beat" together, and one arranges that the beats are as far apart as possible. The analysis of the PEPSIOS vernier method is described by McNutt [1965].

Observations of dayglow sodium emission with tandem etalons are described by Greet *et al.* [1989] and are shown in Figure 5.10. Sodium airglow observations are interesting in that the emission line sits at the bottom of a Fraunhofer line with an absorption depth of about 96%. The four panels in the figure correspond to different viewing directions for a zenith angle of 60°; they are marked E, N, W and S and the differences in the sodium emission rate and the Rayleigh-scattered sky background change drastically in these four directions. The observation of day airglow (dayglow) with ground-based instruments has presented a challenge in parallel with many decades of night airglow observations. Only very specialized instruments with specific operators as described in the previous paragraph have achieved this on a regular basis, and there is no continuous routine set of recorded observations. Bens *et al.* [1965] observed the OI 630 nm dayglow emission with two etalons and an interference filter, but this was a marginal observation. Nevertheless,

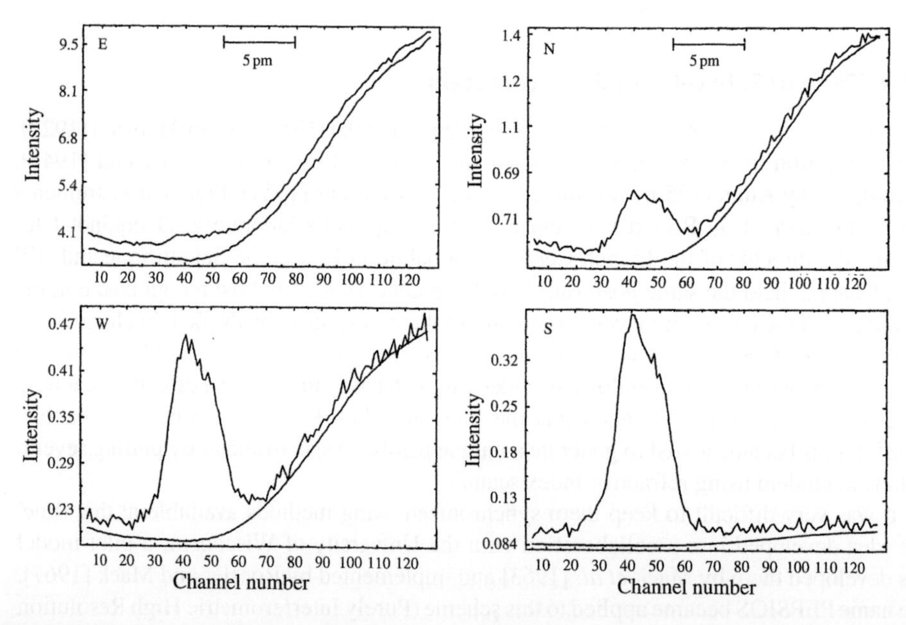

Figure 5.10. Spectra of sodium dayglow in four different viewing directions. After Greet *et al.* [1989].

the measurement suggested that the narrow bandwidth, potentially achievable with tandem Fabry–Perot etalons, was a way to make this measurement. The HRDI instrument, a triple-etalon system with a fully controllable unique passband flown on the Upper Atmosphere Research Satellite is described in a later chapter; it provides a contemporary solution to this problem. Other approaches are best described as modulators, which are covered in Chapter 8.

Another use of tandem etalons is to increase the contrast, the peak to valley transmittance ratio, of a single etalon. For two identical etalons the transmittance between the passband peaks is very much reduced over that obtained with a single etalon. This is important in applications where a weak spectral line is to be observed near a much stronger line, or where there is a continuum to be suppressed. One way to achieve this is simply to double-pass the same etalon, as was done by Hariharan and Sen [1961]. Sandercock [1971] went further in traversing the same etalon five times.

Hernandez [1987] proposed a configuration called a Double-Etalon Modulator (DEM) in which a low resolution (fixed) etalon is combined with a scanning one of high resolution, but where the two are coupled with a pair of telescopes in an afocal system (recall the conservation of étendue in Chapter 3) so that the ring patterns are identical. One etalon acts as a mask for the other, a mask with multiple orders. The number of orders determines the increase in responsivity and, while this places constraints on the quality of the imaging system, gains in responsivity of 100 are thought to be possible. The purpose of the tandem etalons here is not to create a unique passband, but to increase the responsivity of the high-resolution etalon. This system is not currently in use, perhaps because of the availability of the CCD as a Fabry–Perot spectrometer detector, which happened at about the same time.

5.7.5 Stabilized Fabry–Perot Spectrometers

It is important that FPS instruments stay in adjustment. Otherwise the finesse varies during the observations and so does the resolution. For wind measurements the etalon must not only remain parallel, but the spacing must not change either. Wind measurements are described in detail in later chapters but suffice it to say that this is done by measuring the Doppler shift of an emission line where the emitting species is carried by the background gas and thus corresponds to a wind. Ground-based instruments capable of measuring winds began to provide results in 1979 [Hernandez and Roble, 1979; Hays *et al.*, 1979; Sipler *et al.*,1983; Jacka, 1984; Smith *et al.*, 1985]. To measure winds to $3\,\mathrm{m\,s^{-1}}$, a desirable value for the stratosphere and mesosphere, the measurement must be made to one part in 10^8. For a wavelength of 500 nm, that means the measurement of a wavelength shift of 5 fm (femtometre). Since a linewidth is of the order of 0.002 nm, this is 2.5×10^{-3} of the linewidth. Unless the instrument has a similar stability, its variations are interpreted as winds.

Gonzalo Hernandez, author of the authoritative book on Fabry–Perot interferometers [Hernandez, 1986b], has been one of the most consistent practitioners of ground based measurements of winds and temperatures. In order to establish a definitive wind measurement program he developed a system in which the etalons are mechanically positioned by piezoelectric pillars, crystals which change their length when a voltage is applied across them. Independent detectors sense a passband peak generated by a spectral lamp and send a voltage to the piezoelectrics to cause them to set to the desired spacing; this is called servo control. The etalon is then stepped through the spectrum, using this reference control system [Hernandez and Mills, 1973]. An example of wind measurements made at the South Pole, where the instrument must be untended for long periods is shown in Figure 5.11. The winds shown in the figure are obtained from the atomic oxygen 630 nm emission, which occurs over a rather broad range of altitude in the thermosphere, but in this region the winds are known to change slowly with altitude. The authors have assigned an effective altitude of 240 km for these observations. Line-of-sight wind measurements are made in eight equally spaced azimuthal directions (the cardinal directions and the intracardinal directions) as well as at the zenith. Since it is the South Pole in winter, the observations are made during a 24 hour period; in the figure they are converted to magnetic coordinates so the station is not located at the pole in the diagram. The zenithal line of sight winds are averaged over 24 hours and the resulting wind value is taken as the *zero wind value*, and subtracted from all of the azimuthal measurements. This must be done because the wavelength cannot be measured in an absolute sense to the accuracy required, and it is a good assumption that the vertical wind averages to zero, otherwise the atmosphere would disappear. From the horizontal wind values taken in eight directions it is possible to deduce the wind field geomagnetically north

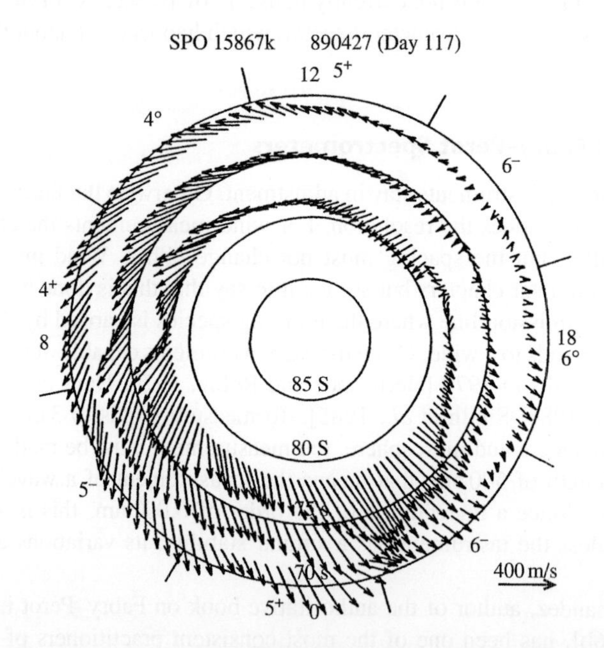

Figure 5.11. Wind measurements from the South Pole for April 27, 1989, using the 15 867 cm^{-1} (630 nm) emission. After Hernandez *et al.* [1990].

of the station, geomagnetically south of the station, and at the station. Thus three rings of wind values are shown in Figure 5.11.

More recently, the company Queensgate [Hicks *et al.*, 1974] developed a servo controlled system based, not on the optical path difference between the plates, but on the capacitance between the ends of two pillars mounted on the plates. It is remarkable that the sensing of capacitance can yield accuracies at levels required for interferometry, a few picometres (pm). While maintaining the separation of the capacitor gap is not the same as controlling the OPD, the physical integrity of the system, with the pillars firmly contacted to the etalons, the effect is just as good, and much simpler in design, avoiding the need for additional light paths through the system. Normally the plates are cemented to three piezoelectric pillars which provide the movement, and the sensing is done with two pairs of pillars, one "horizontal" and the other "vertical".

5.8 APPLICATIONS OF THE FABRY–PEROT IMAGER

5.8.1 Introduction

In Section 5.7 a number of methods of obtaining extended information in the focal plane of the FPS were described. The capability described for those methods is generalized with the use of a two-dimensional array, specifically a CCD imager. However, the additional flexibility offered by a true imager deserves consideration as a different class of instrument, even though the distinction is arbitrary to some extent. As is shown in this section, the introduction of the CCD has further revolutionized the use of the Fabry–Perot etalon. However, all is not as simple as it might seem. The FPS spectral elements are organized into rings, while the pixels of a CCD are rectilinear. Companies have considered the fabrication of a CCD with ring-like pixels and it is considered feasible. However, circumstances have not brought such a device to market, probably because the desired objectives are being achieved in other ways. It turns out that there are many different ways to make use of the 2-D imaging capability, involving different combinations of the space and spectral domains.

5.8.2 PRESTO – A Programmable Etalon Spectrometer for Twilight Observations

Low resolution Fabry–Perot spectrometers provide very high responsivity for applications where the required scanning range is small and high resolution is not required. One application where this is advantageous is the measurement of auroral or airglow emissions in twilight. An instrument was originally built to detect daytime auroras associated with the dayside magnetospheric cusp to support rocket launching at Cape Parry, on the Arctic coastline, when the sun was approximately 2° below the horizon at noon in December. The optical design of the original (photomultiplier) version of PRESTO (PRogrammable Etalon Spectrometer for Twilight Observations) is described by Gault *et al.* [1983]. The etalon system of separation 0.562 mm has plates separated by cemented piezo-electric spacers controlled by a capacitive feedback system that maintains parallelism and allows the separation to be set to a specific value, or changed by precise increments. The etalon system was built by Queensgate Instruments Ltd. in London. The spacing was adjusted manually or controlled through the microcomputer. The étendue of the etalon is matched with that of the interference filter, with a different refractive index, through a pair of transfer lenses L_1 and L_2 according to conservation of étendue.

After moving the instrument back to Toronto it was used for the measurement of the atomic oxygen $O(^3P)$ 844.6 nm twilight emission using a Ga As(cs) photomultiplier to observe both components of the atomic doublet and hence provide a definitive identification of the emission as described by Bahsoun-Hamade *et al.* [1989]. Later the photomultiplier was replaced with a CCD imager [Bahsoun-Hamade *et al.*, 1994], allowing the imaging of Fabry–Perot rings showing the two components of the doublet.

After locating the ring centre, the radial distance of each pixel is computed, and the data are binned according to radial distance. The spectrum is then obtained by plotting the accumulated signal versus radial bin number. For high resolution applications, where the spectral "rings" get very narrow near the edge of the CCD, some other technique is desirable. In addition, at very low light levels, reading out individual pixels to later form ring patterns in the computer adds to the noise, because the readout noise is added to each pixel. Abreu and Skinner [1989] suggested a novel solution, which consists of reading out entire rows at a time, averaging across the entire row. Thus the readout noise is added only to each row, not each pixel. This would appear to destroy the ring pattern, but as they point out, a single row across a circular ring pattern is exactly the same situation as viewing the limb of the earth from a satellite. Thus the same inversion techniques referred to in Chapter 1 can be taken advantage of here so that the signal sums from all of the rows can be inverted to obtain the radial irradiance pattern which is just what is wanted.

5.8.3 MORTI and SATI: Hybrid Spatial and Spectral Instruments

In the previous section both dimensions of the CCD were used for spectral information; this is a way to enhance signal-to-noise ratio. A different approach is to use one dimension for the spatial measurement and the other for the spectral measurement, but because of the way the FPS works, these two dimensions are not rectilinear with this instrument. The radial dimension provides spectral information as has already been described, and the tangential direction is used to image the azimuthal position on the sky. Without fore-optics, the FPS ring pattern is imaged on the sky and so different pixels correspond to different spectral elements and view different locations on the airglow layer in a way that spatial and spectral information are mixed together. By using a conical mirror as shown in Figure 5.12 and a Fresnel lens in front, the MORTI (Mesopause Oxygen Rotational Temperature Imager) instrument projected the interference filter into a circular annular ring on the sky [Wiens *et al.*, 1991]. By dividing the CCD data (in the analysis) into 12 sectors, it was possible to extract a spectrum from each sector, each of which corresponded to a different azimuthal location on the sky. The spectrum contained six rotational lines of the (0, 1) $b^1\Sigma_g^+ O_2$ Atm band at 867.7 nm, from which the rotational temperature was determined to an accuracy of

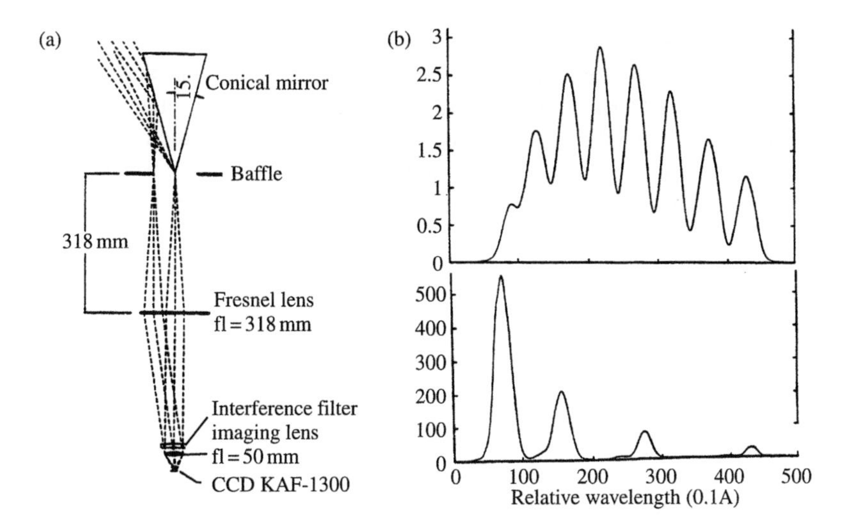

Figure 5.12. (a) Optical diagram for the SATI/MORTI instruments, showing the interference filter projected onto the sky as an annular ring; (b) spectrum of the (0, 1) $b^1\Sigma_g^+ O_2$ Atm band at 867.7 nm (above) showing six rotational lines and of the Meinel (6-2) band Q branch showing four lines. After Wiens *et al.* [1996].

about 2 K. Rotational temperatures are obtained from the relative populations of rotational energy levels; the rotation is usually in thermal equilibrium because of the low energies involved. As a minimum the emission rate ratio for two lines is required, provided they are chosen to be responsive to temperature change. Six lines offer a still more accurate temperature. By detecting phase shifts in emission rate and temperature between sectors, it was possible to determine the direction of propagation of atmospheric gravity waves, as well as the horizontal wavelength and period. Results obtained are described by Zhang *et al.* [1993a, 1993b] and Wang *et al.* [1993].

A later version of the instrument was called SATI (Spectral Airglow Temperature Imager) [Wiens *et al.*, 1996]; it incorporated both O_2 and hydroxyl (836.8 nm) interference filters with spectral widths of 0.23 and 0.18 nm respectively. The latter extracted three lines from the (6-2) Meinel band Q branch. Since, as for MORTI, an interference filter is just a solid FP etalon with all secondary passbands blocked, it is legitimate to consider this an FPS imager. The narrow bandwidth interference filters that are available do overlap what can be achieved with an etalon. Spectra from the O_2 Atm band and the OH Meinel band are shown in Figure 5.12(b).

A different approach giving the same solution was provided by Swenson *et al.* [1990]. They employed a piezoelectrically scanned Fabry–Perot with an etalon spacing of 0.1 mm, yielding a FSR of 50 cm^{-1}, or for a finesse of 20, a FWHM of 2.5 cm^{-1}, or 0.13 nm, which is comparable to an interference filter. They viewed the sky at a zenith angle of 60°, and formed an elliptical ring on the sky by defining a circular ring on the CCD. The FP was scanned to bring one of the two OH lines, $P_1(2)$ at 631.6 nm into the defined circular region and an image was taken. They then scanned further to bring the second line, $P_2(3)$ at 733.0 nm into the same region. By taking the ratio of the emission rates of the two lines at the same place on the CCD and thus at the same place on the sky, the rotational temperature was determined in an annular region. This region was then subdivided in order to study the passage of gravity waves. The authors did not consider this an optimum experiment, rather a demonstration, but it shows the power of using an imaging detector combined with a scanning etalon.

5.8.4 Imaging Low Light Level Application

CCD arrays are not used with the FPS only for imaging; they may also be used for increased responsivity for sources of very low radiance. An excellent example of this is given by Nossal *et al.* [1997] in observations of the Balmer-α line in the night sky. Above 500 km the frequency of collisions between neutral constituents in the Earth's atmosphere is reduced to the point where atoms may escape from the atmosphere without colliding with another constituent, provided that the velocity is sufficiently great to escape the Earth's gravity (the escape velocity). This region is called the exosphere and is dominated by atomic constituents; O, He and H. Of these, the light H atom has the greatest velocity for a given thermal energy, and so can participate in this process – above 1000 km, atomic hydrogen is the dominant constituent. This highly tenuous hydrogen atmosphere is called the Earth's geocorona. After the sun has set, but the exosphere is still illuminated, the absorption of solar Lyman-β populates the $n = 3$ level, giving rise to Balmer-α emission that can be detected at the ground. However, it is very weak, just a few R.

The instrument used by Nossal *et al.* [1997] was a tandem etalon with etalon spacings of 1.49 and 0.524 mm; the origins of this group can be traced back to PEPSIOS. The moderately low resolution was appropriate to the relatively broad Balmer-α line width but two etalons were used to improve contrast and to remove secondary passbands passing through the interference filter. A CCD was used as the detector and the spectra were analysed by organizing the pixels into rings of unit area and equal spectral width, as described earlier. A spectrum is shown in Figure 5.13 where the wavelength scale has been converted to velocity units; a Doppler shift appears to be evident but the situation is complicated by the fact that the line has seven fine structure components. There is a background, most likely from the interstellar medium, although observations were taken only when the galaxy was not in the field of view. From the line widths, effective temperatures in the range 500–1000 K were obtained; these are effective temperatures because of the lack of thermal equilibrium. The complexities of using a CCD imager are accurately described in this paper.

5.8.5 Imaging Winds with an FPS Imager

The term imager is used in many different ways by the practitioners of spectral imaging; the example given in this section is a particularly interesting form of imaging. By combining a CCD imager as in Chapter 4 with an all-sky camera as in Chapter 5 and placing this behind a high resolution Fabry–Perot etalon, a set of FPS rings appears superimposed on an airglow

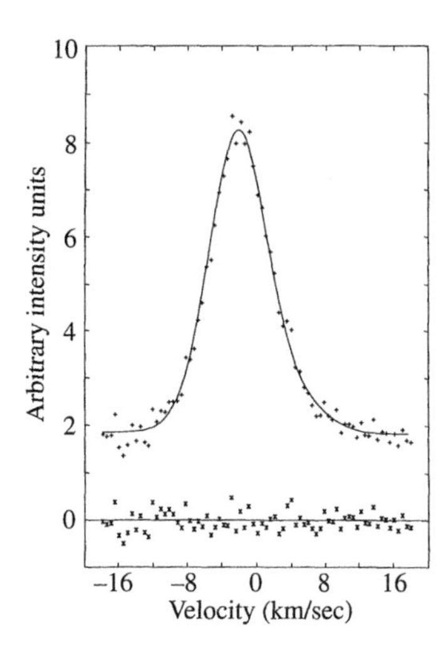

Figure 5.13. The spectrum of Balmer-α from the Earth's geocorona, with the spectral width expressed in velocity coordinates. The emission rate was about 3 R for this measurement. After Nossal *et al.* [1997].

or auroral image of the sky, the kind of image shown in Chapter 1. In principle, the line-of-sight wind can be determined for the altitude of the emission layer from each point on the sky where a ring appears; there is no emission and therefore no information between the rings. The problem is to determine the shift of the ring from its zero-wind position, not an easy problem. However, small-scale wind variations over the sky can be determined from the deformation of the rings. This type of wind imager has been used by Rees *et al.* [1984], Batten and Rees [1990] and Biondi *et al.* [1995]. The instrument itself is very easy to use as it contains a stable etalon of fixed spacing, and the Doppler information is obtained from the CCD, but the fact remains that the data are difficult to interpret.

The concept was dramatically extended by Conde and Smith [1998] by using a capacitance-stabilized piezoelectrically scanned etalon in place of the etalon of fixed spacing. By scanning the etalon through its free spectral range, the spectrum of the line and thus its Doppler shift could be obtained for every pixel, allowing many thousands of line-of-sight wind measurements to be made over the sky. For practical reasons these were grouped into 25 sectors, one overhead, four in the annular ring outside that, eight in the next outward ring and twelve in the outermost ring, corresponding to a zenith angle of 65°. The problem then is, how does one obtain vector winds at various sky locations when all that one has are line-of-sight measurements at each location? The wind field can be constrained by expanding it using first-order Taylor series expansions. Further, the azimuthal variations can be expressed as Fourier series, and inverted. But it turns out that the measured information is insufficient. Thus, drawing on an earlier method by Burnside *et al.* [1981], two sets of sky measurements are taken some interval of time apart. It is assumed that the station rotates with the Earth, under a wind field that is stationary with respect to the sun. This

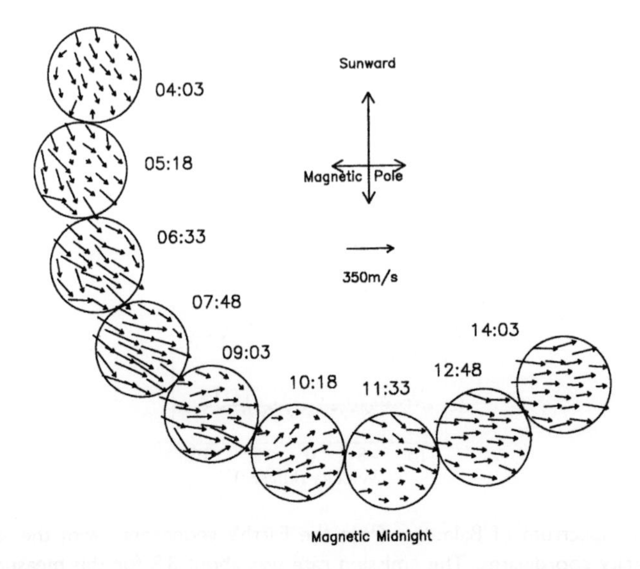

Figure 5.14. The wind field above Alaska, as measured for the night of October 17, 1996. The small circles correspond to the field of view from the ground station, and the circles move according to the universal times given, around the magnetic pole in a coordinate system fixed to the sun. After Conde and Smith [1998].

is a reasonable assumption for the 630 nm line, corresponding to roughly 250 km in the thermosphere, in the auroral zone, where magnetospherically generated electric fields drive the ions, and thus the neutral wind through collisions. With this additional information, the problem is solved. Figure 5.14 shows the variations of winds across the field of view, and throughout the course of the night, where the views are arranged in a clockdial pattern, the Earth rotating within a coordinate system fixed to the sun, consistent with the assumptions. Both the small and large-scale patterns are evident in this presentation.

5.8.6 CLIO (Circle to Line Interferometer Optical System)

The most recent, and perhaps the ultimate in Fabry–Perot configurations is a method that provides a solution to the problem discussed earlier, of matching circular rings to a rectilinear CCD. This method was conceived by Paul Hays [1990], after a career devoted in part to the development of the FPS, and following both the DE FPI and HRDI on UARS, neither of which used CCD imagers. This brilliant idea is as follows. A hollow cone reflector is shown to convert the circular ring (of constant off-axis angle) to a linear pattern in the focal plane as illustrated in Figure 5.15. It is called CLIO (Circle to Line Interferometer Optical system) and it provides a conceptually neat and practical solution to the problem of how to combine the rectilinear CCD with the circular FP rings. An analysis of the instrument has been presented by Wu *et al.* [1994]. This might be considered the ultimate in generating an

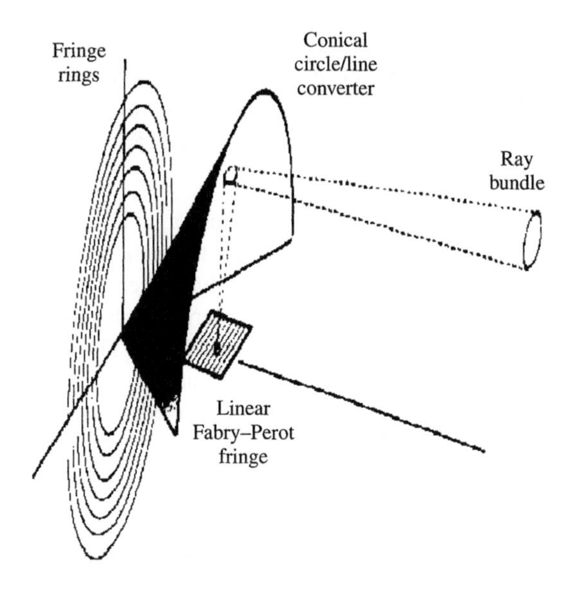

Figure 5.15. Optical ray diagram for a circle to line converter, after Hays [1990].

FPS of maximum responsivity. This method is currently used in the TIDI (Timed Doppler Imager) on the TIMED satellite mission launched in December 2001, and described in Chapter 10. It is also currently being used in a ground-based FPS operating at Resolute Bay, 74° N in Canada [Fisher *et al.*, 1999]. Finally, it is also the basis of a method for the lidar measurement of winds in the troposphere.

5.9 PROBLEMS

5.1. A Fabry–Perot etalon is perfectly polished, but has a spherical error of 0.01 of a wavelength (centre to edge). It is coated with reflective coatings of 90% reflectance. The plates are mounted 2 cm apart. Assume a wavelength of 600 nm. It is used with a lens of 80 cm focal length in front of an aperture and photomultiplier.
(a) Calculate the diameter of the field stop if the aperture spectral width is to be equal to the spectral width of the etalon Airy function.
(b) Numerically compute the shape of the spectral passband function for the overall system and determine its FWHM. Calculate the ratio of this width to that of the perfect etalon, that is, the FWHM of the Airy function.

5.2. It is planned to measure the line width of the OI 630.0 nm line, using a pressure-scanned FPS. The refractive index of air is given by $n = 1 + 2.88 \times 10^{-9}$ P, where P is in Pascals and a pressure change from 1 atm to 1.1 atm is intended to scan the etalon over one order.
(a) Calculate the separation of the air-spaced etalon plates.

(b) Suppose that the etalon is perfectly flat, with high reflectance, and a small aperture is used so that the passband function is dominated by the plate polish, giving a Gaussian passband with a finesse of 20. The measured spectrum yields a FWHM of 3.0 pm (0.003 nm). Assuming that the atomic oxygen line shape is Gaussian, determine its true width, using the results of Problem 2.2.

5.3. The H-β line at 486.1 nm as observed in the aurora, is broadened by the velocity of incoming protons, which become fast neutrals through charge exchange collisions, requiring a spectral range of 0.8 nm, or a somewhat larger FSR of 1.2 nm for its observation. The etalon is highly polished but has a serious spherical error of $\lambda/30$, centre to edge. The aperture has an angular diameter of 2.06°. The reflective (Airy) finesse is 30, so that the spherical error and the aperture function dominate; the Airy function contribution to the instrumental width can be neglected.
(a) Calculate the separation of the air-spaced etalon.
(b) Describe the FPS passband shape and determine its FWHM value.
(c) Calculate the distance one plate must be mechanically moved to scan one order.

5.4. An investigator wishes to measure O_2 rotational temperature from eight lines in the (0, 1) b($^1\Sigma$) Atm band at 864.5 nm, as in Figure 5.12, requiring that the spectrum be measured over an extent of 5 nm around 864.5 nm. A solid etalon is to be used, having an effective refractive index of 2.5 and a HWHM of 0.1 nm. Calculate the maximum off-axis angle required to cover this spectral range.
(a) Using an $f/1$ lens and a CCD of 1 cm × 1 cm having 256 × 256 pixels, determine the required lens focal length. Assume that only complete spectral rings are used, that is, the corners of the CCD are not used.
(b) Determine the aperture spectral width associated with the width of the outermost pixel and compare this with the solid etalon width of 0.1 nm.

5.5. A CCD of dimensions 1 cm × 1 cm has pixels 30 micrometers in size. Calculate how many complete Fabry–Perot rings can be imaged onto this CCD, such that the width of the largest ring is equal to the dimensions of one pixel.

5.6. Consider three tandem etalons which are to be configured using the vernier method. All three air-spaced etalons may be represented by Airy functions with $r_a = 0.95$. For simplicity assume the peak transmittance is unity. Etalon 1 is set to an order of $m = 50\,000$ at $20\,000\ \text{cm}^{-1}$, giving $d_1 = 2.5$ cm.
(a) Determine a d_2 value for Etalon 2. This can be an integral number of orders different from that of Etalon 1, i.e. 50 001, 50 002 etc. Choose an order difference so that, while the primary passbands coincide, the adjacent passbands are separated such that the individual transmittances at the crossover point is only a few percent.
(b) Determine d_3 for Etalon 3 in the same way but on the lower order side, i.e. 49 999, 49 998 etc. such that when the primary passbands coincide, the adjacent passband overlaps that of Etalon 1 by a few percent, but on the opposite side of the Etalon passband than that for Etalon 2.
(c) Plot the transmittance of the tandem etalon system from 20 000 to 20 100 cm^{-1} and comment on the success of this approach towards a unique passband system.

6

THE MICHELSON INTERFEROMETER

6.1 HISTORICAL BACKGROUND

The use of spectrometers in general and the FPS in particular, have now been described in some detail. However, this is not the only method of obtaining a spectrum. An extremely powerful alternative is that in which the Fourier transform of the spectrum is measured, and the spectrum is computed through numerical transformation. Such instruments are called interferometers, and the interferometer most widely used for this purpose is that invented by Michelson.

Albert A. Michelson invented his elegantly simple interferometer in order to measure the effect on the speed of light of the earth's motion through the "luminiferous ether". The experiment was performed in Cleveland in 1887 with the collaboration of E.W. Morley, and as every physics student knows, no effect whatever could be detected. Einstein, impressed with the experimenters' thoroughness, accepted the result and it became one of the experimental cornerstones of the theory of relativity. The negative result puzzled Michelson, however, and it is doubtful that he ever fully accepted it. For many years he seldom mentioned his experiment in public. Even his own students at the Case School of Applied Science in Cleveland did not hear of the Michelson–Morley experiment from Professor Michelson. A history of the many measurements he made is given in the Michelson centennial issue of Physics Today by Swenson [1987].

Michelson went on to make many more important measurements and discoveries with his instrument, including the determination of the length of the metre in terms of wavelengths of light, the discovery of the hyperfine structure of spectral lines and the measurement of the width of many lines. He also measured the speed of light more accurately than anyone before him and then surpassed his own accuracy several times. He invented a stellar interferometer and was the first to measure the diameter of a star, using a principle that later was used in array telescopes and long baseline interferometry. Michelson was the first American scientist to receive a Nobel Prize, but it was in honour of this later work, not for the Michelson–Morley experiment, for which he is best known today.

After Michelson, the interferometer was little used for decades until its revival, chiefly in France, in the 1950s and 60s. Some of the names associated with this revival are P. Jacquinot

(France), P. Fellgett (U.K.), L. Mertz (U.S.A.), H. Gush (Canada) and above all, Pierre and Janine Connes of the Laboratoire Aimé Cotton near Paris. With the advent of sensitive electronic detectors and large digital computers, the Michelson interferometer (MI) became the Fourier Transform Spectrometer (FTS). By the late 1960s, the Connes were producing million-point interferograms and by far the most detailed infrared spectra yet obtained. Instruments of this type have been applied to many types of problems, including the measurement of emission and absorption line profiles in the Earth's atmosphere. Currently, it is possible to buy commercial, long-path interferometers of extremely high resolving power which come complete with data acquisition systems and software to transform the interferograms into spectra.

Other, more specialized applications of the Michelson interferometer have been made. Among these are the field-widened versions (instruments of uniform optical path difference over a wide field) developed at York University and earlier at the University of Saskatchewan in Saskatoon using the Doppler approach for measuring winds and temperatures; a parallel development took place at Service d'Aéronomie du CNRS in France. These were applied to the upper atmosphere using airglow emissions and are now proposed for the middle atmosphere using thermal emission. These instruments are described in later chapters. A history of the development of field-widened Michelson interferometers in Canada has been given by Shepherd *et al.* [1991]. Similar instruments for the observation of solar oscillations were developed for the GONG (Global Oscillations Network Group) ground-based project and for the SOI (Solar Oscillations Investigation) experiment on the SOHO (SOlar and Heliospheric Observatory) satellite, discussed in Chapter 8.

6.2 BASIC CONCEPT

The Michelson interferometer (MI) consists of a plane beam-splitter, which divides an incident beam of light into two beams of equal radiance, and two plane mirrors, which reflect the beams back on themselves to the beam-splitter, where they divide again, and recombine into two output beams as illustrated in Figure 6.1. One mirror is fixed but the other is moved and is shown in two positions in the figure. The one corresponding to the solid rays has an optical path difference (OPD) of zero, and the dashed rays correspond to a finite OPD. As long as the optical surfaces are reasonably flat and well aligned, the recombining

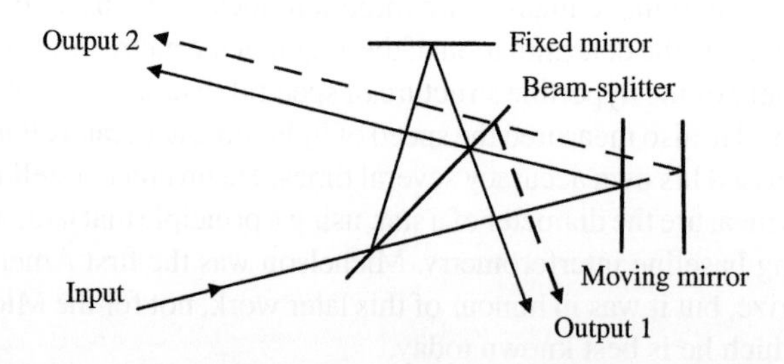

Figure 6.1. Ray diagram for an ordinary Michelson interferometer.

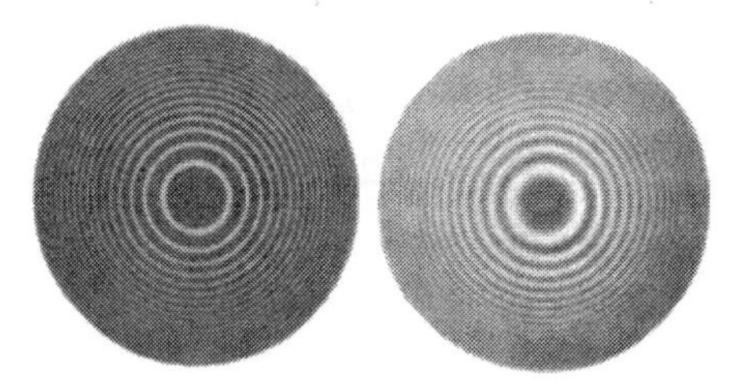

Figure 6.2. A comparison of Fabry–Perot (left) and Michelson (right) interferometer fringes.

beams are sufficiently coherent to produce an interference pattern. The fringes of general interest are those formed at infinity, the Haidinger fringes. These circular bright and dark rings may be studied by looking into the interferometer with the eye relaxed or by forming an image of them with a lens. A photograph of these fringes is shown in Figure 6.2, compared with those obtained from a Fabry–Perot interferometer set to the same interference order. The Michelson fringes are not as sharp as those from the Fabry–Perot instrument for reasons that will become clear.

Suppose that the electric field of the incident monochromatic radiation is represented by $\mathcal{E}_o e^{j2\pi\nu t}$, where $\mathcal{E}_o$ is the amplitude of the electric field, t is time and ν is the frequency. The complex transmittance and reflectance coefficients for the beam-splitter are represented by $t_a e^{j\delta_T}$ and $r_a e^{j\delta_R}$ respectively, where $\sqrt{t_a}$ and $\sqrt{r_a}$ are the amplitude coefficients (= $1/\sqrt{2}$ for an ideal beam-splitter) and δ_T and δ_R are the phase shifts for transmittance and reflectance, respectively. The difference in path length travelled by the beams in the two arms of the interferometer is the OPD, denoted Δ. The MI has two outputs, one where the light comes out to the side (Output 1) and one where it returns back towards the source (Output 2).

Output 2 is usually inaccessible because it is blocked by the input beam, and Output 1 is the one that is commonly used. In some cases, it can be advantageous to use both outputs, and schemes are available for gaining access to the light returning to the source by reducing the size of the input beam and sacrificing some of the instrument's responsivity as compared to that for a single beam. For surfaces that are perfectly flat and perfectly aligned, the electric field, $\mathcal{E}_1$, of the radiation exiting by Output 1 is given by:

$$\mathcal{E}_1 = \mathcal{E}_o\sqrt{t_a}e^{j\delta_T}\sqrt{r_a}e^{j\delta_R}e^{j(2\pi\nu t+2\pi\sigma\Delta)} + \mathcal{E}_o\sqrt{t_a}e^{j\delta_T}\sqrt{r_a}e^{j\delta_R}e^{j2\pi\nu t} \tag{6.1}$$

$$= \mathcal{E}_o e^{j2\pi\nu t}\sqrt{t_a r_a}e^{j(\delta_T+\delta_R)}\left(1+e^{j2\pi\sigma\Delta}\right) \tag{6.2}$$

The radiance is found by multiplying the electric field by its complex conjugate and the result is:

$$L_1 = 2L_o t_a r_a(1+\cos 2\pi\sigma\Delta) \tag{6.3}$$

where L_1 is the radiance from output 1 and L_o is the radiance of the incident beam. When the mirror is moved at constant speed so that Δ increases uniformly with time, the output radiance varies in a cosinusoidal fashion. The average radiance, $\overline{L_1}$ is found by integrating over a complete cycle:

$$\bar{L}_1 = 2L_o r_a t_a \tag{6.4}$$

In a similar way, the electric field at Output 2 can be shown to be

$$\mathcal{E}_2 = \mathcal{E}_o e^{j2\pi\nu t}\left(t_a e^{2j\delta_T} e^{j2\pi\sigma\Delta} + r_a e^{2j\delta_R}\right) \tag{6.5}$$

and the radiance is:

$$L_2 = L_o\left[t_a^2 + r_a^2 + 2r_a t_a \cos(2\pi\sigma\Delta + 2(\delta_T - \delta_R))\right] \tag{6.6}$$

with an average value of

$$\bar{L}_2 = L_o\left(t_a^2 + r_a^2\right) \tag{6.7}$$

Comparing the radiances at the two outputs, the following points may be noted (remembering the assumptions of perfect optics and a single frequency):

1. If $t_a = r_a$, then L_1 and L_2 are the same except for the phase difference of $2(\delta_T - \delta_R)$.
2. L_1 is always fully modulated (i.e. goes to a minimum value of zero), while L_2 is fully modulated only if $t_a = r_a$.
3. If $\delta_T - \delta_R = \pi/2$, then the two outputs are complementary. This is generally true for dielectric beam-splitters, in which the absorption is very small (conservation of energy). It is not generally true for metallic beam-splitters, which do absorb substantially (typically 40% per pass).

Now suppose the spectrum of the incident light is not monochromatic, but is composed of many frequencies, and is represented by $L(\sigma)$. Then the signal at Output 1 is proportional to:

$$I'(\Delta) = \int_{-\infty}^{\infty} L(\sigma)(1 + \cos 2\pi\sigma\Delta)\, d\sigma \tag{6.8}$$

$$= \int_{-\infty}^{\infty} L(\sigma)\, d\sigma + \int_{-\infty}^{\infty} L(\sigma)\cos(2\pi\sigma\Delta)\, d\sigma = \int_{-\infty}^{\infty} L(\sigma)\, d\sigma + I(\Delta) \tag{6.9}$$

(the subscript 1 is dropped from the signal, as all subsequent discussion refers to Output 1). $I'(\Delta)$ is called the interferogram and consists of two parts, one constant and one variable. The variable part, $I(\Delta)$, is the Fourier transform (FT) of the spectrum, and the spectrum can be recovered by performing an inverse Fourier transform on $I(\Delta)$. Ignoring scaling factors, this is:

$$L(\sigma) = \int_{-\infty}^{\infty} I(\Delta)\cos(2\pi\sigma\Delta)\, d\Delta \tag{6.10}$$

As was shown in Chapter 2, the cosine transform alone is adequate for an ideal interferogram that is symmetric about zero path difference, but the complex transform

$$L(\sigma) = \int_{-\infty}^{\infty} I(\Delta)e^{-j2\pi\sigma\Delta}\, d\Delta \tag{6.11}$$

is more generally correct.

From the viewpoint of a linear dynamical system, the response of the interferometer to an impulsive input is a pair of pulses, with a delay of Δ. The FT of this is a cosine, which is the interferogram signal corresponding to a monochromatic input. The other way of thinking about it is that $I(\Delta)$ in the above equations is proportional to the autocorrelation function of the input signal with delay Δ and therefore the power spectrum is the Fourier transform of $I(\Delta)$, as derived above.

6.3 SPECTRAL RESOLUTION

The spectral resolution of an instrument is the minimum separation that two lines of equal radiance can have and still be clearly distinguishable. This is not a precise mathematical definition, and some convention must be adopted so that "resolution" can become a mathematical entity. In all previous discussions involving spectrometers, the spectral resolution was taken as equal to the passband width (FWHM). For the interferometer, a different convention is needed.

Suppose the interferogram of a source consisting of a single frequency, σ_o, is measured. The true interferogram is clearly a consinusoid stretching to infinity, but the recording is terminated at the path difference Δ_m, which is the maximum the interferometer will allow. This is equivalent to multiplying the interferogram by a rectangular function, which equals unity, from 0 to Δ_m and zero after that.

The spectrum derived from this is the FT of the truncated interferogram, which equals the convolution of the spectrum with the FT of the rectangle of width $2\Delta_m$ because both positive and negative sides of the interferogram have to be included. This function has the form:

$$W(\sigma - \sigma_o) = \text{sinc}\, 2\pi(\sigma - \sigma_o)\Delta_m \tag{6.12}$$

and each frequency in the spectrum contributes a function of this shape to the measured spectrum. $W(\sigma - \sigma_o)$ is therefore the instrumental function and its first zero is at an interval $d\sigma = 1/2\Delta_m$ from the central peak.

One sometimes sees $1/2\Delta_m$ used in the literature for the spectral resolution of the Michelson interferometer. It is equal to the full width of the instrumental function at 64% of its peak, and if two such functions are spaced apart by this interval, their sum shows only one broadened peak, so this definition doesn't allow full resolution of two spectral lines. A more practical definition, and the one usually quoted, is $d\sigma = 1/\Delta_m$. The resolving power, $\mathcal{R}$, is then given by $\mathcal{R} = \sigma\Delta_m$.

The side lobes of the instrumental function can be very troublesome in a complicated spectrum and it is usual to reduce the size of the side lobes by weighting the interferogram by a function which reduces the abruptness of the truncation. A spectrum produced in this way is said to be "apodized", as was described for the grating spectrometer in Chapter 2. Two apodizing functions often used are a triangular function:

$$A_1(\Delta) = 1 - |\Delta/\Delta_m| \tag{6.13}$$

and the function

$$A_2(\Delta) = \left[1 - (\Delta/\Delta_m)^2\right]^2 \tag{6.14}$$

A_2 is probably the most commonly used apodizing function. It decreases the size of the first side lobes from 22% of the peak in the unapodized case to 4%. The cost of apodizing, however, is an increase in $d\sigma$ by about 50% in both the above cases.

6.4 FIELD OF VIEW

Consider a ray entering a MI at incident angle i as shown in Figure 6.1. Looking into the beam-splitter, the virtual image of one mirror is separated from the other by a distance t. The difference in path length travelled by the two wavefronts is the same as for the Fabry–Perot etalon where θ is the off-axis angle inside the interferometer:

$$\Delta = 2t\cos\theta \tag{6.15}$$

The path difference at normal incidence is $\Delta_o = 2t$, and so:

$$\Delta = \Delta_o\cos\theta \tag{6.16}$$

By expanding the cosine and using Snell's Law to set $\theta \approx i/n$, the approximation can be written:

$$\Delta - \Delta_o \approx \frac{\Delta_o}{2}\theta^2 \approx \frac{\Delta_o i^2}{2n^2} \tag{6.17}$$

where i has been related to θ through the refractive index n inside the interferometer. For most MI instruments the medium is air, so that $i = \theta$, but other cases are considered in later sections. This describes the variation of path difference across the field of view in the Michelson and explains the appearance of the circular bright and dark rings, whose spacing varies quadratically as for the Fabry–Perot etalon. Thus the locations of the rings are the same as for the Fabry–Perot etalon as shown in Figure 6.2, although the shapes of the individual rings are different, being cosinusoidal rather than described by the Airy function, as for the FPS.

The maximum off-axis angle in the interferometer is determined by the combination of lens and aperture placed at the output of the interferometer. This combination determines the étendue of the instrument. One would like to use a large solid angle in order to have a large responsivity, but the size of the field of view used for measurements must be chosen carefully because of the variation of Δ with angle. As Δ_o increases, and a greater range of path difference is included within the field of view, the contrast of the recorded fringes decreases. This is similar to apodization and the effect is to decrease the spectral resolving power.

The best compromise to use is somewhat a matter of taste and depends on the application. J. Connes plotted a figure of merit against field stop size and found a broad maximum with a field of view having a radius corresponding to $\Delta_o - \Delta = 0.9\lambda$. The effective resolving power with an aperture of this size is reduced to about 80% of its paraxial value. Using (6.17) the radius of the field of view can be written, where $n = 1$ as:

$$i_{max} = \sqrt{\frac{1.8}{\sigma\Delta_o}} = \sqrt{\frac{1.8}{\mathcal{R}}} \tag{6.18}$$

6.5 THE REAL MICHELSON INTERFEROMETER

Michelson used the term "visibility" to describe the contrast of the fringes he observed in his interferometer; it was a quantity that he could measure by eye, and that he used effectively for many spectroscopic measurements. Visibility, V, is defined as the ratio of the amplitude to the average value of the interferogram value, or where I_{max} and I_{min} are the maximum and minimum values of the interferogram at the path difference being considered:

$$V = \frac{I_{max} - I_{min}}{I_{max} + I_{min}} \tag{6.19}$$

Thus, for a quasi-monochromatic spectrum of wavenumber σ_o, the Michelson interferometer equation can be written as:

$$I(\Delta) = I_o(1 + V \cos 2\pi\sigma_o\Delta) \tag{6.20}$$

The visibility is dominantly influenced by the spectrum of the source as described in more detail in Chapter 8. For example, if the spectrum consists of two narrow lines of equal radiance at σ_1 and σ_2, then the visibility goes from unity at $\Delta = 0$ to zero at $\Delta = [2(\sigma_1 - \sigma_2)]^{-1}$ in a repeating pattern, as the two lines beat against one another. Since $V = 0$ is readily identified by eye, Michelson was able to accurately measure the wavenumber difference between two lines in this way, even for very closely spaced lines that were not resolvable by other methods. Using a similar approach it is possible to measure line widths, as discussed in the next chapter. The visibility as determined by the source is normally called *source visibility*.

The discussion so far has assumed perfect optics and ideal alignment, but in a real instrument these ideal conditions are not met. The mirrors can be misaligned, the optical surfaces are not perfectly flat, the glass is not perfectly homogeneous and the transmittances in the two arms might be different. These defects tend to "wash out" the fringes, i.e., reduce their visibility, and a multiplicative visibility factor can be estimated for each defect. This overall *instrument visibility factor*, U, is the product of all these components; it is a slowly varying function of frequency. The interferogram for an emission line measured with this imperfect instrument is then

$$I(\Delta) = I_o(1 + UV \cos 2\pi\sigma_o\Delta) \tag{6.21}$$

In practice, it is usually not necessary to measure U accurately, but the closer to unity it is, the more complete is the modulation of a spectral line. The instrument is calibrated by measuring a source of known radiance. In cases where it is necessary to know U, such as in the measurement of line width by the visibility method, it can be determined by observing a spectral line of negligible width, such as a stabilized laser line, and it is not necessary to estimate each visibility component separately.

6.6 SAMPLING THE INTERFEROGRAM

To provide a data set for transformation, the measured interferogram, $I(\Delta)$, has to be sampled at regular intervals. If the sampling interval is g, the number of samples in the interferogram is $N = \Delta_m/g + 1$, going from $\Delta = 0$ to Δ_m. From Chapter 2, the Fourier transform of such a discretely sampled finite interferogram is a function which repeats periodically at frequency intervals of $1/g$. Within each $1/g$ interval are two mirror images of the spectrum. The reflection occurs at $\sigma_m = 1/2g$, and to avoid aliasing (overlapping the orders), the sampling interval must be chosen to be no greater than

$$g = \frac{1}{2(\sigma_{max} - \sigma_{min})} \tag{6.22}$$

where $\sigma_{max} - \sigma_{min}$ covers the complete range of frequencies present in $I(\Delta)$. The sampling theorem also states that no more spectral information is gained by using a smaller value of g. In order to save computer time, N is kept near the minimum and an interpolation procedure is used to produce a smooth spectrum for plotting.

In practice, several methods are available for sampling the interferogram. They all involve the use of a reference beam of a frequency outside the spectral region being measured (usually at a higher frequency), in order to ensure that the sampling is done at a constant interval. One cannot rely on the accuracy of a mechanical drive alone.

The path difference can be stepped, held constant for a brief integration time, stepped again, and so on. The size of the steps must be accurately controlled by means of a feedback system operating from the reference detector. Alternatively, a continuous scan can be used with integration triggered by the reference fringes. This involves a slight loss of resolution because integration is not done at a unique path difference. Another method is to digitize both the interferogram and the reference fringes at constant time intervals and then use the computer to determine the path difference intervals from the reference fringes and do the sampling from the interferogram recording. The latter method involves considerably more computing than the other two. Some extra effort put into an automatic sampling system in the instrument is bound to make the data analysis easier.

6.7 SUPERIORITY OF THE MICHELSON INTERFEROMETER

Equation (6.17) gave the variation of OPD with off-axis angle and is repeated below for $n = 1$.

$$\Delta - \Delta_o \approx \frac{\Delta_o}{2}\theta^2 = \frac{\Delta_o i^2}{2n^2} \tag{6.23}$$

For the Fabry–Perot spectrometer the same equation holds, with Δ replaced by σ. Thus the two instruments are governed by the same equation, but since Δ and σ are not the same thing it is necessary to calculate the superiority, using Eq. (6.23) and $d\sigma = 1/\Delta_m$ where Δ_m is replaced by Δ_o. This is appropriate because the aperture size (maximum off-axis angle) is determined at the maximum OPD. In addition the assumption is made in Eq. (6.17) that an aperture of $\Delta - \Delta_o$ equal to λ can be tolerated, which is only slightly different from Eq. (6.18) where a value of 0.9λ was used.

$$S_{MI} = \Omega\mathfrak{R} = \pi i^2 \frac{\sigma}{d\sigma} = \pi \frac{2(\Delta - \Delta_o)}{\Delta_o}\frac{\sigma}{d\sigma} = 2\pi\lambda\, d\sigma \frac{\sigma}{d\sigma} = 2\pi \tag{6.24}$$

which is exactly the same as for the Fabry–Perot spectrometer.

Fellgett [1951] pointed out that in addition to having the same superiority as the FPS, the MI has a multiplex advantage. The Michelson interferometer records the signal from each frequency in the spectrum all the time, unlike a scanning spectrometer, which records successive frequencies sequentially. The dominant source of noise in an infrared detector is noise generated within the detector itself. Suppose a spectrometer takes a certain time to record a certain frequency in the spectrum with a certain signal-to-noise ratio. A Michelson interferometer with the same throughput takes the same time to achieve the same SNR, but during that time it can do the same thing at all other frequencies in the spectrum. Or, for equal scanning times of the same spectral region at the same resolution, the SNR advantage of the MI over the spectrometer is $\sqrt{N}$, where N is the number of spectral elements (instrumental function elements). This is often referred to as the Fellgett advantage. In the visible region, where photon counting techniques can be used, the MI loses its Fellgett advantage because the dominant source of noise is shot noise from the random arrival of photons, and extending the exposure to a spectral element by a factor of N increases the noise by a factor of $\sqrt{N}$ which just cancels the multiplex gain.

Thus the MI has two advantages over a diffraction grating spectrometer, the multiplex advantage and the superiority advantage. When the contemporary MI first burst onto the scene there was some confusion about whether these two advantages were distinct. Once that was resolved they were often referred to as the Fellgett and Jacquinot advantages respectively.

Since the superiority of the MI equals that of the FPS there is nothing to distinguish them in that regard. In the infrared region of the spectrum the MI will enjoy the Fellgett advantage as well, and is therefore a better choice. In the visible region they are again equal in this sense. However, the MI is able to cover large spectral regions in a way that is extremely difficult to manage for an FPS and so, where this is a requirement, the MI is again the better choice.

Early in the days of Fourier transform spectroscopy the main disadvantage of the MI was the necessity to perform a lot of computation, including a Fourier transform on the data in order to get the desired spectrum. But in recent decades this effort has become trivial and is more than offset by the amount of control the computations give over the final spectrum. The spectrum can be recomputed from the interferogram over and over again until the desired result is obtained.

6.8 SCANNING METHODS FOR THE ORDINARY MICHELSON INTERFEROMETER

6.8.1 Overview

There is an incredible variety of methods that have been tried for the scanning of the ordinary Michelson interferometer. The works by Vanasse [1977, 1983] are recommended for comprehensive descriptions of many of them. In this section the emphasis is on the concepts behind the scanning methods, not the technology. For the Fabry–Perot spectrometer the possible methods were spatial scanning (variation of off-axis angle), refractive index scanning and mechanical scanning. The first of these produces only a small variation of optical path difference, which is inadequate for a Michelson interferometer. Refractive index scanning was proposed by Hirshberg [1974] and it is feasible for low resolution interferometers. However, it is generally true now that all Michelson interferometer users are interested in moderately high, if not very high, resolution, so that this method is not in current use. Thus mechanical scanning through the movement of one or more mirrors, or sometimes the beam-splitter, is the only method employed for atmospheric measurements. In this field, high resolution means a resolution of 0.002 cm^{-1}, which corresponds to a maximum optical path difference of 5 m, and a primary challenge in that is to maintain the interferometer in alignment during this long displacement of the optics. There are three main methods by which this has been achieved.

6.8.2 Cube Corner Reflectors

A cube corner consists of three mutually perpendicular mirror surfaces; literally the corner of a cube. It can be hollow, with mirror surfaces joined as at the corner of a room, at the floor or ceiling, or it can be a solid glass prism. Such devices are retro-reflectors, which return an incident ray precisely parallel to that of the incident ray, though displaced laterally, and thus they may be used to replace the mirrors in a Michelson interferometer. Since rotation of the cube corner about its vertex does not change the direction of the returned ray, the scanning mechanism need only be concerned with accurate displacement and not rotations. There are many other applications, for example, there are cube corner reflectors on the moon, which return laser beams accurately to the same point on the Earth. This allows precise determination of the distance to this astronomical body.

The properties of this retro-reflector were first worked out long ago by Peck [1948]. Peter Fellgett used corner cubes in his first interferometer [Fellgett, 1951] but they have not been much used in scanning instruments until recently, as is seen in Chapter 10. One exception is Ward [1988] but that was in a field-widened Doppler Michelson interferometer, as is discussed later.

6.8.3 Cat's Eye Retro-Reflector

The cat's eye retro-reflector consists of a concave spherical (or parabolic) primary mirror with a secondary mirror such that the focal point of the primary reflector falls on the secondary mirror. It then follows that the device acts as a retroreflector as shown in Figure 6.3. The detailed theory has been worked out by Beer and Marjaniemi [1966]. Pierre Connes was an early advocate of the cat's eye system, and a detailed description of a complex long-path ($\Delta_m = 1$ m) cat's eye system for astronomical use is described by Connes and Michel [1975]. As is evident in Figure 6.3, the emerging ray is not closely coincident with the incident ray, as in Figure 6.1, but comes out on the opposite side. This can be useful in applications where both exits from the interferometer are to be used, i.e. a double-output system.

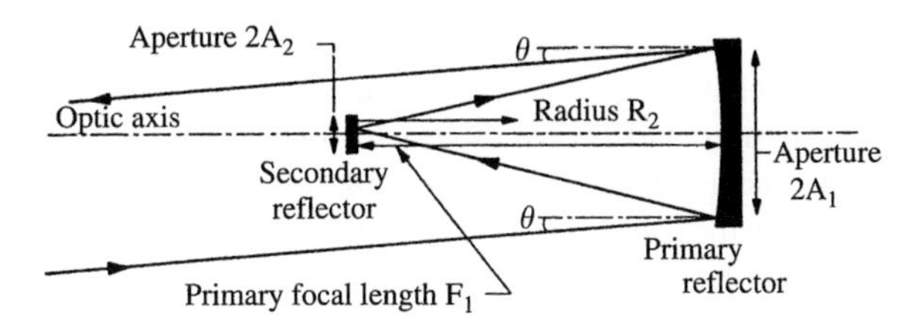

Figure 6.3. The cat's eye retro reflector. After Beer and Marjaniemi [1966].

6.8.4 The Dynamic Alignment System

Although retro-reflectors have the advantage of making it possible to separate the two outputs from the MI, the primary reason for their use is to maintain alignment during the scan. Another approach to this was developed by Henry Buijs, which he called dynamic alignment. In principle this is similar to the servo stabilized Fabry–Perot spectrometers described in the previous chapter. However, the FPS involves only very small movements while a long-path MI must first find the adjustment and then maintain it accurately during the rapid motion. This principle has been successfully implemented in the commercial ABB Bomem series of interferometers. The moving mirror moves vertically in a tower, allowing a movement of 2.5 m, a Δ_m of 5 m and thus a resolution of 0.002 cm^{-1}. There is an output optics selector, allowing different outputs to be used. By changing the beam-splitter and detector an enormous spectral range, from 200 nm to 1 mm may be covered, which is remarkable for one instrument.

6.9 SOME ATMOSPHERIC APPLICATIONS OF THE MICHELSON INTERFEROMETER

The first measurements of the 1.2–2.0 μm region of the near infrared spectrum of the upper atmosphere were made by Vallance Jones and Gush [1953] and Gush and Vallance Jones [1955] using a diffraction grating spectrometer. At that time Herbert Gush and the author were M.Sc. students at the University of Saskatchewan in Saskatoon. Subsequently both were PhD students of Harry Welsh at the University of Toronto, following which Gush moved to Jacquinot's laboratory for a postdoctoral fellowship. During that time he collaborated with Janine Connes in obtaining the first Michelson interferometer spectrum of the near infrared airglow [Connes and Gush, 1959, 1960], demonstrating the superiority of the Fourier transform method over that of the grating spectrometer used earlier. Both during his time in France and following his return to the University of Toronto, Gush was a key source of advice to the author and others on interferometric methods. Henry Buijs, one of the founders of the company now known as ABB Bomem, was his graduate student. Gush also encouraged the Canadian astronomical community to undertake the concept of using two spaced radio antennas as a Michelson interferometer, to measure the diameter of astronomical sources [Gush, 1988]. Instead of combining the signals on a beam-splitter, they are independently recorded, and then correlated afterwards when the two recordings are brought to the same site, using highly accurate time bases to accomplish the correlation. Thus Canadian astronomers executed the first *long baseline interferometry* experiment [Broten *et al.*, 1967], now widely used and known as *very long baseline interferometry* (VLBI).

One of Gush's first projects was to build a balloon-borne Michelson interferometer [Gush and Buijs, 1964] to observe the spectral region from 1.2 to 2.5 μm, part of which is absorbed in ground-based observations. A drawing of the instrument is shown in Figure 6.4. The spectrum obtained during a flight on Sept. 20, 1962 is shown in Figure 6.5. It was obtained from an altitude of 24 km, and the spectrum was calculated from 12 interferograms obtained over an interval of 144 min. The spectrum is dominated by the Meinel bands of hydroxyl; the vibrational quantum numbers are shown below the spectrum, covering a range for the vibrational level of the upper state from 2 to 9. One other feature, that of the $O_2(^1\Delta_g)$ (0, 0) band is evident. Another example of results obtained from a balloon-borne platform with a scanning MI, but at the opposite end of the infrared spectrum, was obtained by Naylor *et al.* [1981] and is shown in Figure 6.6. It covers the spectral range from 30 to 80 cm^{-1}, approaching the millimetre region. Unlike the example of Figure 6.5 which is an emission

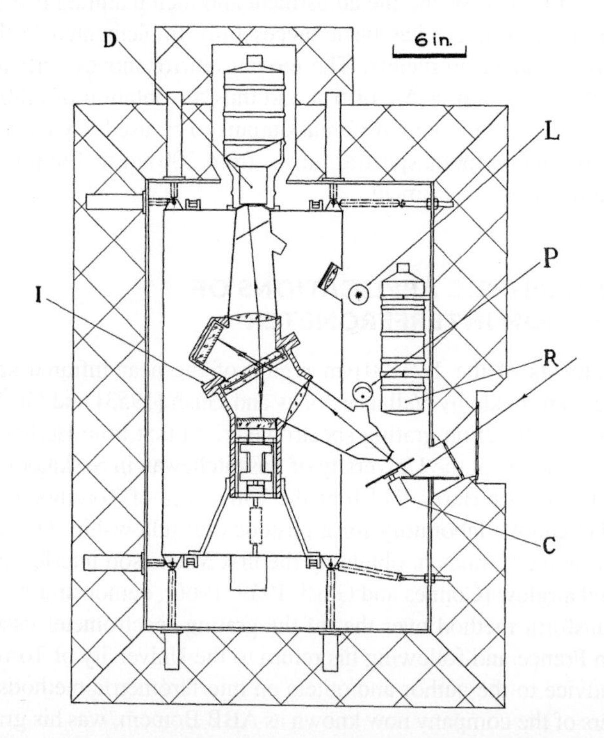

Figure 6.4. Cross-section of a balloon-borne infrared spectrometer. I, interferometer; D, cooled infrared detector; L, helium discharge lamp; P, photomultiplier; C, reflective chopper; R, cooled black body reference source. The incoming ray is seen just under the letter "R". After Gush and Buijs [1964].

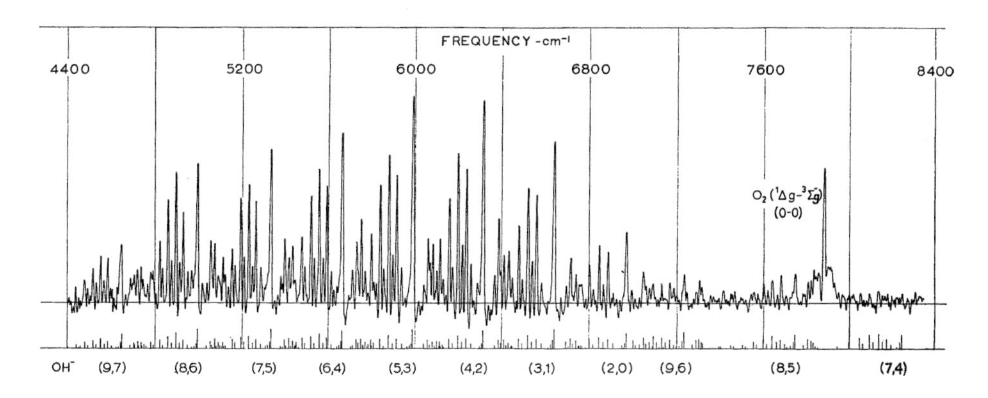

Figure 6.5. Spectrum of the 1.2–2.5 μm region taken from 24 km with a Michelson interferometer. The vibrational levels for the Meinel hydroxyl bands are shown. After Gush and Buijs [1964].

spectrum, the far-infrared spectrum is taken in absorption by viewing the sun, and yields the absorption spectrum of the stratosphere. Extensive measurements in the mid-IR have been made from balloons by Murcray *et al.* [1969].

This has become a very powerful method of satellite observation, in which the instrument views the sun at the Earth's limb as it rises and sets over the horizon; the apparent vertical motion of the sun caused by the motion of the satellite as it moves from daylight to darkness, or the reverse, generates a vertical profile of absorption measurements. This method was used by the ATMOS MI flown on the space shuttle; for which a sample of a spectrum was shown in Chapter 1.

The earliest spacecraft measurements were of emission spectra; an instrument called IRIS (Infra Red Interferometer Spectrometer) was developed and described by Hanel *et al.* [1970]. It was flown on Nimbus 3 and Nimbus 4. The latter instrument had a resolution of 2.8 cm^{-1} and a range of 6–25 μm. Similar instruments flew on the Mariner-9 and Voyager planetary missions [Hanel, 1983].

While measurement of the cosmic background radiation (CBR) is not an atmospheric measurement, it was one of the most remarkable observations of "sky" radiation made with a MI, for which the measured spectrum was shown in Chapter 1 [Gush *et al.*, 1990]. Because the wavelength is around 1 mm, this is on the boundary between optical and radio methods. The optical version required a very special all-reflective approach with no windows, as described by Gush and Halpern [1992]. The concept is shown in Figure 6.7. There are two inputs and two detectors. One input is from the sky where the field-of-view is defined by a reflective horn, H1. The other input is a 2.7 K black body B, radiating into horn H2. Light from both inputs is then reflected into a polarizing beam-splitter constructed of gold-coated tungsten wires stretched on an invar frame, an effective technique for this wavelength region. The Michelson mirrors are cones which ride on a common carriage that executes the OPD scan; the output radiation is brought to two detectors which record the difference between the interferograms of the two sources. That is, if the CBR had exactly the same spectrum as the blackbody, the recorded interferogram would be zero. In order to reduce the

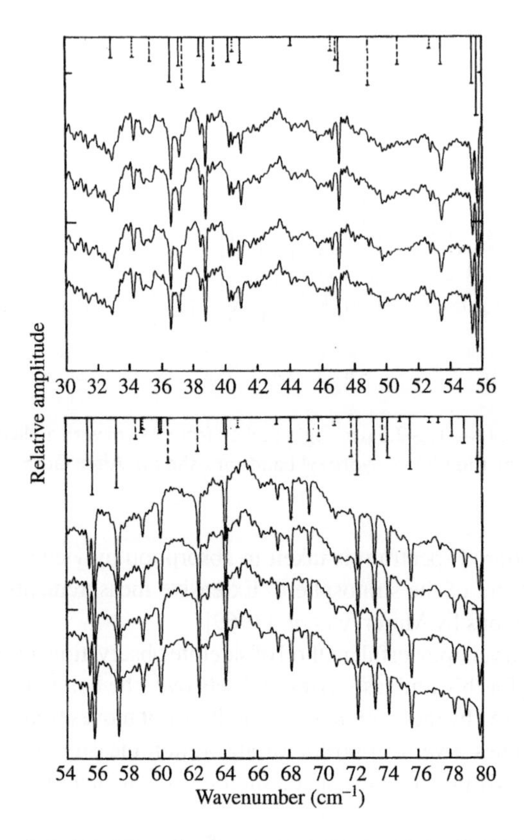

Figure 6.6. Spectrum of the 30–80 cm^{-1} region taken with a balloon-borne MI through the absorption of solar radiation. Vertical lines indicate positions of the main absorption features, with H_2O the solid lines, the Q branches of O_3 the dotted lines, and O_2 the dashed lines. After Naylor *et al.* [1981].

radiation from the instrument to acceptable levels it was cooled to 2 K with liquid helium. The detectors had to be cooled to an even lower value, 0.29 K and this was done using liquid ^3He. For the many interesting details of this remarkable sounding rocket experiment the reader is referred to the authors' description.

6.10 FIELD WIDENING

One interesting feature of the Michelson interferometer is that it is possible, at least in principle, to overcome the restriction on the field of view described in Section 6.4 and to make an instrument with a very large central fringes, and therefore large étendue, even at large resolving power. This is called field widening or field compensation. It may be explained simply in the following way. In Figure 6.1 the moving mirror is shown in two positions. In the first position the two mirrors are equidistant from the beam-splitter so that

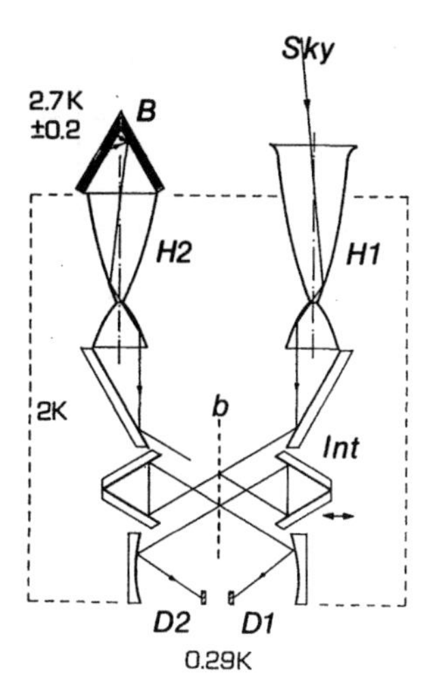

Figure 6.7. Concept for a rocket-borne Michelson interferometer to measure the Cosmic Background Radiation. H1 and H2 are reflective horns, while B is a blackbody. "Int" is for interferometer within which b denotes the beam-splitter and D1, D2 the detectors. The temperatures of the interferometer and detectors are shown. After Gush *et al.* [1990].

the two returning rays meet at the same point on the beam-splitter and emerge co-linearly. This corresponds to zero OPD. Since the ray paths form identical triangles in the two arms of the interferometer and will do so for all incident angles, the OPD is independent of incident angle and the instrument could collect a full hemisphere of radiation, all with the same OPD. However, since the OPD is equal to zero, no useful information is obtained. When the movable mirror is translated as shown in Figure 6.1 the symmetry is destroyed, the triangles are no longer identical and the off-axis angle is limited as described in Section 6.4.

The introduction of an OPD can be achieved in a different way, by introducing a refractive plate, with a mirrored surface on the back, into one arm of the interferometer as shown in Figure 6.8. When this is done a virtual mirror appears inside the plate, at a depth of t/n, where t is the thickness of the plate and n is the refractive index. This is the "depth of the swimming pool" effect as is shown in any elementary text; here it will be derived later. Since the ray in this arm appears to reflect from this virtual mirror, the mirror in the other arm can be made conjugate to it and the emerging rays are co-linear as before. Thus the OPD becomes independent of angle to a large extent but now there is a finite optical path difference because in the air arm the travel distance is $2t/n$ and in the refractive plate it is $2nt$, the difference being $\Delta = 2t(n - 1/n)$, which is, of course, the OPD.

The configuration of Figure 6.8 was first employed for airglow measurements by Hilliard and Shepherd [1966a] at the University of Saskatchewan and called by them a Wide Angle Michelson Interferometer (WAMI). Figure 6.9 shows a comparison of the size of the fringes for the field widened and conventional cases at a resolving power of 8×10^4. The relative location of M_1 and M_2 is critical, as a shift of a few microns can change the appearance of the fringes when the mirrors are near the wide-angle condition.

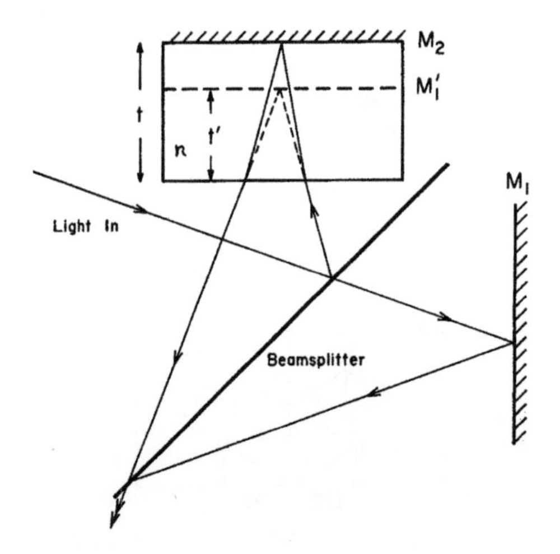

Figure 6.8. A Wide Angle Michelson Interferometer (WAMI), in which a refractive plate is used in one arm of the interferometer. After Hilliard and Shepherd [1966a].

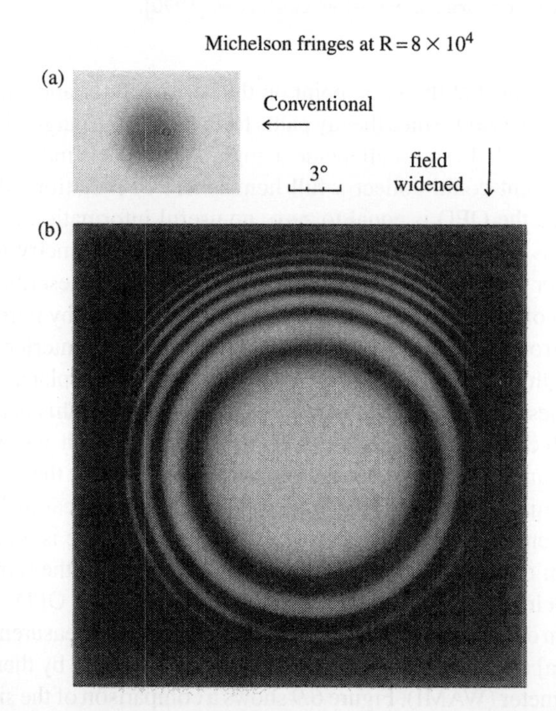

Figure 6.9. Michelson interferometer fringes (a) for an ordinary Michelson interferometer and (b) for a WAMI, with a resolving power of 80 000 for both configurations. These photos were taken by W.A. Gault, and are reproduced after Shepherd *et al.* [1991].

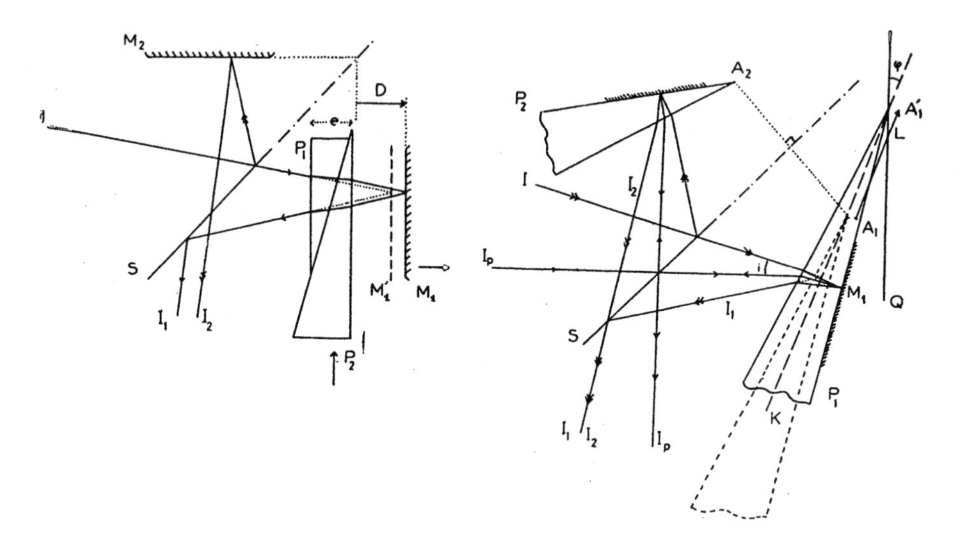

Figure 6.10. Configurations for field-widened Michelson interferometers after Mertz (left) and after Bouchareine and Connes [1963] (right), from which both figures are taken.

The first report of field-widening was by Hansen [1955] and Hansen and Kinder [1958], but the first spectral instrument making use of field widening was described by Connes [1956]. However, he did not use refractive plates to image one mirror on the other; he used an afocal telescope in each arm of the interferometer to perform the imaging. To scan, a mirror was moved in one arm and the telescope in the other arm was moved synchronously. Subsequent versions using prisms were described by Mertz [1959] and by Bouchareine and Connes [1963]. These are shown in Figure 6.10; Mertz [1959] used two prisms in one arm, and Bouchareine and Connes [1963] used one in each arm of the Michelson interferometer.

For a quantitative derivation, it is convenient to use the configuration of Figure 6.8, for which the two arms are superimposed in Figure 6.11. A beam of light approaches at incident angle i (in air) and enters the glass block with refractive angle θ. The beams reflect in mirror M_1 in one arm and in M_2 in the other arm. The difference in optical path travelled by the two wavefronts is as shown in Figure 6.11.

$$\Delta = 2na - 2b - s, \quad \text{where } a = \frac{t}{\cos\theta}, \quad b = \frac{t'}{\cos i} \tag{6.25}$$

and

$$s = 2\sin i(t\tan\theta - t'\tan i) \tag{6.26}$$

so that

$$\Delta = 2t\left(\frac{n}{\cos\theta} - \tan\theta\sin i\right) - 2t'\left(\frac{1}{\cos i} - \tan i\sin i\right) \tag{6.27}$$

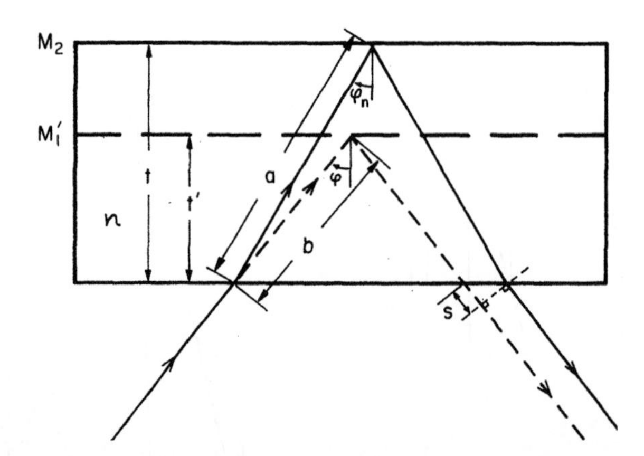

Figure 6.11. WAMI geometry for the calculation of optical path difference. Note that φ and φ_n are used in place of the i and θ employed in the text. After Hilliard and Shepherd [1966a].

Application of Snell's law gives $\Delta = 2nt\cos\theta - 2t'\cos i$ which can also be written in the following form:

$$\Delta = 2nt\left(1 - \frac{\sin^2 i}{n^2}\right)^{1/2} - \sqrt{2t'(1-\sin^2 i)} \tag{6.28}$$

and binomial expansion yields

$$\Delta = 2(nt - t') - \left(\frac{t}{n} - t'\right)\sin^2 i - \frac{1}{4}\left(\frac{t}{n^3} - t'\right)\sin^4 i \cdots \tag{6.29}$$

The $\sin^2 i$ dependence of Δ on angle can be eliminated by choosing $t' = t/n$, i.e., by locating mirror M_2 at the virtual image of M_1 as was done in the original simple explanation. Under this condition, the path difference at normal incidence becomes:

$$\Delta_o = 2t\left(n - \frac{1}{n}\right) \tag{6.30}$$

and the variation of OPD with off-axis angle can be written:

$$\Delta_o - \Delta = \frac{\Delta_o}{8n^2}\sin^4 i \approx \frac{\Delta_o}{8n^2} i^4 \tag{6.31}$$

Compared with the expression for the conventional MI in (6.17), it is seen that the quadratic dispersion has been replaced by one of fourth order. The dependence on incident angle has been greatly reduced, resulting in a large central fringe as shown in Figure 6.9. In fact, the larger the path difference, the greater the solid angle advantage over the conventional Michelson. Using $d\sigma = 1/\Delta_m$ and Eq. (6.31), and setting $\Delta_o - \Delta = \lambda$ as before, the superiority of a wide angle Michelson interferometer (WAMI) is given by:

$$S_{WAMI} = \Omega \mathcal{R} = \pi i^2 \mathcal{R} = \pi \sqrt{\frac{\Delta - \Delta_o}{\Delta}}\, 2\sqrt{2}\, n \frac{\sigma}{d\sigma} \tag{6.32}$$

$$S_{WAMI} = 2\sqrt{2}\pi n \sqrt{\lambda d\sigma} \frac{\sigma}{d\sigma} = 2\pi n \sqrt{2\mathcal{R}} \tag{6.33}$$

The superiority gain over the ordinary MI in air is thus:

$$\frac{S_{WAMI}}{S_{MI}} = n\sqrt{2\mathcal{R}} \tag{6.34}$$

This gain increases with resolving power and for $\mathcal{R} = 10^5$ and $n = 1.5$ for a typical glass, the gain is 670.

The above discussion refers to a fixed path difference given by Eq. (6.30); it applies directly to the two-prism systems shown in Figure 6.10 which is a way to vary the thickness of the glass plate. These methods require moving large prisms very smoothly and very accurately and, in some cases, keeping two moving components synchronized very precisely. The engineering problems are formidable. A bewildering variety of instruments have been suggested, some using prisms as shown above and others using focussing with confocal optics. These have been thoroughly reviewed by Ring and Schofield [1972] and by Baker [1977]. There are many suggested methods described in the literature and few reports of success. Elsworth *et al.* [1974] had some success in the visible region where the enhanced superiority is a particular advantage. With the same motivation a scanning wide angle Michelson interferometer (SWAMI) for the visible region was built at York University based on the dynamic alignment principle [Lu *et al.*, 1988a, 1988b].While this system shown in Figure 6.12 looks complex, this is because it involved a new design for the desensitizing of response to component alignment as well as a computer-controlled dynamic alignment system. The objective was to use commercially available positioners under computer control. It performed extremely well, although it was not built to the engineering level required for field use.

The most notable success has been obtained at Utah State University in the near infrared, where tolerances are somewhat less severe than in the visible [Baker *et al.*, 1981], but remarkable engineering skills had to be employed to make this feasible. A gas-lubricated precision slide and drive was specially developed; it should be noted that the drive distances are larger in a field-widened interferometer than for the conventional instrument of the same resolving power. A sequence of cryogenic instruments was built and flown in rockets as described by Baker *et al.* [1981]. Liquid nitrogen was used to cool the calcium fluoride optics and a liquid helium-cooled bismuth-doped silicon detector was used to cover the spectral range from 2 to 8 μm at a resolution of 2.7 cm^{-1}. To match the linear expansion of the calcium fluoride optics the interferometer structure was machined from a special alloy of high-carbon nickel manganese iron that was custom fabricated for this instrument. The étendue was 0.5 cm^2 sr, an enormous value, allowing airglow spectra to be obtained in 1.3 s during a rocket flight. A photograph of the instrument is shown in Figure 6.13. The detailed historical development is described by Baker *et al.* [1981] along with much of the historical development of the Michelson interferometer for atmospheric applications in general, and the wide-angle instrument in particular.

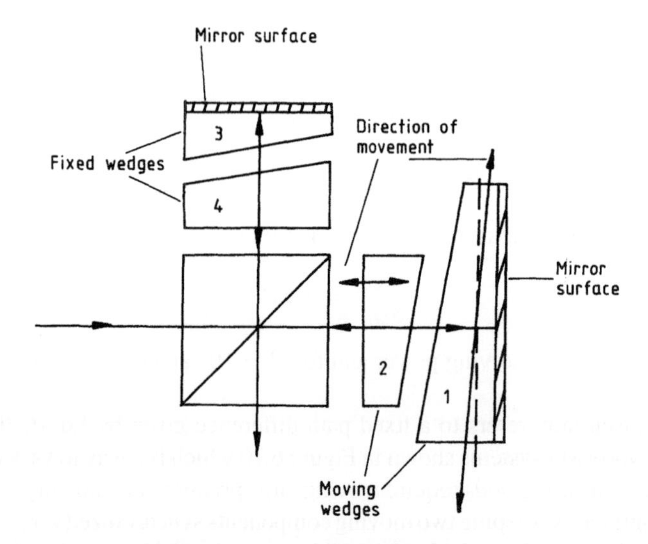

Figure 6.12. A scanning field-widened Michelson interferometer by Lu [1988a, 1988b]. As wedge 1 is moved laterally to change the path difference, wedge 2 moves parallel to the optical axis to maintain compensation against astigmatism by keeping the diagonal gap equal in the two arms.

Figure 6.13. USU rocketborne liquid-helium cooled field-widened Michelson interferometer (RBFWI) flown aboard a Sergeant rocket at Poker Flat, Alaska, on April 13, 1983, obtained high resolution LWIR spectra of an aurora. Courtesy of Doran Baker (Utah State University).

6.11 PROBLEMS

6.1. Referring to $d\sigma = 1/2\Delta_m$, where the instrumental function width is given in terms of Δ_m, confirm the statement that this resolution element corresponds to the full width of the passband at 64% of its peak value.

6.2. In Section 6.4, J. Connes' criterion for aperture size is presented. Consider an alternative criterion, where the figure of merit is the solid angle, multiplied by the visibility. Proceed by constructing a series of apertures of angular radius such that $\Delta_o - \Delta = 0.01, 0.02, \ldots, 1.0$ wavelengths. For each of these elementary annular apertures the solid angle is the same, as was the case for the Fabry–Perot spectrometer. But, for a circular aperture containg $1, 2, 3, \ldots$ of these elementary annuli, the solid angle increases, increasing the signal level in proportion, but the visibility decreases, because of the phase shift from one aperture element to the next. Perform the sum of the cosine signals using vectors to represent the phases, and multiply by the solid angle for the corresponding aperture. Plot this figure of merit versus $\Delta_o - \Delta_{max}$, where Δ_{max} corresponds to the outside of the aperture, and identify the value of $\Delta_o - \Delta_{max}$ giving the largest figure of merit.

6.3. Consider a wide-angle Michelson interferometer operated at 500 nm with an optical path difference of 4.0 cm. Using a glass plate of refractive index 1.5, compute t and t' for conventional field widening. Plot $\Delta - \Delta_o$ versus i for this interferometer and determine the value of i for which $\Delta - \Delta_o = 0.5\lambda$. Now let $t/n - t' = \xi$, and let ξ take on the values $-0.002, -0.001, 0.000, 0.001$ and 0.002. Plot $\Delta - \Delta_o$ versus i for each of these values, and determine the value of ξ that gives the largest value of i for $\Delta - \Delta_o = 0.5\lambda$. This enhancement of the field-widening is called defocussing.

6.4. Consider an interferogram given by:

$$I(\Delta) = \cos 2\pi(19\,700\Delta) + 5\cos 2\pi(20\,100\Delta) + 3\cos 2\pi(20\,500)\Delta$$

(a) Simulate the output of a Michelson interferometer by calculating $I(\Delta)$ over the range $\Delta = 0$ to $\Delta_{max} = 0.04$ cm at an interval of 2×10^{-5} cm; that is, with 2000 points. Simulate the effect of the finite path difference by multiplying by an apodizing function that is unity from points 0 to 800, and from 1200 to 1999 and zero everywhere else. By using a symmetrical function, the spectrum will be real. Calculate and plot the FT of this interferogram, which corresponds to a Δ_{max} of 0.016 cm.
(b) Repeat with $\Delta_{max} = 0.008$ cm and discuss the differences between the computed spectra of (a) and (b).

6.5. (a) Repeat the instructions of Problem 6.4(a) but apodize the interferogram before computing the FT, using the apodization function of Eq. (6.13) for a Δ_{max} of 0.016 cm.
(b) Repeat (a) but with the apodizing function of Eq. (6.14) and the same Δ_{max}. Compare the lineshapes of (a) and (b), and with the lineshape for the unapodized function.

6.6. Extract the spectral line of Problem 6.4 at 20 500 cm^{-1} from the interferogram by filtering, which for the interferogram means convolution by $G(\Delta)$ as given by:

$$G(\Delta) = \frac{\sin \pi(\sigma_{max} - \sigma_{min})\Delta}{\pi(\sigma_{max} - \sigma_{min})\Delta} \cos 2\pi(\sigma_{max} + \sigma_{min})\Delta$$

where the sinc function is the FT of the square filter function, and the cosine term results from the shift of this function from zero to σ. Use $\sigma_{min} = 20\,400$ and $\sigma_{max} = 20\,600\ \text{cm}^{-1}$. Calculate and plot the spectrum of the filtered function.

7

MULTIPLEXERS AND MODULATORS

7.1 SPECTRAL OPERATING MODES

Of the devices discussed so far, the diffraction grating and Fabry–Perot devices produce spectra and are therefore called spectrometers, while the Michelson instrument produces an interferogram and for that reason is called an interferometer. While the Fabry–Perot and Michelson instruments have the same superiority, since they both enjoy the "Jacquinot advantage", the Michelson interferometer has an additional advantage, through multiplexing. In this chapter other instruments are described which have a multiplex advantage. These instruments are neither spectrometers nor interferometers, but take different forms.

The multiplex advantage arises from the fact that each wavelength measured is observed for the whole time of the observation, unlike the conventional spectrometer, where the wavelengths are observed sequentially in time. For example, the MI interferogram is a superposition of cosinusoids, each one corresponding to one wavelength element, and all are simultaneously observed. Thus if there are P spectral elements, the signal for each element is increased by a factor of P, giving an improvement in the signal-to-noise ratio of $\sqrt{P}$. This advantage thus increases with the number of spectral elements viewed, which tends to increase with the resolving power, though not necessarily. If one wishes to measure the width of a single line, for example, only a few spectral elements are involved, even though the resolution is large.

As mentioned in the last chapter, the multiplex advantage is a clear advantage for detectors which are limited by thermal noise, since for them the noise level is the same whether the detector is exposed to light from one spectral element or P spectral elements simultaneously. For a visible region detector, such as a photomultiplier which is shot noise limited, there is a difference because the noise is determined by the signal level, being equal to $\sqrt{N}$ for a signal of N photoelectrons. Thus for a multiplexed system of P spectral elements, the signal level is enhanced to PN, and the noise increased to $\sqrt{PN}$, an increase of $\sqrt{P}$, a factor which is exactly offset by the multiplex gain. But this is true just for a continuous spectrum with equal signal in all spectral elements. This is the situation where one observes an absorption spectrum, for example, which is often the case in the infrared. For an emission line spectrum in the visible region having a few (say P') spectral lines scattered sparsely across a wide spectral range, then the increase in noise level is $\sqrt{P'}$ rather

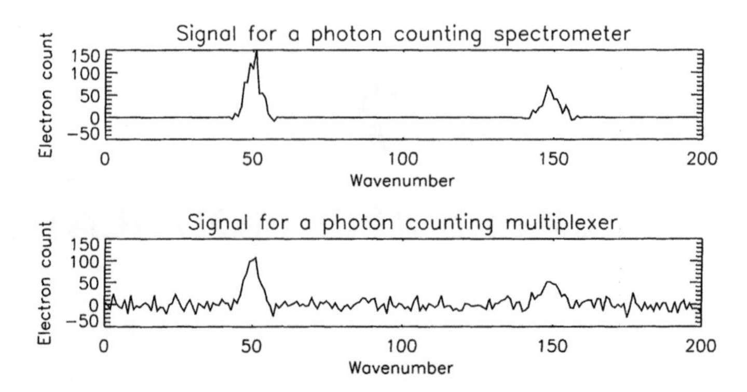

Figure 7.1. Noise distribution for a spectrometer (above) and a multiplexer (below).

than $\sqrt{P}$, and there is a multiplex advantage. Thus while it is often said that in the visible region the Fellgett advantage becomes a disadvantage, this is not so in general. At worst, it is a neutral advantage, and for a sparse spectrum it is a positive advantage.

The only situation in which multiplexing is a disadvantage is where one wishes to measure a weak emission line in the presence of strong ones. Then the shot noise from the strong features is imposed on the weak feature, which is definitely a disadvantage. Further thought about this leads to a realization that the true distinction between the noises in a spectrometer and a multiplexer is how the noise is distributed across the spectrum. As illustrated in Figure 7.1, for a spectrometer, the noise follows the signal level, so is large at the top of a spectral line, but small (zero, apart from some thermal noise) in between spectral lines. For the multiplexer, the noise is uniformly distributed across the spectrum, so is the same at the spectral peak as it is between the spectral lines, as is shown in the same figure.

7.2 MULTIPLEXERS

7.2.1 Introduction

The Michelson interferometer, presented in the previous chapter, employs a very subtle form of multiplexing. All of the light entering the device reaches the detector, but the instrument codes the different wavelengths in a way that they can be distinguished one from the other, even though they are superimposed into one output signal. Each wavelength in the spectrum is coded with the recorded frequency in the interferogram. The spectrum is then recovered through Fourier transformation; but other codes, such as Hadamard codes, may be used. Jacques Hadamard is the renowned French mathematician responsible for this orthogonal transformation (like the Fourier transform), that is binary in nature.

7.2.2 The Hadamard Spectrometer

The concept is described for a diffraction grating spectrometer in which the coding is carried out spatially, in the dispersion direction of the spectrometer. Suppose that the exit slit is replaced by a strip mask of N segments, each segment corresponding to one spectral element.

Each segment is either closed (opaque) or open (transmitting) so that the transmission area A_v of the vth segment is 0 or 1 respectively. Half of the segments are closed and half are open. If the radiance at the exit slit of the vth element is L_v and all of the light transmitted by the mask is measured with a single detector then the interferogram signal measured by the detector is:

$$I = \sum_{v=1}^{N} A_v L_v \tag{7.1}$$

Suppose then that the mask is replaced by one mask after another, with N different masks altogether. This yields N values of I_u as follows.

$$I_u = \sum_{v=1}^{N} A_{uv} L_v \tag{7.2}$$

Since there are N measured values of I_u and N unknown values of L_v, it is possible in principle to solve the set of simultaneous equations and obtain the desired set of L_v, the spectrum. The gain arises in that each spectral element is observed $N/2$ times, yielding a corresponding gain of $\sqrt{N/2}$ in signal-to-noise ratio. Nelson and Fredman [1970] have shown that, if Hadamard codes are used to construct the mask, this recovery is not only possible, it is optimum as regards the noise.

This technique, known as Hadamard spectroscopy became widely known through the work of Decker [1969, 1971a, 1971b] and his colleagues. A detailed review is given by Decker [1977]. The relation between I_u and the L_v can be written in matrix form.

$$I = AL \tag{7.3}$$

$$L = A^{-1} I \tag{7.4}$$

Below are shown the matrices $\mathbf{A}$ and $\mathbf{A}^{-1}$ for a seven-step code. A description of how to construct Hadamard matrices has been given by Baumert [1964] and Nelson and Fredman [1970].

$$A = \begin{bmatrix} 1 & 1 & 0 & 1 & 0 & 0 & 1 \\ 0 & 1 & 1 & 1 & 1 & 0 & 0 \\ 1 & 0 & 1 & 1 & 0 & 1 & 0 \\ 0 & 0 & 0 & 1 & 1 & 1 & 1 \\ 1 & 1 & 0 & 0 & 1 & 1 & 0 \\ 0 & 1 & 1 & 0 & 0 & 1 & 1 \\ 1 & 0 & 1 & 0 & 1 & 0 & 1 \end{bmatrix}$$

$$A^{-1} = \left[\frac{2}{M+1}\right] \begin{bmatrix} 1 & -1 & 1 & -1 & 1 & -1 & 1 \\ 1 & 1 & -1 & -1 & 1 & 1 & -1 \\ -1 & 1 & 1 & -1 & -1 & 1 & 1 \\ 1 & 1 & 1 & 1 & -1 & -1 & -1 \\ -1 & 1 & -1 & 1 & 1 & -1 & 1 \\ -1 & -1 & 1 & 1 & 1 & 1 & -1 \\ 1 & -1 & -1 & 1 & -1 & 1 & 1 \end{bmatrix} \tag{7.5}$$

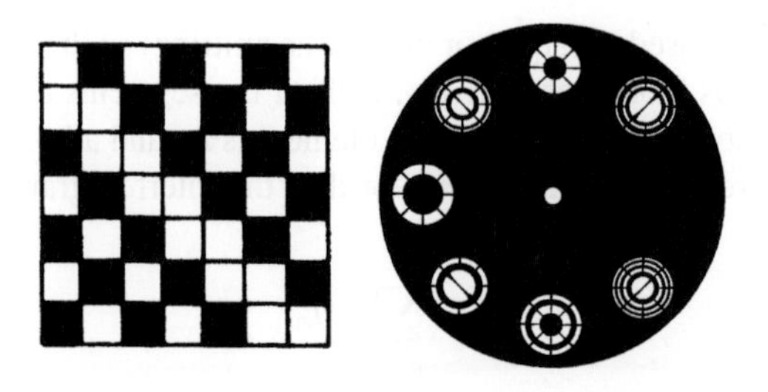

Figure 7.2. Hadamard coded rectangle (left) and disk (right). After Shepherd *et al.* [1978].

Seven-step codes implemented as a rectangular grid, and as a stepped disk are shown in Figure 7.2. Note that the rectangular code is cyclic, that is in going down one row, the pattern simply shifts to the right by one element (and wraps around from the right column to the left column). This simplifies the operation considerably as the fabricated mask then need not be a square array, but simply two rows, one after the other, forming a single strip. To move from measurement 1 to measurement 2, the mask is simply shifted to the right by one spectral segment. For N segments the mask requires a strip of $2N$ segments to allow for N shifts, rather than consisting of a mask of N^2 segments, and is much simpler to use. For the circular pattern required for an FPS, or interference filter photometer, it is necessary to have N segments, each of N elements as shown in Figure 7.2. This is the rotating mask used by Deans and Shepherd [1978] for the rocket measurements described in Chapter 1. The instrument concept is described by Shepherd *et al.* [1978]; earlier results were presented by Neo and Shepherd [1972] for ground-based airglow photometry.

7.2.3 Grating Spectrometers with Array Detectors

With the advent of imaging detectors, a simpler kind of multiplexing is made possible. Since the rays corresponding to different wavelengths are dispersed inside the instrument, they exit at different angles and therefore arrive at different locations on the imaging detector, placed at the focal point of an exit collimator. Thus for a diffraction grating instrument, for example, different wavelengths fall on different elements of the detector and all wavelengths are recorded simultaneously. This multiplexed signal may be recorded with a linear (one-dimensional) array detector. If a two-dimensional (array) detector such as a CCD is used, then spatial information may be obtained in the direction parallel to the slit. In other words, one CCD dimension may be used to record the spectrum and the other to obtain a one-dimensional image. Such devices are effectively multiplexers, although that is not the conventional classification. Specific configurations are discussed in detail in Chapter 11.

7.3 MODULATORS

7.3.1 The SISAM

One approach to the problem of achieving the simplicity of the grating spectrometer with the advantages of the MI was proposed by Pierre Connes [1957, 1958b], in which the

Michelson mirrors were replaced by diffraction gratings. SISAM stands for Spectromètre Interférential a Sélection par l'Amplitude de Modulation. Connes invented this device with the idea of creating a diffraction grating instrument having the étendue of a Michelson or Fabry–Perot interferometer. Jacquinot [1960] makes the following distinction between the SISAM and the Michelson interferometer; calling the former an instrument that works by "amplitude of interferometric modulation", while the Michelson interferometer works by "frequency of interferometric modulation". Here the approach is taken that the Michelson interferometer is a multiplexer, while the SISAM is the first example of a modulator.

Jacquinot's [1960] description begins with a Michelson interferometer in which each mirror is tilted by an angle $\varepsilon/2$, causing a rotation by angle ε at each mirror, and thus causing interference at an angle of intersection 2ε. Across the width D of the mirror, this angular difference corresponds to $D\varepsilon$ in distance and $2D\varepsilon$ in Δ. The result is $2D\varepsilon/\lambda$ straight-line fringes across the aperture, separated in distance by $\lambda/2\varepsilon$. For rectangular mirrors the interferogram has, for monochromatic light, a visibility given by the FT of a rectangular function of width $2D\varepsilon$:

$$V = \frac{\sin 2\pi\sigma D\varepsilon}{2\pi\sigma D\varepsilon} \tag{7.6}$$

where the denominator is adjusted to normalize $V = 1$ at $\varepsilon = 0$.

Now consider the SISAM, in which the Michelson mirrors are replaced by diffraction gratings as shown in Figure 7.3. The gratings are set at an angle such that one particular wavelength of light is returned to the beamsplitter along the same path, for both of the gratings. For that wavelength, and that wavelength only, the device behaves as a Michelson interferometer, and produces proper fringes. The output samples an interferogram at that wavenumber, say σ_a. For wavenumber σ, the light is not returned along the same path, so that the two outgoing beams are rotated in opposite directions. The angle is obtained from Eq. (3.17), $m\lambda = 2t \sin i$; or t sin i here, because only the diffracted angle changes. So $m\,d\lambda = t \cos i\, di$ and recalling that $d\lambda/\lambda = d\sigma/\sigma$ and setting $di = \varepsilon$ and $d\sigma = (\sigma - \sigma_a)$ the result is:

$$\varepsilon = 2 \tan i_a \frac{(\sigma - \sigma_a)}{\sigma} \tag{7.7}$$

By putting $\Delta_m = 2D \tan i_a$ (maximum value of the path difference) in Eq. (7.7), the following visibility is obtained from Eq. (7.6):

$$V = \frac{\sin 2\pi(\sigma - \sigma_a)\Delta}{2\pi(\sigma - \sigma_a)\Delta} \tag{7.8}$$

Since the visibility is the modulation of the interferogram, it is the extracted signal for this modulator. This equation is also the equation for the instrumental function, and since the first zero of the visibility is at $(\sigma - \sigma_a) = 1/2\Delta$, this is also the resolution. However, Eq. (7.8) is not a desirable instrumental function, so the instrument would be apodized, using a diamond-shaped mask, as described earlier. The resolution is the same as the maximum possible resolution of the grating spectrometer.

The instrument is operated by rapidly oscillating the compensating plate through a small angle to produce a modulation of the Michelson fringes at a given wavenumber, and by synchronously rotating the two gratings to scan through the spectrum. The degree of

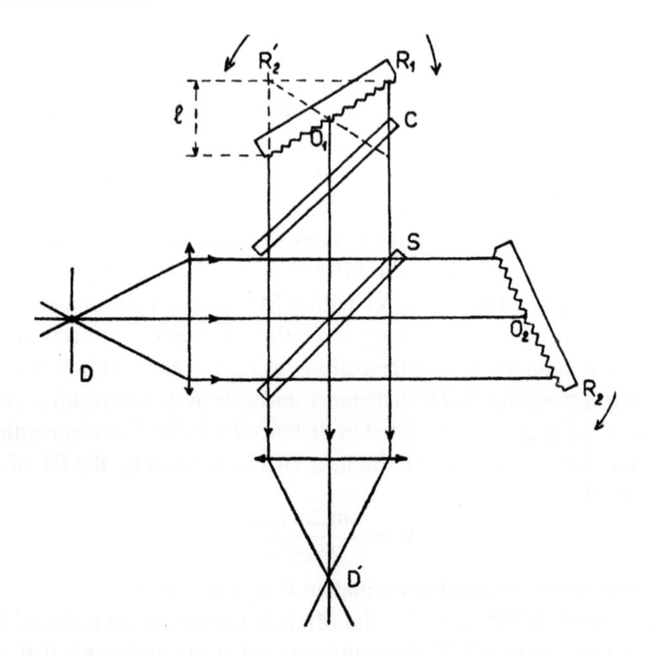

Figure 7.3. The SISAM of Pierre Connes. After Connes [1958b].

modulation of the output light is proportional to the signal at that wavelength. As far as is known, Connes is the only person to build such a device. He did this because at that time, the computation of the Fourier transform was still a major obstacle in using the MI in its simple form. With the SISAM, the mechanical problem of synchronously rotating the gratings at a uniform rate was extremely difficult. Another problem was that the resolution of SISAM could not be adjusted after construction, it was fixed to be the limiting resolution of the chosen gratings. In summary, the SISAM had a resolution fixed by the grating spectrometer, and an angular acceptance governed by the Michelson equation. Altogether, it was an ingenious but very inflexible instrument.

7.3.2 The Birefringent Photometer

The birefringent filter was invented by Lyot [1944] and has been used in many later applications. Several designs have been described by Evans [1949]. The principles of operation are described by one of its practitioners in the early years of contemporary upper atmospheric observations, Donald M. Hunten [Hunten *et al.*, 1967]. The device is fundamentally a two-beam interferometer with close similarities to the Michelson interferometer discussed in the previous chapter; but since the interferogram is used only to modulate the spectrum at a single wavelength, it is properly classed as a modulator.

Diagrams of both the ordinary and "split" birefringent photometer are shown in Figure 7.4. A linear polarizer at the entrance, labelled simply as "Polaroid", generates

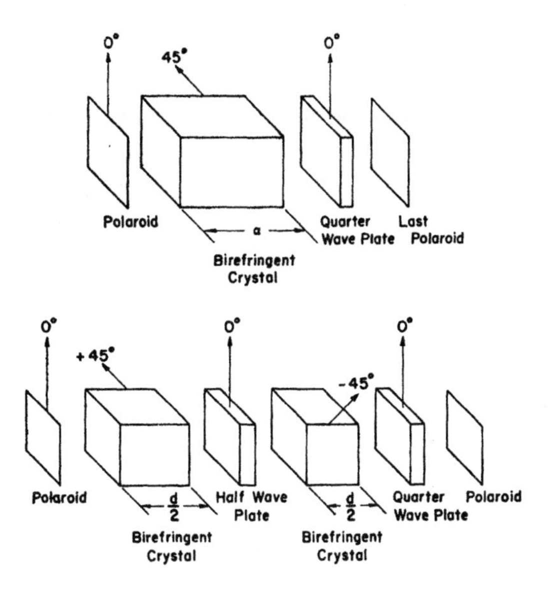

Figure 7.4. The optical system of the birefringent photometer showing the ordinary configuration above and the split configuration below. After Hunten et *al.* [1967].

linearly polarized light that enters the birefringent crystal and is divided into two beams, one polarized along the direction of the optical axis of the crystal, and the other polarized perpendicularly to it. The refractive indices are different for the two so that an optical path difference develops as the beams traverse the crystal. The beams are then mixed in the quarter wave plate which follows the crystal; the plate generating circularly polarized light from each beam, with the vectors for each rotating in opposite directions. The sum of these two counter-rotations is a single beam of linearly polarized light making an angle that depends on the phase difference between the two beams, or thus on the wavelength. This polarized component is extracted by detecting the modulated signal produced by rotating the "Last Polaroid" following the quarter wave plate. A more detailed explanation of this process is given in the next chapter, in the description of a polarizing Michelson interferometer.

The measurement problem to which Hunten and his students applied this method was the detection of weak atomic emissions in twilight. Following sunset, metallic species such as sodium, potassium or lithium, which are deposited through the impact of meteorites, fluoresce in the solar-illuminated upper atmosphere, and can be detected at the ground where the lower atmosphere is no longer in sunlight. The Rayleigh scattered background from the middle atmosphere is still considerable, but this background is not modulated by the birefringent photometer, which is set up for the specific wavelength of the metallic line to be detected. The signal generated at the detector contains the integrated background light from the scattered spectrum contained in a limited band formed with an interference filter, and a cosinusoidal component with a specific phase; this is extracted electronically using a *phase-sensitive detector*, with the reference phase generated by some transducer on the rotating polaroid.

The superiority of the birefringent photometer is determined by the variation of optical path difference with off-axis angle. Some of this variation can be cancelled by splitting the birefringent crystal and placing a half-wave plate in between. The superiority of this split system is given below by Hunten *et al.* [1967]. In this equation n_o and n_e are the ordinary and extraordinary indices of refraction and Δ_m is the maximum difference of optical path allowed between a ray at normal incidence and a ray with maximum incident angle.

$$S_{BP} = \Omega\Re = \left[\frac{\Delta_m}{\lambda}\right] 4\pi n_o n_e \frac{2n_o}{n_o - n_e} \tag{7.9}$$

If $\Delta_m \approx (\lambda/2n_o n_e) \approx \lambda/5$ then the superiority can be written as:

$$S_{BP} = 2\pi \frac{2n_o}{n_o - n_e} \tag{7.10}$$

In this equation, the factor of 2π is the superiority of the the single element design of Figure 7.4 (above), which is the same as for an ordinary Fabry–Perot spectrometer or Michelson interferometer and the factor following is the gain of the split design over that with the single element. For a wavelength of 590 nm, this factor is 20 for calcite, 65 for ammonium dihydrogen phosphate (ADP) and 340 for quartz. Thus the split system is field-widened, or is a higher order device, very similar to the Doppler Michelson interferometers described in the following chapter, except that it is not feasible to construct birefringent photometers with large optical path differences.

7.3.3 The Grille Spectrometer

This device is simply a spectrometer, with the single slit replaced by a grille that admits much more radiation than a single slit, but modulates one wavelength at a time. The early principles were described by Golay [1951]. This concept was later implemented by Girard [1963] for middle atmospheric measurements and by Tinsley [1966] for airglow measurements. An example of the circularly symmetric grille employed by Tinsley [1966] is shown in Figure 7.5.

Identical grilles are located at the locations of the entrance and exit slits. In Figure 7.5 one grille pattern represents the image of the entrance grille for one particular wavelength, and the other is the exit grille itself. For this particular wavelength the two grille patterns are not superposed, but displaced. This produces what is called a Moiré pattern of light and dark regions. Some fraction of the light, corresponding to the total area of the white spaces, is collected by a lens and focussed onto a detector. For a second wavelength, adjacent to the first one, the entrance image is shifted, and the overlap pattern changed, but the integrated radiance reaching the detector remains the same. However, for the one wavelength for which the patterns are superposed, much more light reaches the detector, about twice as much. Since the radiance for this one wavelength produces a signal component that is different from all the other wavelengths, the signal may be said to be modulated. The manufacture of

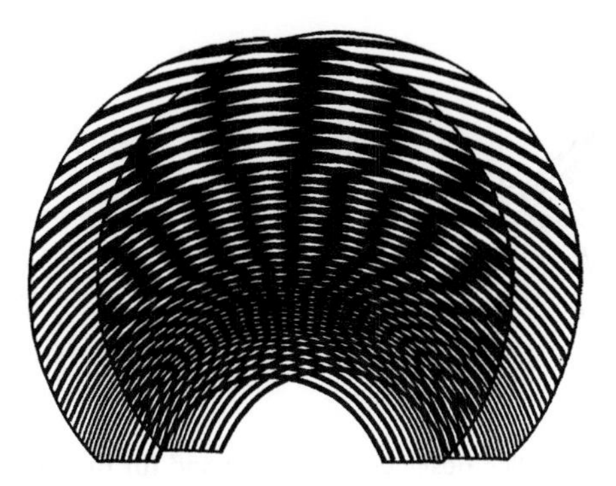

Figure 7.5. A circularly symmetric grill, with the input and output grilles overlaid. Adapted from the single grille pattern shown by Tinsley [1966].

the grilles is no simple matter, as great accuracy is required. In fact, this approach is used as a basis for the testing of optical elements, lenses and mirrors. If a grille is re-imaged upon itself with the optical element, and the coincidence is not exact, Moiré fringes appear and these can be used to diagnose the imperfections in the optical element. This type of measurement was highly developed by Ronchi [1964] and rectangular grilles are often called Ronchi gratings.

Of course, modulation with the grille spectrometer occurs only if there is light at the particular wavelength that is modulated. Thus in practice the spectrometer is scanned by rotating the diffraction grating in the same way as a conventional spectrometer. As the spectrometer scans, the signal remains constant (corresponding to the integrated radiance from all wavelengths reaching the detector), and it departs from that value whenever a line in the source spectrum corresponds to the superposed grilles. Thus the signal looks like that from an ordinary spectrometer. It truly is a spectrometer but, because of this different mode of operation, must be considered a modulating spectrometer.

For a real instrument one must consider the distortion produced by the grating optics, or by the grating itself [Laws, 1962]. For this reason each grille has to be made specifically for each spectrometer and even for each wavelength band. The entrance grille is created first, and then positioned on the spectrometer. Monochromatic light is used to form an image at the exit location, and a photograph taken. This photograph becomes the exit grille.

The instrumental passband is the autocorrelation between the two grilles in the direction in which they are offset; the result depends on the actual grille employed. Tinsley chose the circularly symmetrical grille for an Ebert–Fastie spectrometer for which circular slits are used to minimize distortion. Girard [1963] used parabolic patterns arranged in quadrants as shown in Figure 7.6(a), with the corresponding instrumental passbands in Figure 7.6(b). This instrument was flown by Girard *et al.* [1988] in space shuttle missions for atmospheric remote sensing applications.

Tinsley's [1966] instrument was applied in the visible region and performed very well. However, because of the photon (photomultiplier) detector, the noise went as the square root of the photon count of the entire signal, not just the signal associated with the spectral

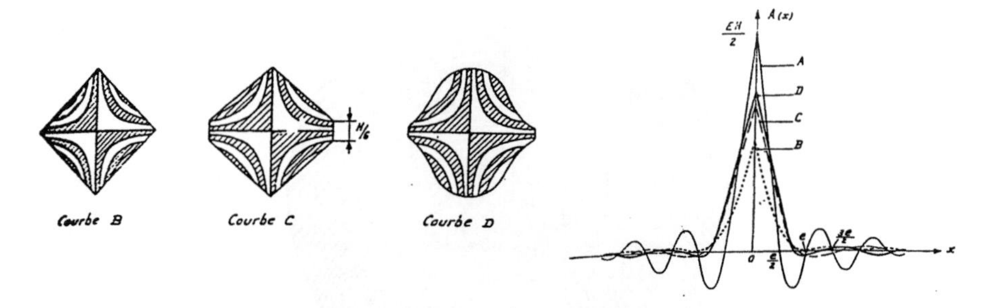

Figure 7.6. (a) Examples of grilles designed by Girard; and (b) their corresponding instrumental passbands. After Girard [1963].

line being detected at a given time. Thus the instrument was used in a band-limited region, defined by an interference filter. In the infrared, where the instrument was applied by Girard, the noise is limited by thermal considerations and this is not a problem. Thus modulators are more effective in the infrared.

7.3.4 The Correlation Spectrometer

This is a interesting variation on the grille grating spectrometers in which a mask is used that is a replica of the spectral pattern one wishes to detect, say the spectrum of SO_2, for example. Any spectral content of SO_2 will correlate with this pattern, while other spectral contributions will not correlate, and so will not contribute to the observed signal. The idea appears to have originated with the renowned optical experimentalist John Strong. In a paper concerning a balloon-borne infrared telescope for astronomical observations, Strong [1967] describes a spectrometer with a focal plane mask consisting of a number of slits. He called them Benedictine slits in recognition of William S. Benedict, who made the original suggestion. Correlation spectroscopy is also discussed by Richard Goody [1968], but instead of using slits, he used a pressure modulated radiometer, as is described below. The author is grateful to Bob Dick of Barringer Research for providing these historical details. However, the application which survives was patented by A.R. (Tony) Barringer [1969] of Barringer Research Ltd. in Toronto for an instrument called COSPEC. This is a very specialized approach that is being widely used to detect SO_2 in the plumes of volcanoes such as Mt. St. Helens, and Mt. Etna. In current applications, the COSPEC instrument is flown past the plume in a way that the absorption of solar radiation in the plume is mapped. For an example of results see Bluth *et al.* [1992].

7.3.5 The Pressure Modulator Radiometer (PMR)

The Pressure Modulator Radiometer (PMR) is a powerful method that has contributed a great deal to our knowledge of the middle atmosphere; its development is described by Taylor [1983]. The instrument is a correlation spectrometer that is essentially a photometer (usually called *radiometer* in the infrared) with a cell in the optical path containing the same gas as is observed in the atmosphere, CO_2, for example. Because some of the atmospheric radiation is absorbed in the cell, there is a difference between the signals observed with and without the gas; by switching back and forth between these two conditions, a modulated signal is generated. In the first such instruments the optical path was switched (chopped) between a cell containing gas, and an empty cell. These instruments were called Selective Chopper Radiometers (SCR) and were flown on the Nimbus 4 and 5 missions [Abel *et al.*, 1970].

Subsequently the concept was improved by Taylor *et al.* [1972] using only one gas cell, but varying the pressure within it; the first application was on Nimbus 6 [Curtis *et al.*, 1974]. The technique takes advantage of the way in which the lineshape varies with pressure, both in the atmosphere and in the cell, as shown in Figure 7.7. In (a) the lineshapes are shown as a function of wavenumber for various pressures, indicated by the labels on the curves which denote "$-\ln p$" with p in atmospheres. In (b) the transmittance of a cell of 1 cm length is shown for various pressures, along with the differences between different pressure curves, indicating the extent of the modulation as a function of wavenumber. All of the instruments mentioned so far were nadir viewing, where a large field of view was used, yielding a large responsivity without the need for cooled detectors. With CO_2 as a gas of known mixing ratio, the radiance yielded the temperature and, using lines of different absorption coefficient, the temperature at several layers in the atmosphere was determined as described in Chapter 1.

The instrument matured as SAMS (Stratosphere And Mesosphere Sounder) which flew on Nimbus 7 [Drummond *et al.*, 1980] as a limb-viewing instrument. Limb viewing requires a small field of view and cooled detectors. A further improved version called ISAMS (Improved Stratosphere And Mesosphere Sounder) was launched on NASA's Upper

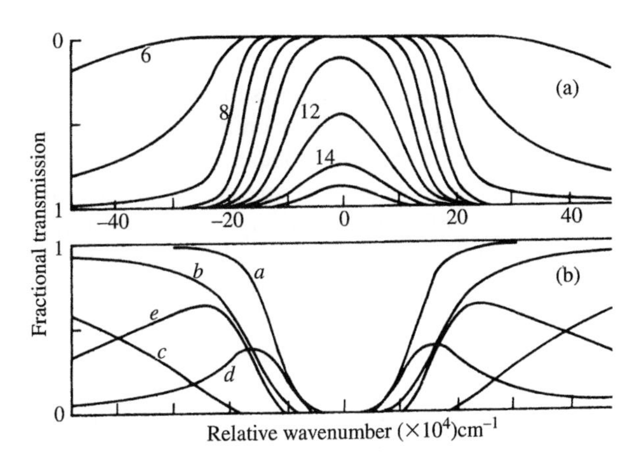

Figure 7.7. (a) The spectral profiles of atmospheric lines for various pressures p, where the numbers on the curves indicate $-\ln p$ atmospheres; (b) transmittance of a gas cell of 1 cm length with the cell pressures indicated as $a = 2$ hPa, $b = 5$ hPa, $c = 14$ hPa, $d = a - b$ and $e = b - c$. After Taylor [1983].

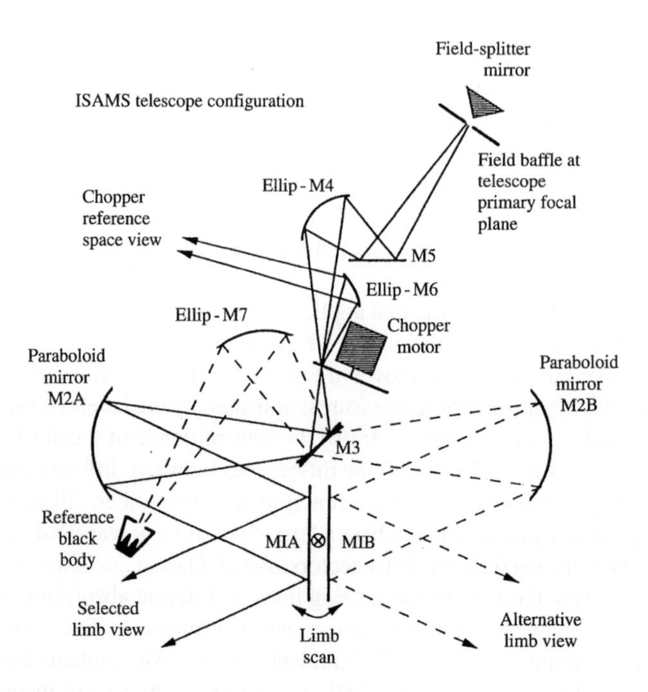

Figure 7.8. The ISAMS input optical system, with limb selector, scanner and chopper. After Taylor *et al.* [1993].

Atmosphere Research Satellite in 1991 [Taylor *et al.*, 1993]. The instrument field of view is rapidly (1 kHz) alternated between a view of the atmosphere and a view of cold space. This is done with a rotating *chopper*, a segmented wheel which alternately presents a mirror and an opening to the instrument. This process subtracts the background signal, corresponding to dark space, from the signal containing both the emission and the background signal.

ISAMS viewed the Earth's limb in order to obtain vertical profiles. To extend the local time (and latitude) coverage observable from the spacecraft, the instrument viewed on both sides of the spacecraft; that is, at 90° from the velocity vector on both the right and left-hand sides of the spacecraft. The input optics which provided for this are shown in Figure 7.8. Flipping the mirror M3 caused the instrument to view on one side, or the other. The limb mirrors for the two views were placed back to back, and the limb scan took place through the rotation of this mirror unit. The optics involving the chopper are also shown, as well as the capability of viewing a black body calibration source.

The spectral resolution is essentially that of the spectral linewidth. This pressure modulated signal arises near the centres of the lines where the absorption is greatest. Other emission, not associated with a cell spectral line, is not modulated by this pressure change. The detection optics for a single channel are shown in Figure 7.9.

The ISAMS system included seven gases; CO_2, CO, NH_4, NO, N_2O, NO_2 and H_2O; the cells were modulated at a frequency of 30 Hz. Interference filters were used as band-limiters

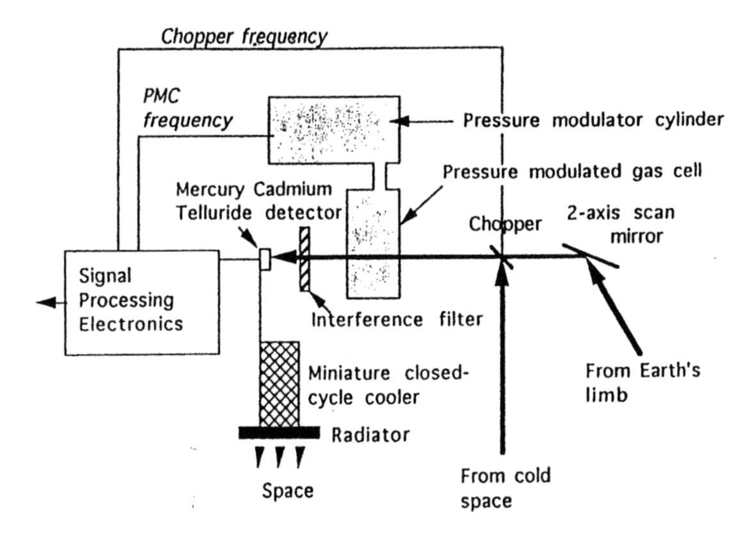

Figure 7.9. The ISAMS spectral selection system for one channel showing the interference filter, pressure cell, closed-cycle cooler and detector. After Taylor *et al.* [1993].

for each spectral region, chosen to correspond to the vibration-rotation band of the species under investigation. The pressure modulation of roughly a factor of two was produced by an oscillating piston suspended on diaphragm springs. The design was changed little from its origins on Nimbus 6, but for ISAMS the unbalanced momentum was eliminated using two pistons running in antiphase. The pistons were made of titanium alloy, suspended by stainless steel suspension springs, with a 75 μm clearance between the piston and the liner.

Depending on the gas, the cells were made out of stainless steel with germanium windows, or aluminum with calcium fluoride windows. The gas pressure is made as close as possible to that observed in the atmosphere, accomplished by using molecular sieves, reservoirs which hold gas in such a way that the pressure is controlled by the sieve temperature. Thus the mean gas pressure was controlled from the ground by varying the temperature, for two reasons. The first was to compensate for any small gas leakage and the second was to fine-tune the weighting functions to optimize the performance. One novel aspect of the ISAMS-type of radiometer was the use of closed cycle coolers based on the gas piston technology (refrigerators) that reached temperatures of about 80 K. Avoiding cryogenic fluids removes an otherwise serious restriction on the lifetime in orbit.

Because ISAMS uses both space and spectral discrimination, it is a double-modulator. As noted, this concept has been highly successful because of the inherently high resolution. Because there is no significant change of absorption with off-axis angle in the cell, the étendue is limited only by the optical design, and a theoretical superiority cannot be defined. In fact, because these measurements are made in the infrared, between 4.5 and 16.3 μm, the detectors are the limiting factor in the responsivity. These radiometers have the advantage that they operate not on a single line, but on all the lines in a band, enhancing the signal significantly. Because of the linewidth resolution, the Doppler shift of the atmospheric line induced by the motion of the spacecraft is important. This can be eliminated by looking at 90° to the spacecraft velocity vector, but in fact the instrument views slightly off this angle in order to compensate for the velocity of the Earth rotation at the look-point on the atmosphere. The instrument has the potential to measure winds, but this has not been accomplished to date.

7.3.6 Instruments for Dayglow Observations

In Chapter 5, the use of tandem Fabry–Perot etalons for the observation of daytime airglow was described. Chakrabarti [1998] has recently reviewed all of the ground based observations of sunlit airglow and aurora. The problem is to observe a signal of a few kR against a background of Rayleigh scattered light of about 30 MR nm^{-1}. The high resolution tandem FPS solves this problem by using a unique passband that is comparable with the linewidth. An alternative approach is through the use of modulators which can detect extremely small differences between the "dayglow + background signal" and that of the background alone. In a review paper on day airglow, Noxon [1968] describes his approach using a grating spectrometer and a chopper technique based on polarization. The dayglow is unpolarized but the background Rayleigh scattered light is strongly polarized if viewed at 90° to the sun. The system had two input beams with orthogonal polarizations, each of which viewed the same point on the sky. A rotating chopper switched the spectrometer input from one input to the other. The two beams were balanced to give the same signal on the detector when viewing the background Rayleigh-scattered signal, but when the spectrometer scanned across the $O(^1D)$ 630 nm unpolarized line, the radiances became unbalanced, generating a signal which was the spectrum of the emission line. The observation of day airglow is complicated by another phenomenon, which Noxon called the *Ring effect*, after James Ring. Grainger and Ring [1962] had noticed that Fraunhofer lines observed in the reflected light from the moon had a different depth from those observed in day-sky spectra. This was originally thought to be a fluorescence effect in the atmosphere, but was later shown to be the result of the rotational Raman shifts in wavenumber. The Raman effect is a process of inelastic scattering, in which the molecule that emits the photon is in a different energy level from the one in which it absorbed the original photon; the result is a wavenumber shift of the scattered light. When the atmosphere is illuminated with monochromatic radiation, as with a laser (lidar), the scattered light contains not only the Rayleigh component at the same wavelength, but lines that are shifted by an amount corresponding to the differences between rotational energy levels. For solar illumination, near a Fraunhofer absorption line, the continuum radiation adjacent to the line is shifted to the line wavelength, partially filling in the absorption line. Radiation at the line wavelength is also shifted out, but since the radiance is less at the bottom of the absorption line, less is shifted out than is shifted in. The net effect is that the absorption line is less deep than would otherwise be the case. This is a problem for day airglow measurements, but also for the detection of species through the absorption as seen in scattered light. Thus radiative transfer models of scattering in the atmosphere must include the Ring effect. For example, see Vountas *et al.* [1998].

More recently, Narayanan *et al.* [1989] developed a somewhat similar method, using a single Fabry–Perot etalon and an interference filter, not using the polarization method, but instead a specially designed chopper to chop between the dayglow + background spectrum and the background as observed with a diffuser. Their rotating mask, shown in Figure 7.10 exposes first the central portion of the FPS ring pattern containing the dayglow+background, and then an outer ring which contains only background. For one particular dayglow emission, atomic sodium, a technique somewhat akin to the gas cell technology of pressure modulator radiometry, has been used, in several different forms. All methods use a heated cell containing sodium vapour. Blamont and Donahue [1961] allowed day-sky light to be scattered from the cell, so that the resonance fluorescence scattering contains only the sodium lines and the solar background at the bottom of the Fraunhofer lines, amounting to about 5% of the full day-sky value. In order to separate the sodium radiation from the background, the Zeeman effect of a magnetic field was used to shift the wavelengths of the sodium lines in the cell from those in the atmosphere.

Scrimger and Hunten [1957] also used a sodium cell, but simply used a grating spectrometer of 0.1 nm resolution to detect the sodium lines scattered from it. In this case the sun was the source, and what was observed was the absorption in the sodium layer, near 85 km. Their optical set-up is shown in Figure 7.11. Light from the sun is focussed into the sodium cell, forming an image of the sun just behind the window, to reduce the effects of absorption

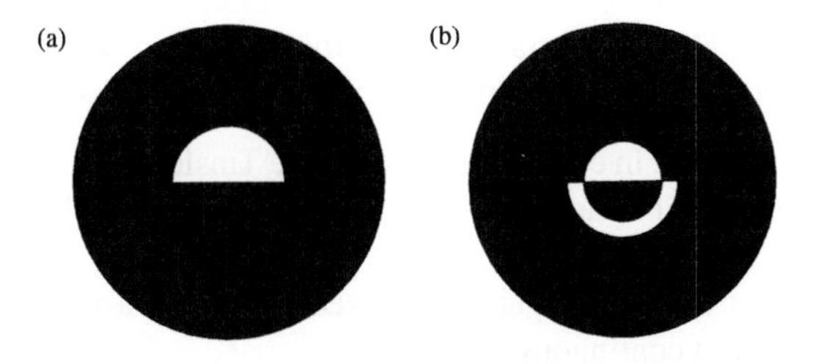

Figure 7.10. A dayglow chopper that alternates between the central FPS ring and an outer ring. (a) is the rotating mask, and (b) the stationary one. After Narayanan *et al.* [1989].

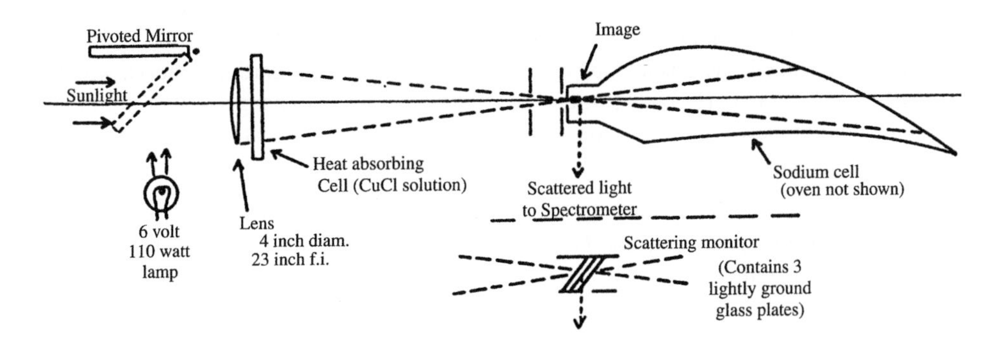

Figure 7.11. A schematic diagram of the sodium dayglow experiment, not to scale. After Scrimger and Hunten [1968].

in the cell. The light scattered at 90° is observed with the spectrometer. The cell is replaced by a scattering monitor in order to observe the sky spectrum without the sodium emission. This instrument is not really a modulator but is included here with the other instruments that use gas cells and are modulators.

7.4 PROBLEMS

7.1. Consider a visible region emission spectrum consisting of a series of lines 2.0 nm apart with their radiances enclosed by a Gaussian envelope of FWHM 10 nm; one of the lines is located at the peak of the envelope.

(a) Using a conventional sequential scanning spectrometer, the line at the Gaussian peak yields a count of 1000 electrons per integration period. Plot the measured spectrum and plot the noise in the measured spectrum as a function of wavelength; that is, calculate the noise for each line and plot it as a spectrum.

(b) Suppose a multiplexer of the same responsivity is used instead of the spectrometer so that all lines simultaneously contribute to the signal. The noise is now determined by the sum of the signals for all of the lines. Plot the noise as a function of wavelength for the multiplexer.

7.2. Consider a seven-element spectrum with relative radiances given by [0, 1, 3, 15, 0, 7, 2]. Compute the Hadamard transform of this spectrum using the formulation of Eq. (7.5), and retrieve the original spectrum using this formulation.

7.3. (a) Estimate the increase in étendue offered by the Tinsley grille shown in Figure 7.5, based on reasonable assumptions about the geometry. Assume that a single slit would correspond to the innermost ring.

(b) Explain to what extent this gain is realized for the measurement of: (i) a single spectral line; and (ii) a continuous spectrum.

7.4. In the description of ISAMS it was noted that, by looking at 90° to the velocity vector, the Doppler shift associated with the spacecraft motion is eliminated.

(a) For a spacecraft velocity of 7 km, calculate the accuracy with which this angle must be achieved, for a lineshift expressed as a velocity error of 5 m s^{-1}.

(b) Also as noted, there is a Doppler shift associated with the Earth's rotation, which can be accommodated by changing the 90° angle. Calculate the change in angle required as the field of view crosses the equator, travelling along a track that makes an angle of 57° with the equator (the orbit inclination).

7.5. The atomic oxygen green line in the dayglow (Figure 10.11) has a vertically integrated emission rate of typically 3.5 kR. The day-time sky background as seen from the ground is about 30 MR nm^{-1}.

(a) Calculate the bandwidth that would be required to make the background signal equal to that of the spectral line. Comment on the feasibility of such a measurement.

(b) For a rectangular instrumental passband of width 0.01 nm, and an integrated measurement of line plus background of 10^4 electrons, calculate the signal-to-noise ratio.

7.6. A measurement of sodium dayglow emission sitting in the bottom of the sodium solar Fraunhofer line, is shown in Figure 5.10. The background at the location of the line is reduced to about 5% of the value quoted in Problem 7.5. Take the emission rate of the sodium line, at 589 nm, as 5.0 kR.

(a) Repeat the calculation of Problem 7.5(b) for the measurement of sodium dayglow.

(b) One interesting aspect of the sodium dayglow problem is the Doppler shift associated with the motion of the observing point on the Earth with respect to the sun because the sodium emission is in the reference frame of the Earth, while the Fraunhofer absorption line is in the reference frame of the sun. The largest component of this is caused by the Earth's rotation. Calculate the Doppler shift in wavelength between sunrise and sunset caused by Earth rotation, where the Doppler shift is given by $\lambda = \lambda_o(1 + w/c)$ where w is the velocity responsible for the shift and c is the velocity of light. Comment on whether this calculated shift is consistent with Figure 5.10.

8

DOPPLER MICHELSON INTERFEROMETRY

8.1 THE MEASUREMENT OF DOPPLER TEMPERATURE

The Michelson interferometer described in Chapter 6 is often called a *Fourier Transform Spectrometer* (FTS) since it is scanned in optical path difference from zero to some finite value, yielding an interferogram that is Fourier transformed to generate the spectrum. In this chapter a specialized configuration of the Michelson interferometer is described. While it can be regarded as a subset of the FTS, it is used in a very different way, and applied to a very specific type of atmospheric measurement, the determination of atmospheric linewidths and lineshifts. The basis of the linewidth method goes back to Michelson's [1927] observation of fringe visibility which was recast by Hilliard and Shepherd [1966a] (see also Section 6.10) in its contemporary form. In this method a single spectral line is observed with a field-widened MI that is set to a specific optical path difference that is then scanned over one fringe to determine the visibility and phase at that value of OPD; the instrument is called a Doppler Michelson Interferometer (DMI).

The atomic oxygen green line airglow at 557.7 nm has been the subject of many linewidth investigations, beginning with the FPS observations described in Chapter 6. This emission is produced by a transition from the $O(^1S)$ to the $O(^1D)$ level. This is a *forbidden* transition because of the change in orbital angular momentum (S to D); consequently it has a lifetime of about 1 s compared with about 10^{-8} s for an *allowed* transition. At night the $O(^1S)$ population is dominantly produced by the recombination of atomic oxygen in the altitude range from 90 to 110 km as described in Chapter 1; during the 1 s lifetime the excited atom collides with the surrounding atmospheric molecules and at 100 km altitude makes roughly 500 collisions before radiating. This ensures that the radiating atoms are in thermal equilibrium with the surrounding atmosphere and that their velocity distribution reflects the temperature of that background atmosphere. Since the natural width of the line is small, its observed shape arises solely from the Doppler shifts associated with the Maxwellian velocity distribution. The $O(^1S)$ transition is a singlet, the nuclear spin is zero and as well there are no significant isotopes so this line consists of a single component whose spectral radiance can be represented by a Gaussian which mirrors the Maxwellian velocity

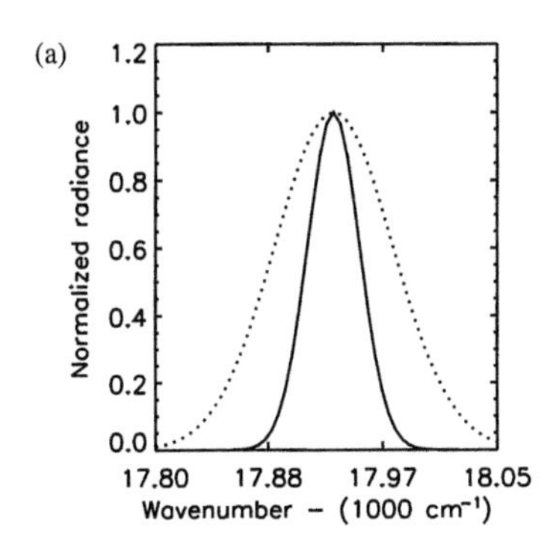

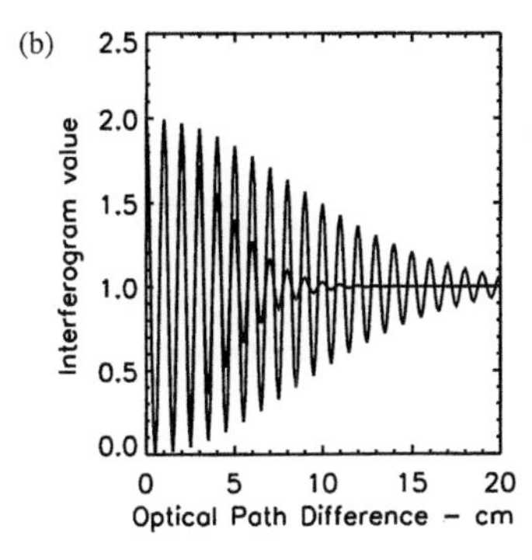

Figure 8.1. The spectrum of the 557.7 nm airglow green line for two temperatures: dotted – 1000 K, and solid – 200 K. (b) The interferograms for these temperatures; the trace which extends to larger OPD values corresponds to 200 K. The oscillation period of the interferogram is exaggerated to make it visible.

distribution, where η is the full width at half maximum of the spectral line as shown in Figure 8.1(a) for two temperatures:

$$L(\sigma) = L(\sigma_o)\, e^{-(4\ln 2)(\sigma-\sigma_o)^2/\eta^2} \tag{8.1}$$

The interferogram of this Gaussian line spectrum is its Fourier transform, which is also a Gaussian. Here the constant term I_o is added to the variable part of the interferogram to give the total signal and the frequency-shifting Property 4 is employed from Chapter 2, with the following result:

$$I = I_o\left(1 + e^{-(\pi\eta\Delta)^2/(4\ln 2)}\cos 2\pi\sigma_o\Delta\right) \tag{8.2}$$

This interferogram is a damped cosinusoid with a Gaussian envelope, as shown in Figure 8.1(b). The two different curves are identified by their temperatures (200 K and 1000 K) rather than their widths as is explained shortly. The cosine oscillation represents the individual fringes with a wavenumber corresponding to the line centre at σ_o, and the envelope is the Gaussian which is the FT of the spectral line. Michelson defined the fringe visibility, as already presented in Chapter 6 as:

$$V(\Delta) = \frac{I_{max} - I_{min}}{I_{max} + I_{min}} \tag{8.3}$$

where I_{max} and I_{min} are the maximum and minimum values of the interferogram over one fringe in the vicinity of the optical path difference Δ. Substituting (8.2) into (8.3) the result is:

$$V = e^{-(\pi\eta\Delta)^2/(4\ln 2)} \tag{8.4}$$

so that:

$$I(\Delta) = I_o(1 + V(\Delta)\cos 2\pi\sigma_o\Delta) \tag{8.5}$$

where I_o, the mean value of the signal, corresponds to the integrated radiance of the emission line. Michelson [1927] did not record the interferogram; he simply estimated the visibility

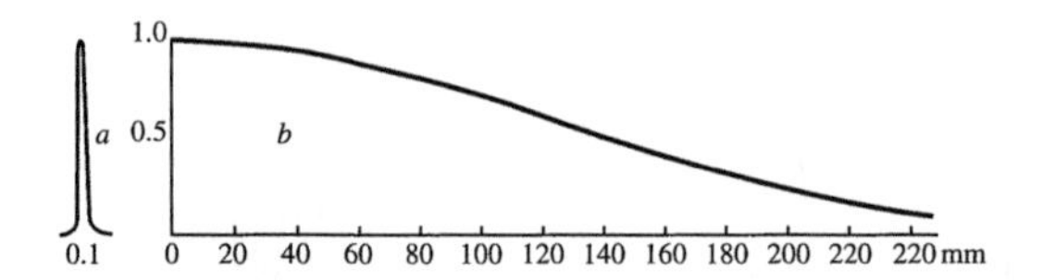

Figure 8.2. The precursor to Figure 8.1 as drawn by A.A. Michelson: (a) the spectrum of a narrow line; and (b) the resulting visibility as a function of the OPD in mm. After Michelson [1927].

by eye. For example, in Figure 8.1 if the fringes were observed to disappear when the MI was adjusted to about 10 cm OPD then the temperature of the emission was 1000 K, while if the fringes were visible to a little beyond 20 cm then the temperature was 200 K (this is for oxygen atoms; for a species of different molecular weight the numbers are different).

If the upper envelope of the oscillations in Figure 8.1(b) is denoted $I(\Delta)_{env}$ then $I(\Delta)_{env} = I_o(1 + V(\Delta))$, or the visibility is given simply by $V(\Delta) = I(\Delta)_{env} - I_o$, where in Figure 8.1, $I_o = 1$. In other words, the visibility is just the envelope measured above the mean value, and it is just the FT of the spectral line centred at zero wavenumber.

Figure 8.2 is taken from the work by Michelson [1927]; it is essentially identical to Figure 8.1 except that, since it is the visibility that is measured and is plotted, the cosinusoidal oscillation does not appear. Eqs. (8.3) and (8.4) both appear in Michelson's (1927) book, following a description of what he calls "the interferometer", namely the MI.

Hilliard and Shepherd [1966a] noted that if the visibility is determined at a single value of Δ then the temperature may be calculated if the lineshape has a known analytical form; Gaussian in this case. Since as explained above, the linewidth reflects the velocity distribution of the excited species, the visibility can be directly related to the atmospheric temperature for those species, which is the same as that of the background gas. The following equations apply:

$$\frac{\eta}{\sigma_o} = 7.16 \times 10^{-7}\sqrt{\frac{T}{M}} \tag{8.6}$$

and therefore:

$$V = e^{-QT\Delta^2} \tag{8.7}$$

where

$$Q = \frac{(7.16 \times 10^{-7})^2}{4 \ln 2}\frac{\pi^2\sigma_o^2}{M} \tag{8.8}$$

and M is the molecular weight. The same principle works for other known lineshapes. It can be shown that dV/dT reaches a maximum when $\Delta = (QT)^{-1/2}$ and $V = 1/e$. This can be taken as an indication of the optimum value of Δ to use for measuring temperature by this method, but for any particular case it is best to calculate the temperature error as a function of Δ, including noise in the calculation.

There is a second broader layer of night airglow green line emission formed in the ionosphere around 250 km. It originates in the recombination of O_2^+ ions with ambient electrons. Since the ionization energy made available in this process exceeds the dissociation energy of the O_2 molecule, the recombining molecule flies apart, with the excess energy going into the excitation of the dissociated atoms. This may create one $O(^1S)$ and one ground state O atom, with 2.79 eV of energy left over as in Eq. (8.9). Another possibility is that one $O(^1S)$ and one $O(^1D)$ metastable are produced with 0.83 eV of excess energy as in Eq. (8.10):

$$O_2^+ + e \rightarrow O(^1S) + O + 2.79\,\text{eV} \tag{8.9}$$

$$O_2^+ + e \rightarrow O(^1S) + O(^1D) + 0.83\,\text{eV} \tag{8.10}$$

The excess energy goes into kinetic energy of the excited and neutral atoms so that their velocity distribution is far from that of the background atmosphere and, since at these altitudes the time between collisions is more than one second, the majority of the $O(^1S)$ metastable atoms radiate before they are able to lose their excess energy through collisions. Hays and Walker (1966) investigated this process and predicted the lineshape that would be observed in the night airglow, as shown in Figure 8.3. Note the unusual shapes and widths of these lines shown in (a) corresponding to Eqs. (8.9) and (8.10), while in Figure 8.3(b) the predicted visibility values (the interferograms) are shown for these two cases. Paul Hays and Jim Walker made the MI visibility calculation because they overlapped with the author as visiting scientists at the Queen's University, Belfast, where the late Professor Sir David Bates led a group which pioneered in the understanding of upper atmospheric processes. These calculations presented an observational challenge which has never been fully met. The observation of the F-region non-thermal emission is difficult from the ground as the "thermal" lower thermospheric emission near 100 km is roughly 200 R while the F-region non-thermal emission varies is about 10 R and is difficult to distinguish from the stronger lower-altitude component. Here the word "thermal" has a somewhat different meaning from the way it has previously been used. In this case the emission is produced through photochemical non-thermal processes, but the species comes into thermal equilibrium with the surrounding atmospheric constituents through collisions, prior to radiation. The best way to make the measurement would be to make the visibility observations for not just one OPD, but several; with this configuration the easiest way would be to build separate WAMIs

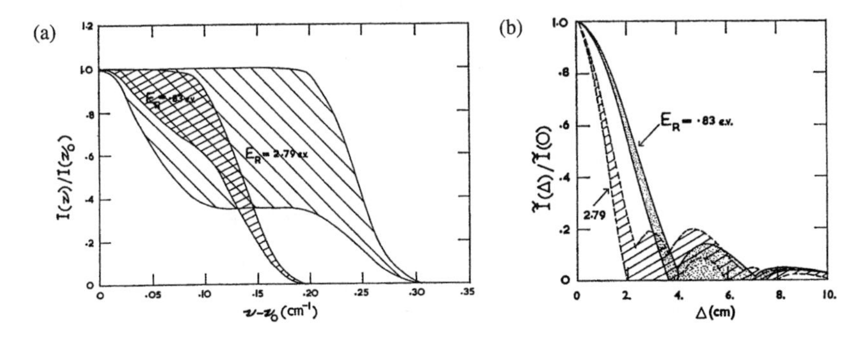

Figure 8.3. (a)The calculated spectrum of the non-thermal green line airglow for an ambient temperature of 800 K. (b) The corresponding interferogram envelope computed from (a); in both cases the cross-hatched areas correspond to variation of the collision cross-section from zero to 10^{-15} cm^2. After Hays and Walker [1966].

for each OPD. An instrument of this type, having six different optical path differences is described in the Appendix of a thesis by Marie-Louise Duboin [1974]. While the instrument was built, the experiment was never completed. For satellite viewing of the emission from above, the measurement is easier; some results are described in the next chapter.

8.2 THE MEASUREMENT OF DOPPLER WIND

In the previous section only the visibility was extracted from the measured interferogram and the phase information was ignored. Figure 8.4 shows what happens to the interferogram for a single line if the wavelength of the line is shifted by $d\lambda$ ($\delta\lambda$ in the figure). The interferogram is stretched, so that the phase of a fringe at some large OPD value (shown as 10^5 fringes in the figure) is shifted. Thus a measurement of the phase at a single location in the interferogram yields a value of the wavelength shift, without executing a Fourier transform. If the wavelength shift is the Doppler shift of a line in the atmosphere, then the measurement is that of a wind; since the radiating atoms are being carried by the motion of the background atmosphere, then the measurement of Doppler shift yields the line-of-sight wind. It is not necessary that the atoms be in thermal equilibrium, only that their mean velocity be the same as that of the background gas.

At first sight this is a difficult measurement, since if winds are to be measured to an accuracy of $10\,\mathrm{m\,s^{-1}}$ this is a measurement to one part in 3×10^7 of the velocity of light. However, the relative phase shift of *w/c* takes place for each fringe, so for an OPD of 10^5 fringes, the measurement needs to be made to $10^5\, w/c$, i.e., only one part in 300, which is possible.

The Doppler shift in wavenumber is:

$$\sigma = \sigma_o \left[1 + \frac{w}{c}\right] \tag{8.11}$$

where a positive velocity w denotes motion towards the observer, and c is the velocity of light. However, as first pointed out by Thuillier and Hersé [1988], because of the refractive

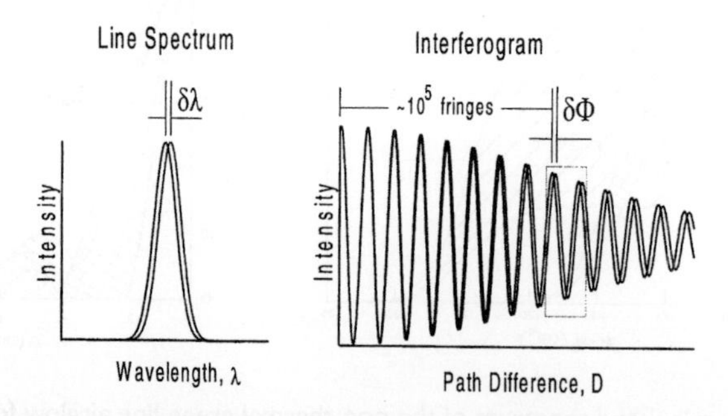

Figure 8.4. The effect on the interferogram (right) of a Doppler shift in wavelength (left). The size of the fringes is exaggerated to make them visible on the graph. Courtesy of W. Gault (York University).

element in the DMI, the change of wavenumber causes a change in Δ which must be taken into account. This change is written as:

$$\Delta(\sigma) = \Delta(\sigma_o) + (\sigma - \sigma_o)\frac{d\Delta}{d\sigma} \tag{8.12}$$

The equation for the absolute interferometer phase may then be written as:

$$\phi_a = 2\pi\sigma\Delta = 2\pi\sigma_o\left(1 + \frac{w}{c}\right)\left[\Delta(\sigma_o) + \left(\sigma_o\left(1 + \frac{w}{c}\right) - \sigma_o\right)\frac{d\Delta}{d\sigma}\right] \tag{8.13}$$

Multiplying this out, but neglecting the term in $(w/c)^2$, the result is:

$$\phi_a = 2\pi\sigma_o\Delta_o + 2\pi\sigma_o\left[\Delta(\sigma_o) + \sigma_o\frac{d\Delta}{d\sigma}\right]\frac{w}{c} \tag{8.14}$$

$$= 2\pi\sigma_o\Delta_o + 2\pi\sigma_o\Delta_{eff}\frac{w}{c} = \Phi + \phi \tag{8.15}$$

The term in square brackets was named by Thuillier and Hersé [1988] the *effective optical path difference*, Δ_{eff}, because it is this value and not Δ that determines the phase shift. The interferometer equation, Eq. (8.5), is thus written with two phase terms in Eq. (8.15), the second corresponding to the phase of the wind, ϕ; while the Φ is the absolute phase of the signal for zero wind, which for a given spectral line depends only on Δ, that is to say, only on the instrument. The interferogram equation is then written:

$$I = I_o(1 + V\cos[\Phi + \phi]) \tag{8.16}$$

For a wavelength of 500 nm, an optical path difference of 5 cm and a velocity of 10 m s^{-1} (a reasonable requirement for wind accuracy), the phase shift ϕ is 1.2°. Thus accurate phase shift measurements are required in order to make useful measurements of atmospheric winds but this turns out to be feasible.

8.3 PHASE STEPPING INTERFEROMETRY

From Figure 8.1 and Figure 8.4 it is clear that measurements need be made over only one fringe. This is further illustrated in Figure 8.5, where a single interferogram fringe is shown, with four samples indicated. The question immediately arises as to the optimum number of samples required for a determination of fringe phase. This consideration is a topic in *phase-stepping interferometry*, as discussed by Hariharan [1987, 1989], among others. A particularly simple and elegant approach as described by Shepherd *et al.* [1985] results from the four samples illustrated in Figure 8.5. In order to sample the four points in the figure, the interferometer OPD is stepped by amounts that correspond to phase (Φ) intervals of 90°; thus $\Phi_u = 0, \pi/2, \pi, 3\pi/2$, where $u = 1, 2, 3, 4$. The result is the four

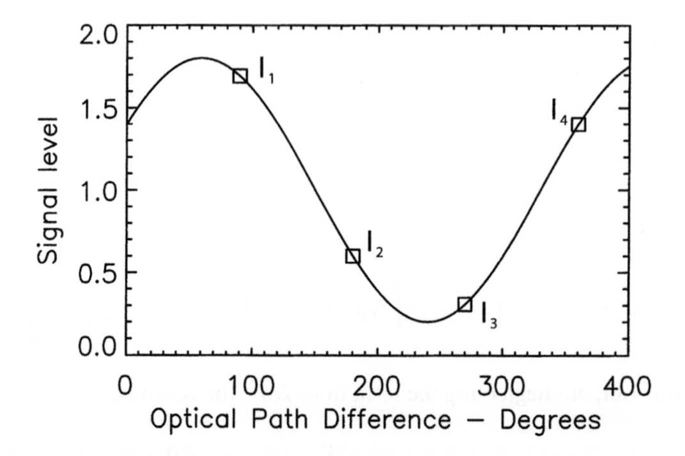

Figure 8.5. Illustrating the sampling of four points on a fringe; the four-point algorithm.

following equations:

$$I_1 = I_o[1 + V\cos(0 + \phi)] = I_o[1 + V\cos\phi] \tag{8.17}$$

$$I_2 = I_o\left[1 + V\cos\left(\frac{\pi}{2} + \phi\right)\right] = I_o[1 - V\sin\phi] \tag{8.18}$$

$$I_3 = I_o[1 + V\cos(\pi + \phi)] = I_o[1 - V\cos\phi] \tag{8.19}$$

$$I_4 = I_o\left[1 + V\cos\left(\frac{3\pi}{2} + \phi\right)\right] = I_o[1 + V\sin\phi] \tag{8.20}$$

Combining (8.17)–(8.20) the result is:

$$\tan\phi = \frac{I_4 - I_2}{I_1 - I_3} \tag{8.21}$$

The mean value, I_o, is obtained as follows:

$$I_o = \frac{I_1 + I_2 + I_3 + I_4}{4} \tag{8.22}$$

while V is obtained from:

$$V = \frac{\sqrt{(I_1 - I_3)^2 + (I_4 - I_2)^2}}{2I_o} \tag{8.23}$$

Equations (8.21), (8.22) and (8.23) constitute the *four-point algorithm*. This kind of phase-stepping is crucial to the method of optical Doppler imaging. While this algorithm is very convenient and practical to use under ideal conditions, it is restrictive in requiring accurate 90° phase steps. A more general approach is to calculate the Fourier coefficients. Suppose that N interferometer steps are taken with $\Delta_u - \Delta_o = ud$, where $u = 0, 1, \ldots, N$, over an integral number of fringes. Multiplying (8.16) by 1, $\cos(2\pi ud/\lambda)$ and $\sin(2\pi ud/\lambda)$ in turn, and summing over all phase steps in each case, one obtains:

$$J_1 = \frac{1}{N}\sum_{u=1}^{N} I_u = I_o \tag{8.24}$$

$$J_2 = \frac{2}{N}\sum_{u=1}^{N} I_u \cos\left(2\pi u\frac{d}{\lambda}\right) = I_o V \cos\phi \tag{8.25}$$

$$J_3 = -\frac{2}{N}\sum_{u=1}^{N} I_u \sin\left(2\pi u\frac{d}{\lambda}\right) = I_o V \sin\phi \tag{8.26}$$

The quantities J_1, J_2 and J_3 contain the desired quantities of radiance, phase (wind) and visibility (temperature); the application of this method is described in the next chapter.

8.4 THE WIDE-ANGLE MICHELSON INTERFEROMETER

The principle of field-widening was introduced in Chapter 6, and while it has been little used in FTS devices it has had an enormous impact on Doppler Michelson interferometers. In 1961, during an extended visit to Jacquinot's laboratory, the Laboratoire de Bellevue, also known as Laboratoire Aimé-Cotton, just outside Paris, Pierre Connes showed the author a field-widened MI similar to that of Bouchareine and Connes [1963], described in Chapter 6. This was being constructed by Ové Harang, a Norwegian scientist working in collaboration with Connes; they planned to use this instrument for the observation of auroral spectra. Harang's clear explanation of the operating principle, already described in Chapter 6, made it immediately evident to the author that this could provide a much more powerful method of measuring upper atmospheric temperature than that provided by the Fabry–Perot spectrometer already operating in his laboratory. This is feasible because, for the DMI, the visibility is needed at only one path difference, as has been described.

Ronald Hilliard was the graduate student who built the first DMI instrument, described by Hilliard and Shepherd [1966a], and given by him the name WAMI (Wide-Angle Michelson Interferometer). The layout is shown in Figure 8.6(a), with the interferometer optics at the upper left. This consists simply of a cubical beamsplitter with a glass plate in one arm of the interferometer (above) and an air space with a moving mirror in the other (left). The air space is adjusted to satisfy the wide-angle condition. It is interesting that in Michelson's [1927] original interferometer he inserted a *compensating plate* to ensure the same amount of glass in both arms. This is because he wanted to be able to see fringes in white light which is possible only if all wavelengths have the same phase at one OPD value, the one for which Δ is said to be zero. The WAMI is deliberately designed to be uncompensated in order to be field widened, but because this is now considered to be a desirable effect, the field-widened version is said later also to be compensated, though in a different sense. At this time the goal was to measure upper atmospheric temperatures from the linewidths of auroral and airglow emissions. The instrument shown in Figure 8.6(a) has

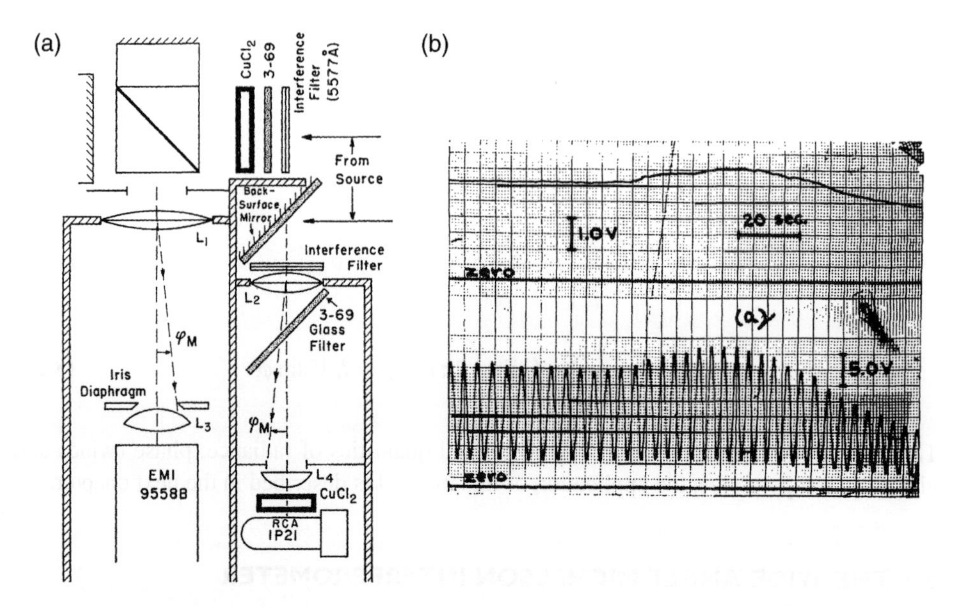

Figure 8.6. The WAMI instrument (left) and a WAMI scan across an auroral form (right). After Hilliard and Shepherd [1966b].

two channels. Light entering from the upper source beam passes through some filters and enters the MI, following which the exit light for a defined solid angle is measured with the EMI photomultiplier. The lower input beam is filtered to accept background radiation, but with no MI; it measures the background signal which must be subtracted from the interferogram. Some raw data are shown in Figure 8.6(b) with the background signal shown above and the interferometer signal below, for a scan across an auroral form; extensive results were obtained and presented by Hilliard and Shepherd [1966b].

8.5 CUBE CORNER DOPPLER MICHELSON INTERFEROMETER

Following the Hilliard and Shepherd [1966b] DMI, other configurations were experimented with by the author's graduate students, but building a truly stable instrument for field use that could be used at more than one wavelength proved difficult. The last one constructed before the development of the solid fully compensated instruments, to be described shortly, was that by Ward [1988], shown in Figure 8.7. Cube corners provided the desired stability and the MI elements were made so that they could be removed in transit and replaced in a reproducible location. The piezoelectric driver did not have a servo positioning system, and exhibited hysteresis in that the change in length was not linear with the applied voltage, although it was repeatable. Reference lamp fringes were measured simultaneously with the atmospheric observations so that this non-linearity could be removed and the phase of the atmospheric emission correctly determined.

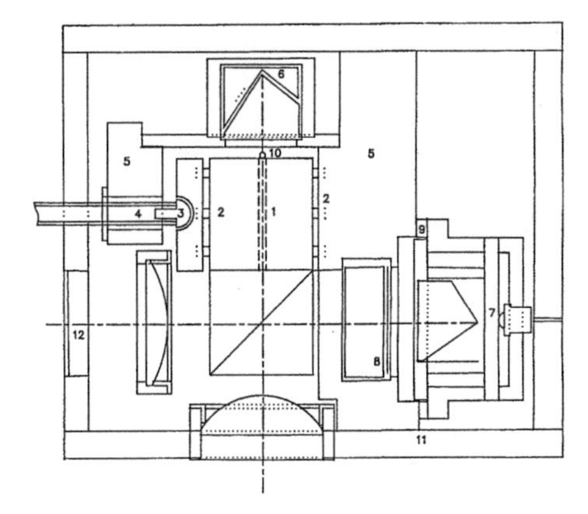

Figure 8.7. A Doppler Michelson interferometer using a fixed cube corner (6) and a movable one (7), driven by a piezoelectric driver. Light enters from the bottom, passes into the beamsplitter, and vertically upward into a glass block held by a knife edge (1), invar positioning pins (2) and a teflon pad (3). To achieve the field-widened condition, a second block (9) is required in the second arm. The detector is located behind the window (12). After Ward [1988].

8.6 ACHROMATIZING A FIELD-WIDENED MICHELSON INTERFEROMETER

From Chapter 6, the field-widening condition for a Michelson interferometer with a plate of thickness t_1 and refractive index $n(\lambda)$ in one arm of the interferometer and a balancing length t_2 of air in the other arm was given as: $t_1 = t_2 n(\lambda)$.

Since the refractive index is a function of λ, the wide-angle condition is effective at only one wavelength, and since t_1 is fixed, t_2 is adjusted to satisfy the equation. To observe at another wavelength, t_2 must be changed, requiring that the moving mirror must be precisely moved by some calculated amount, and that is a complication. Title and Ramsey [1980] showed that an achromatic system could be fabricated by using two different glass plates, one in each arm. Suppose that these have thicknesses and refractive indices of t_1, n_1, and t_2, n_2 respectively, and let the departure from the wide-angle condition be ξ. Then:

$$\xi = \frac{t_1}{n_1} - \frac{t_2}{n_2} \tag{8.27}$$

Taking the derivative with respect to wavelength:

$$\frac{d\xi}{d\lambda} = -\frac{t_1}{n_1^2}\frac{dn_1}{d\lambda} + \frac{t_2}{n_2^2}\frac{dn_2}{d\lambda} \tag{8.28}$$

Setting $d\xi/d\lambda = 0$ and $\xi = 0$, the requirement for achromaticity becomes:

$$\frac{1}{n_1}\frac{dn_1}{d\lambda} = \frac{1}{n_2}\frac{dn_2}{d\lambda} \tag{8.29}$$

Fortunately there is a very wide variety of optical glasses available and by systematically searching an optical glass data base it is possible to find pairs of glasses satisfying this condition.

8.7 THERMALLY STABILIZING A SOLID MICHELSON INTERFEROMETER

Although it has been little discussed to this point, the stability of an interferometer is critically important to its practical use and this is particularly so when the phase is measured because a phase drift in the instrument cannot be distinguished from a wind. Regular phase calibration is required in any case but it is desirable to reduce the requirement for this. Fabry–Perot etalons can be made extremely stable by using spacers of low expansion coefficient between the plates, quartz or zerodur, for example. Mounting the elements of a wide-angle Michelson interferometer presents a mechanical problem, but the thermal problems involved in using glass plates are even more horrifying to the designer. This can be put in perspective by writing the equation for the derivative with respect to temperature of the path difference for the "one-glass" system of Hilliard and Shepherd [1966a], of thickness t_1 and refractive index n, as follows:

$$\frac{d\Delta_o}{dT} = 2t_1 \frac{\partial n}{\partial T}\left(1 + \frac{1}{n}\right) + 2t_1\left(1 - \frac{1}{n}\right)\frac{1}{t_1}\frac{\partial t_1}{\partial T} = t_1\left(3.33\frac{\partial n}{\partial T} + 1.66\alpha\right) \tag{8.30}$$

Here the reasonable value of $n = 1.5$ has been used for simplicity and α is the linear expansion coefficient, $\alpha = (1/t)(\delta t/\delta T)$. Using typical values for the constants; $\delta n/\delta T = 3 \times 10^{-6}\,\mathrm{K}^{-1}$, $\alpha = 6 \times 10^{-6}\,\mathrm{K}^{-1}$, and $t_1 = 3\,\mathrm{cm}$, one obtains a derivative of $6 \times 10^{-5}\,\mathrm{cm\,K}^{-1}$, or for a wavelength of 600 nm, a value of one wavelength per K, or in terms of phase, $360°\,\mathrm{K}^{-1}$. This means that to maintain a stability of better than 10 m/s (1.2° of phase from Section 8.2) for a wind measurement system, the temperature must be controlled to 1.2/360 K.

Title and Ramsey [1980] showed that temperature stabilization was possible using a two-glass system. The optical path difference for a two-glass system is given by:

$$\Delta = 2[n_1 t_1 - n_2 t_2] \tag{8.31}$$

The derivative with temperature then takes the form:

$$\frac{d\Delta}{dT} = 2\left(n_1\frac{\partial t_1}{\partial T} + t_1\frac{\partial n_1}{\partial T} - n_2\frac{\partial t_2}{\partial T} - t_2\frac{\partial n_2}{\partial T}\right) \tag{8.32}$$

Using the wide-angle condition this equation can be written as:

$$\frac{d\Delta}{dT} = 2n_1 t_1\left[\frac{1}{t_1}\frac{\partial t_1}{\partial T} + \frac{1}{n_1}\frac{\partial n_1}{\partial T} - \left(\frac{n_2}{n_1}\right)^2\left(\frac{1}{t_2}\frac{\partial t_2}{\partial T} + \frac{1}{n_2}\frac{\partial n_2}{\partial T}\right)\right] \tag{8.33}$$

This derivative can be set equal to zero by satisfying the equality:

$$n_1^2\left(\frac{1}{t_1}\frac{\partial t_1}{\partial T} + \frac{1}{n_1}\frac{\partial n_1}{\partial T}\right) = n_2^2\left(\frac{1}{t_2}\frac{\partial t_2}{\partial T} + \frac{1}{n_2}\frac{\partial n_2}{\partial T}\right) \tag{8.34}$$

Since the linear expansion coefficient $\alpha = (1/t)(\partial t/\partial T)$ this further simplifies to:

$$n_1^2\left(\alpha_1 + \frac{1}{n_1}\frac{\partial n_1}{\partial T}\right) = n_2^2\left(\alpha_2 + \frac{1}{n_2}\frac{\partial n_2}{\partial T}\right) \tag{8.35}$$

Thus, as before, one can search a database of glass constants in order to find glass pairs that satisfy this equation. Title and Ramsey [1980] demonstrated that there were glasses satisfying this condition that could, at the same time, satisfy the achromaticity condition, and this has been confirmed with further experience by the author's colleagues. Specific details for other instruments are provided in the next chapter.

8.8 A FULLY COMPENSATED SOLID DOPPLER MICHELSON INTERFEROMETER

Thuillier and Shepherd [1985] proposed that in addition to the above requirements that the condition for thermal stabilization should be made achromatic; that is, that $\partial^2\Delta/\partial\lambda\partial T$ should also be considered. The additional degree of freedom required to accomplish this can be provided by making use of the linear expansion coefficients of the pillars that hold the moveable mirror. The air space also is a parameter that can be adjusted. Thus for a *fully compensated* Michelson interferometer there are five conditions to be met:

- The required optical path difference.
- The field widening condition.
- The achromaticity condition.
- The thermal stabilization condition.
- The achromaticity of the thermal stabilization.

Descriptions of implementations of such systems are described by Thuillier and Hersé [1991], and by Shepherd *et al.* [1993]. Measured thermal drift values for a system designed for minimal thermal drift yield values of about 30° K^{-1}, a factor of ten better than an unstabilized system. The limitation here is likely to be the knowledge of the relevant glass constants, so further improvement is possible. The specification of the stability of a wind-measurement system is relative to the time taken between calibrations. If the drift in Δ_o is at a constant rate between calibrations, then the measured phase may be corrected. For example, for the WINDII instrument [Shepherd *et al.*, 1993] on the Upper Atmosphere Research Satellite, phase calibrations were made every 20 min, although this turned out to be a very conservative approach, because of the very large heat capacity of the solid interferometer, preventing rapid temperature changes.

The two-glass design implemented by Shepherd *et al.* [1993] not only solves the thermal drift problem but it results in a solid instrument, which solves the mechanical stability problem as well. Thus the concept for a fully-compensated phase-stepping Doppler Michelson interferometer suitable for space flight is achieved. The configuration of the WAMDII (Wide-Angle Michelson Doppler Imaging Interferometer) instrument [Shepherd *et al.*, 1985] which is fabricated according to these principles, is shown in Figure 8.8.

(a)
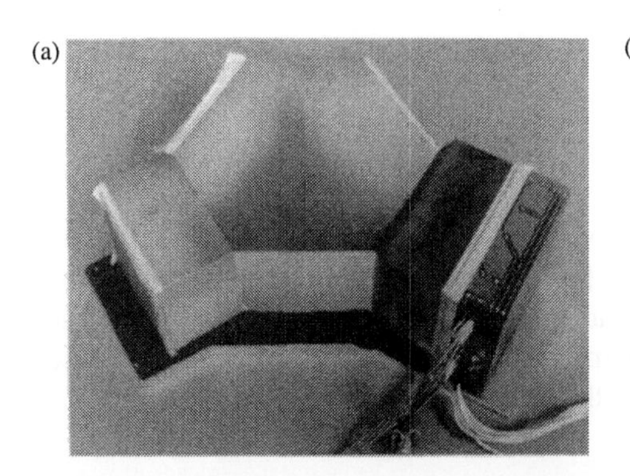

(b)
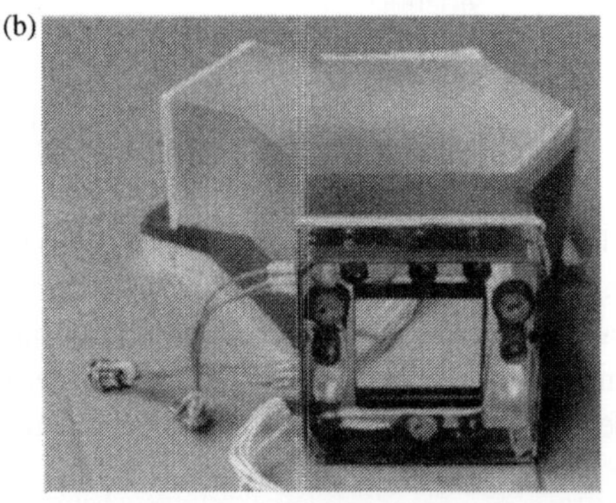

Figure 8.8. Photographs showing the configuration of the solid WAMDII Michelson interferometer. (a) Showing the beam-splitter hexagon in the centre, which is split vertically to allow the deposition of the dielectric beam-splitter layers, the fixed glass arm on the left and the stepped arm on the right. (b) A different view showing the stepped mirror, with the three piezoelectric posts and four capacitive sensors. The masking tape is for protective purposes in the laboratory only!

A hexagonal configuration is used in order to generate a 30° angle of incidence on the beam-splitter in order to reduce polarization of the beam [Dobrowolski *et al.*, 1985]. The stepped mirror is moved with piezoelectric drivers of the same type described for the FPS in Chapter 5, along with the capacitive sensors used to accurately sense the mirror position. WAMDII was planned for a flight on the space shuttle, and while it made wind measurements from the ground [Wiens *et al.*, 1988] it never did fly. The interferometer is not easy to fabricate, because the glass elements must be cemented together while maintaining interferometric alignment of the system. This is done using ultraviolet curing cements which are quickly hardened with UV radiation only after the fabricator is satisfied that everything is in alignment [Wimperis and Johnston, 1984].

8.9 DEFOCUSING A WIDE-ANGLE MICHELSON INTERFEROMETER

In the earlier discussions of field widening, and the compensated Michelson interferometer, it was implied that the optimum setting of the instrument was that of the wide-angle condition as stated. Hilliard and Shepherd [1966a] showed that this was not necessarily the case, and Zwick and Shepherd [1971] explored this further. In fact a small departure from this condition can bring significant improvement. In Chapter 7 an expression for the value of Δ as a function of off-axis angle i to fourth order for a two-glass system was given; here it is extended to three glasses:

$$\frac{\Delta}{2} = n_3t_3 + n_2t_2 - n_1t_1 - \frac{\sin^2 i}{2}\left(\frac{t_3}{n_3} + \frac{t_2}{n_2} - \frac{t_1}{n_1}\right) - \frac{\sin^4 i}{8}\left(\frac{t_3}{n_3^3} + \frac{t_2}{n_2^3} - \frac{t_1}{n_1^3}\right) \quad (8.36)$$

The wide-angle condition as previously stated makes the term in $\sin^2 i$ equal to zero, but it is evident that this leaves a fourth-order term (as well as higher order terms not shown).

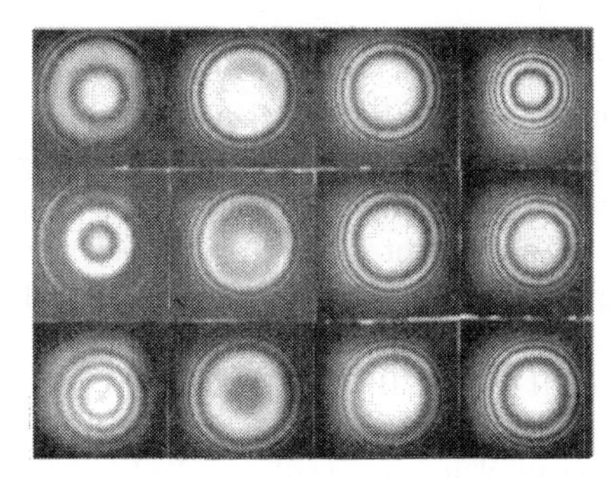

Figure 8.9. Wide-angle Michelson interferometer fringes photographed at a sequence of positions near the one satisfying the wide-angle condition. After Zwick and Shepherd [1971].

In principle it is possible to balance off the second and fourth-order terms in $\sin i$, over a useful range of off-axis angles. The images shown in Figure 8.9 were photographed as the Michelson interferometer was moved over a distance of about 210 μm. The fringe pattern changes rapidly as the $\sin^2 i$ and $\sin^4 i$ terms change drastically as a function of i; some patterns resemble conventional MI interference patterns, while others are very different. Zwick and Shepherd [1971] referred to this positioning of the mirror away from the wide-angle condition as defocusing. Gault *et al.* [1985] showed that fringes could be obtained with an angular diameter as large as 50°, a condition that is called "superwidening". The DMI has many aspects yet to be explored; a history of the Canadian experience is given by Shepherd *et al.* [1991].

8.10 POLARIZING DOPPLER MICHELSON INTERFEROMETERS

8.10.1 Introduction

The polarizing MI is a relative of the birefringent photometer introduced in Chapter 7, as described by Evans [1949]. The birefringent photometer is an MI with only one arm, in which the phase difference is created between the two beams by birefringence. The polarizing MI uses a polarizing beam-splitter but has conventional glasses in the two arms as described by Title and Ramsey [1980]; this led to the SOHO/SOI and GONG instruments described in Section 8.10.3. The Polarizing Atmospheric Michelson Interferometer (PAMI) combines the polarization tuning technique of Title and Ramsey [1980] with the Shepherd *et al.* [1985] technique of deriving winds and temperatures from emission rate measurements at sequential path differences. Thus it too is a phase-stepping interferometer. This scanning system used by PAMI is much simpler than the moving mirror system of WAMDII.

Polarization interferometers are discussed in a general context by Françon and Mallick [1971]. In particular, they introduce polarized light as background to birefringent beam-splitters, and polarization interference microscopes. In addition, they discuss testing of optical surfaces and other applications. Other introductions to interferometry are provided by Mertz [1965], Steel [1983] and Hariharan [1991].

8.10.2 PAMI Polarization States

The geometrical configuration for the PAMI Michelson optics is shown in Figure 8.10, taken from Bird *et al.* [1995]. Entering the instrument, the light first encounters a 45° linear polarizer, which creates equal amplitude components of horizontally and vertically polarized light. Because the input light is unpolarized, half the light is absorbed in this polarizer. The polarizing beam-splitter (PBS) transmits all of the horizontally polarized light and reflects all of the vertically polarized light. Thus the two components are then sent down different arms of the MI by the PBS.

This horizontally polarized light then traverses the 45° quarter-wave plate, QWP, at the end of the arm. A quarter-wave plate consists of so-called *birefringent* material, a material in which the refractive index depends on the orientation of the electric field vector. This means that the two different polarization components travel at different speeds in the crystal (birefringent material is normally crystalline). The orientation within the crystal for which the wave travels fastest is called the fast axis, and this axis is rotated to 45° from the horizontal. Thus the horizontally polarized light is resolved into components parallel to and perpendicular to the fast axis. In traversing the QWP, a phase difference of one quarter wave (90°) develops between the fast and slow components, so the sum of the two creates a rotating electric field vector, a condition called *circularly polarized light*. Upon reflection from the mirror behind the QWP, the QWP is traversed again, introducing a further phase difference of 90° for a total of 180° between the two components. This phase shift reverses the direction of the vector which restores the light to linear polarization again, but now the electric field is oriented vertically, rather than horizontally. Thus when this beam arrives again at the beam-splitter it is reflected, emerging from the exit rather than the input face of the beam-splitter. Since the two arms of the interferometer are of different lengths, an optical

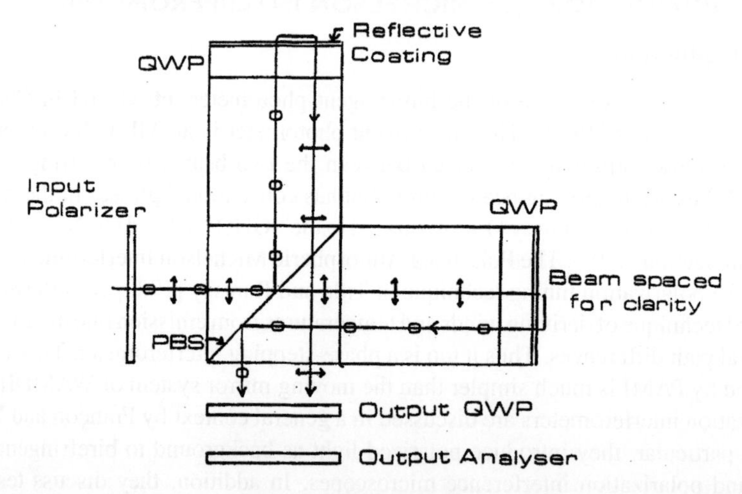

Figure 8.10. Configuration of a Doppler polarizing Michelson interferometer. After Bird *et al.* [1995].

path difference Δ is introduced between the two beams as they arrive at the beam-splitter. On leaving the beam-splitter the two beams traverse another QWP which converts each beam to circular polarization again, one left-handed, the other right. The sum of these is a linearly polarized beam having a rotation angle that depends on Δ. A rotating output analyser (a linear polarizer) transmits maximum radiance when the polarization and polarizing angles coincide, and a minimum radiance when they are perpendicular. If the rotation angle of the analyser is ψ, the output radiance is given by:

$$I(\psi) = \frac{I_o}{2}\left[1 + \cos\left(\frac{2\pi\,\Delta}{\lambda} - 2\psi\right)\right] \tag{8.37}$$

where $2\pi\,\Delta/\lambda$ is the phase difference introduced by the MI. This is precisely the MI equation derived in Chapter 6, except for the factor of 1/2 which arises because of the loss of radiance at the entrance polarizer. Thus rotation of the output polarizer by 180° allows a phase scan of 360°. To apply the four-point algorithm, the polarizer is rotated to four steps of 45° in angle of rotation. It is remarkable that the stepped rotation of a polarizer outside the MI where the interference takes place can create phase stepping. The rotation of a polaroid is mechanically much simpler than moving the mirror to high accuracy because the polarizer need not be oriented with interferometric accuracy. The polarizing Michelson interferometer has another important advantage and that is that no light emerges from the MI entrance face. For non-polarizing MIs the light rejected from the beam-splitter is directed toward the input, where it may reflect from the optics, perhaps an interference filter, and re-enter the system [Ward *et al.*, 1985]. If this happens, the MI equations no longer apply and erroneous results are obtained. Steps must be taken to avoid this and these are described for non-polarizing systems in the next chapter.

8.10.3 The SOHO SOI (Solar Oscillations Investigation) and GONG Instruments

These instruments were applied to measurements of the solar atmosphere rather than the Earth's. In 1960 a five-minute oscillation was found in the vertical motion of the sun's photosphere and a decade later it was found that this was the result of acoustic gravity waves trapped below the photosphere [Scherrer *et al.*, 1995a]. It was soon realized that these trapped waves could be used to probe the interior structure of the sun, giving rise to the field of helioseismology. An instrument containing two polarizing imaging field-widened Michelson interferometers and a Lyot filter was developed to study these oscillations. The MIs were called MDIs (Michelson Doppler Interferometers) and were similar to the PAMI instrument just described. A Lyot filter is a birefringent filter, as described in Chapter 7. For analysis of the data, long sequences were required, and so the concept was developed of having a ring of observatories around the Earth, at roughly six different longitudes, to allow continuous recording of the oscillations. This is known as the GONG (Global Oscillations Network Group), which has been described by Harvey *et al.* [1988].

A spacecraft instrument of a similar design was developed for the SOHO (SOlar and Heliospheric Observatory), a joint ESA and NASA mission that was launched on

Dec. 2, 1995. The two MDIs (the same instrument as the author's DMI) were operated in a mode similar to tandem FP etalons, with inter-order spacings of 37.7 pm (10^{-12} m) and 18.9 pm, but the field-widening allowed imaging of the oscillations over the solar disk. In conjunction with the MDIs were fixed filters, including the Lyot filter, with a combined pass-band width of 45.5 pm. This allowed tuning on and off the Fraunhofer absorption line as shown in Figure 8.11. Thus the instrument is used as a spectrometer rather than an interferometer. The MDI's were a solid design, similar to WAMDII, and were also fabricated by GSI Lumonics (then Interoptics). A novel method was developed by Jeff Wimperis for the thermal compensation, based on the fact that copper has a linear expansion coefficient that balances the *dn/dT* for the BK7 glass used as the refractive element. However, this linear expansion coefficient is very different from that for the BK7, so the two could not be cemented together as a solid element. Instead the MI mirror was cemented to copper pillars which were in turn cemented to the cubical beam-splitter. The dissimilarity in expansion coefficients over the diameter of a single pillar did not break the cement bond. This ring of pillars became known as the Stonehenge design. Space does not permit a full description

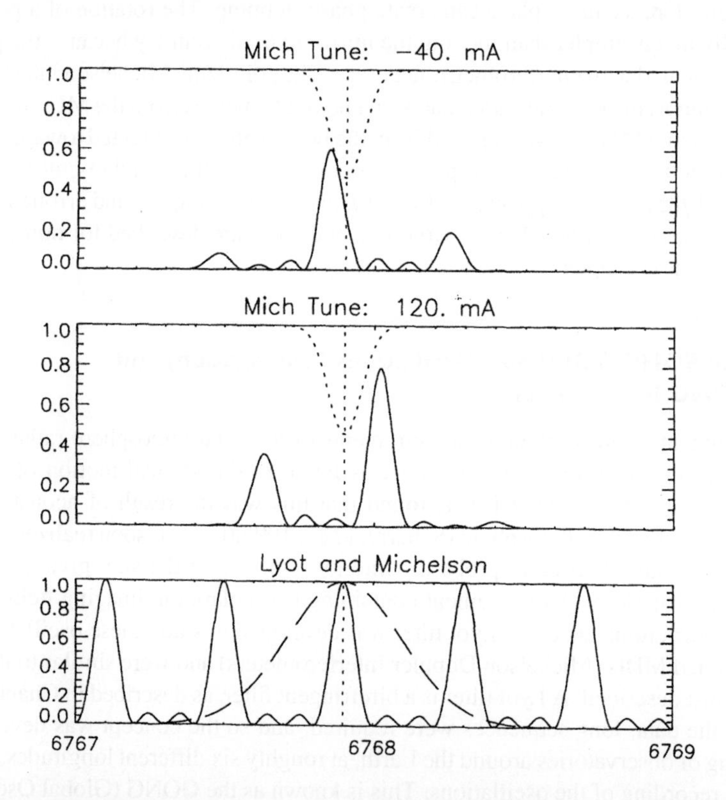

Figure 8.11. The SOI principle of operation, showing the individual passbands of the Lyot filter (dashed line) and the joint passband of the two MIs (solid line). The two panels show two different positions for the Doppler tunings. After Scherrer *et al.* [1995a, b].

of the elegant SOI instrument, and the reader is referred to the very detailed Solar Physics publication by Scherrer *et al.* [1995a], or the book in which these results are reprinted [Scherrer *et al.*, 1995b].

8.11 THE PHASE QUADRATURE MICHELSON INTERFEROMETER

8.11.1 Concept

This interferometer arose with the motivation of making the four phase steps of a Doppler Michelson interferometer simultaneously, rather than sequentially. One apparently easy way to accomplish this is to make one of the mirrors in four sectors, each with an OPD different from the next by $\lambda/4$. In Figure 8.12, three methods of incorporating the phase steps between the segments are illustrated, where the segments are labelled A, B, C and D. In the first method, on the left, the segments are stepped in steps of $\lambda/8$ in order to achieve the desired OPD steps of $\lambda/4$. This will certainly work, but only for a given value of λ. If the instrument is required to observe more than one wavelength, this configuration is not adequate. In method 2, shown in the middle panel, each segment is physically moved in order to change wavelengths. This will also work, but is mechanically very complex, as very precise motions are required. In the third method, shown on the right, the desired steps are produced using thin film multilayers.

8.11.2 Phase-Shifting with Optical Thin Film Multilayers

An optical thin film is a layer of an element or inorganic compound deposited on a substrate, whose thickness is comparable to the wavelength of interest. It is characterized by its thickness t, index of refraction n and absorption coefficient α, as well as t, n, and α for the surrounding incident medium and substrate. A thin-film multilayer is a stack of thin films on a substrate. Whereas a "thin-film multilayer" refers to a number of thin film layers (more than 1) a "coating" can refer to either one or more thin films.

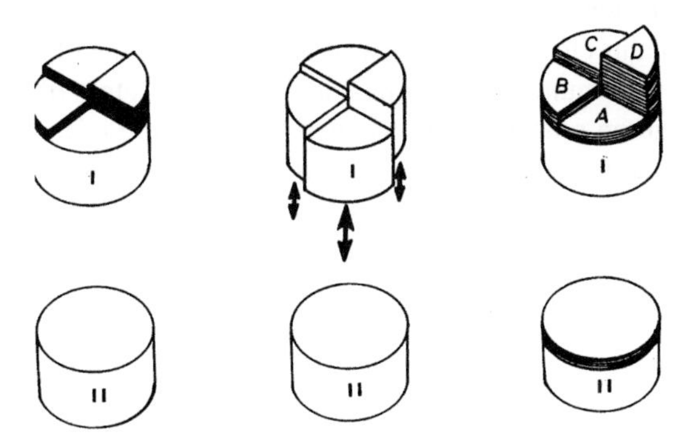

Figure 8.12. A comparison of three different methods of producing phase differences between the mirror segments of a Michelson interferometer. Courtesy of S. McCall (Stellar Optics).

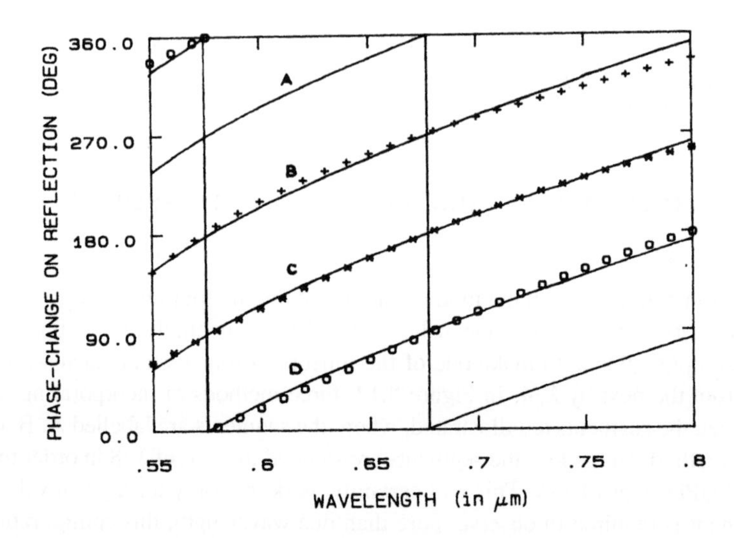

Figure 8.13. Superimposed calculated curves of phase change on reflection for multilayers A, B, C and D in the wavelength region bounded by the design wavelengths, shown as vertical lines. After Piotrowski McCall *et al.* [1989].

The multilayers for the PQI were designed by Susan Piotrowski McCall using program FILTER [Dobrowolski, 1981], an extremely powerful thin-film design research tool, written by J.A. (George) Dobrowolski at the National Research Council of Canada (NRCC). The designs define a unique set of four coatings which have high reflectance and a very special phase relationship between them. The phase change on reflection between each multilayer, is 90° not just at one wavelength, but over a range of wavelengths in the visible region. The superimposed curves of phase change on reflection for sectors A, B, C, and D are shown in Figure 8.13 as presented by Piotrowski McCall *et al.* [1989]. The design wavelengths were 0.5577, 0.6328, 0.8661, and 0.8920 μm. It can be seen that, although the phase shifts from one of these wavelengths to the next, the phase difference between sectors for any given wavelength remains close to 90°. The end of each arm of the PQI has a different fixed array of optical thin film multilayer coatings, and the four sectors of each array induce a different phase change on reflection (determined by the nature of the multilayer), so that after recombination, the interference produces four adjacent outgoing sectors of radiation which differ in phase by 90° from each other. This enables a Michelson interferometer to produce all four phase-shifted images simultaneously (as opposed to sequentially), as well as for any desired wavelength and without mirror motions.

In the generic Michelson interferometer, the ends of the two arms are usually coated with a reflective metal such as aluminum or silver. Since each mirror has the same coating, the phase change experienced by the incident light upon reflection is also the same from each arm. The definition of phase change on reflection adopted here is the one which gives a value of π radians for reflection at an air–glass interface, and 0.8π radians for an air–silver interface for visible light. The phase change on reflection at a surface can be any value between 0 and 2π radians (modulus 2π). It is dependent on the wavelength, polarization, angle of incidence, and on the nature of the surface, the incident and emergent media. In theory, any phase change on reflection is possible to achieve.

Following the development of the MI equation in Chapter 6, if one arm introduces a phase shift of ε_1 and the other a phase shift of ε_2, and the reflectances r_a in the two arms are allowed to be different, then the MI equation is given by:

$$I = r_{a1} + r_{a2} + 2\sqrt{r_{a1}r_{a2}}\cos[(\varepsilon_2 - \varepsilon_1) + 2\pi\sigma\Delta] \tag{8.38}$$

When the reflectances and phase shifts are equal, the ordinary MI equation results. The PQI was implemented in the laboratory and fully evaluated. It performed in accordance with the predictions and so fully met requirements. To date, an operational PQI of this type has not been built, but a monochromatic version of it is described in Chapter 10.

8.12 OPTIMIZED REFLECTIVE WIDE-ANGLE PHASE-STEPPING MI

A reflective DMI is of interest in extending the spectral range over which Doppler measurements of wind and temperature may be made. The design of such an instrument based on cat's eye reflectors has been explored by Wang *et al.* [2000]. This follows on a number of other studies which are cited in this paper, and have also been reviewed by Baker [1977]. One of particular relevance to this discussion is that by Cuisenier and Pinard [1967], in which the secondary mirror is made of a thin diaphragm of stainless steel which is deformed as the interferometer is scanned to maintain the wide-angle condition. The diaphragm is displaced at the edges just enough to compensate for the phase variation over the field to the first order. Another reflective configuration is by Hirshberg [1974] which used a spherical mirror in each interferometer arm without a secondary mirror, forming a laser cavity-like configuration with a beam-splitting cube. Steel [1974] corrected the design by putting a spherical concave surface on the cube. A modified design was described by Hirshberg and Cornwall [1979] using a planar beam-splitter and off-axis paraboloids. His idea was to use high-pressure (6 atm) scanning to achieve significant path differences using a high-index gas such as sulphur hexafluoride in one arm, and a low-index gas such as helium in the other, both to have the same pressure. However, there are no reports on the operational use of such devices.

Previous studies for ordinary MIs have tended to favour parabolic primary mirrors [Beer and Marjaniemi, 1966; Ezhevskaya and Sinitsa, 1978], but Wang *et al.* [2000] show that, while these provide better performance at small off-axis angles, for the larger angles required for a DMI off-axis aberration becomes increasingly significant and it is better to use spherical surfaces. The layout of the cat's eye in one arm of the interferometer is shown in Figure 8.14. The quantities d_1 and d_2 are the distances of mirrors M1 and M2 respectively from the aperture stop (AS), and the mirrors have focal lengths f_1 and f_2, also respectively. In order to follow the same notation as those authors, the notation in this section is not consistent with that otherwise followed throughout this work; P is the total path in the interferometer arm. Wang *et al.* [2000] derive what they call the cat's eye system equations, for which the interferometer is field-widened if Eq. (8.39) is satisfied for both cat's eyes

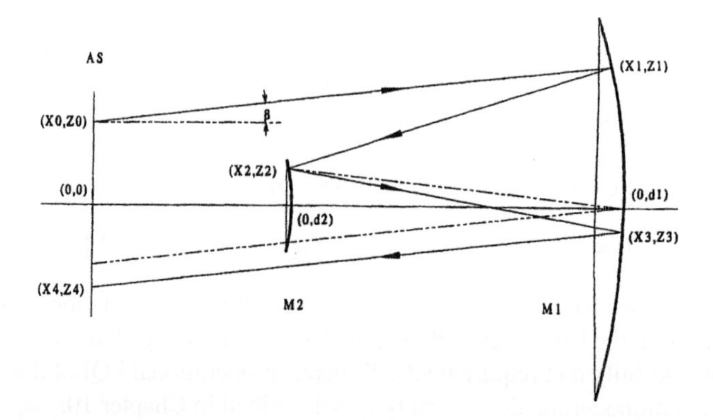

Figure 8.14. Schematic diagram of a cat's eye retroreflector for calculation of the first-order condition of field widening. After Wang *et al.* [2000].

and if their exit pupils are placed symmetrically about the beam-splitter. These are:

$$d_1 = f_1 \left(1 - \frac{f_1}{2f_2}\right) \tag{8.39}$$

$$P = 2d_1 + 2f_1 \tag{8.40}$$

It is of interest that these equations do not depend on any assumptions about the conic figure of the primary and secondary surfaces, and that the coordinates of the ray do not appear in Eq. (8.40). Wang *et al.* [2000] show that, if the aberration integration of each individual cat's eye is minimized, then the integration of the wave aberration difference between the two cat's eyes has its minimum value. Two different configurations were studied, and the results are shown in Figure 8.15.

The results in Figure 8.15 show that a cat's eye MI consisting entirely of spherical mirrors can be highly optimized. For configuration 1 (left) only one cat's eye was optimized and while the aberrations are small they are different for different incident angles. For configuration 2 (right) where mirrors in both cat's eyes were adjusted, the wave aberration is independent of incident angle, and has a small value, 0.04 μm maximum difference from the mean, which is 0.1λ at 400 nm.

The system has two limitations. One is that a refractive beam-splitter is used, meaning that there are some limitations on the wavelength range, although there are a number of optical materials available (such as CaF_2 and MgF_2) throughout the vacuum-uv (> 120 nm). The other is that small focal ratios are not possible without sacrificing some temporal resolution by using a smaller path difference, making the overall instrument comparatively large, which is a limitation for some situations, e.g. a small satellite. The possibility of achieving smaller focal ratios through the use of aspheric correctors could be investigated and the lengths of the MI arms could possibly be reduced with fold mirrors. In any case,

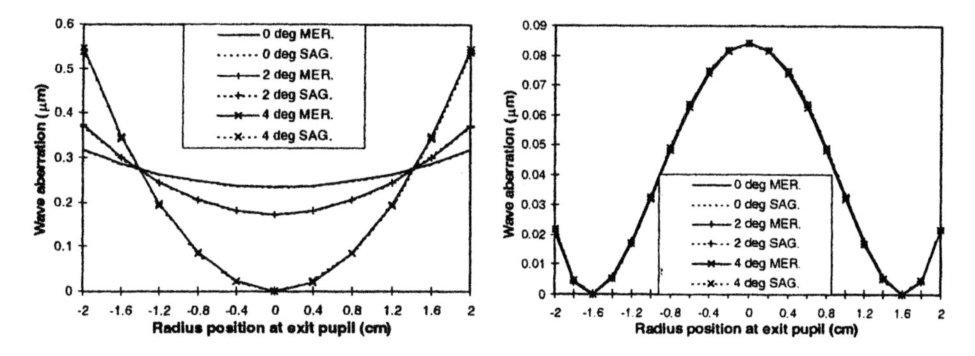

Figure 8.15. Optical path difference variation for two different configurations, as a function of radius position at the exit pupil. For configuration 1 (left), only one cat's eye was optimized while for configuration 2 (right), both cats' eyes were optimized. After Wang *et al.* [2000].

the reflective instrument provides an alternative configuration to the "solid" DMIs that may well play a role in future atmospheric observations.

8.13 PROBLEMS

8.1. (a) Calculate the FWHM in wavenumbers of a spectral line of O_2 at 763 nm, for a temperature of 200 K.
(b) Plot the visibility versus OPD for this line.

8.2. Approximate the Hays and Walker [1966] spectrum in Figure 9.3 for $E_R = 2.79\,\text{eV}$ with a rectangular function, using the outer envelope of the shaded area.
(a) From this approximate spectrum, plot the visibility as a function of OPD and compare with Figure 9.3(b). Make use of the fact that the visibility is just the FT of the line centred at zero wavenumber, but note that the visibility at some Δ values is negative. This just means that the "upper" envelope has become the "lower envelope" of the oscillations, so you may take the absolute value as the visibility. This is the visibility of the F-region $O(^1S)$ 557.7 nm airglow.
(b) Calculate and plot the visibility for the $O(^1S)$ 557.7 nm E-region airglow, with a temperature of 200 K.
(c) Add components (a) and (b) to form the simulated visibility that would be measured, looking upward from the ground. The E-region emission rate is ten times that of the non-thermal component in (a), so multiply the E-region visibility by 10 before adding. Comment on the feasibility of retrieving the non-thermal component from the measured total visibility.

8.3. From the plot shown in Figure 8.4 for the $O(^1S)$ 557.7 nm emission:
(a) Calculate the OPD corresponding to 10^5 fringes.
(b) Estimate the wind velocity appropriate to the figure from the $d\phi$ shown.
(c) Estimate the temperature for the interferogram.

8.4. (a) Calculate the phase shift ϕ in degrees for a spacecraft velocity w of 7 km s^{-1}, for a fixed OPD of 5.0 cm, using the OI 630.0 nm line. By phase shift is meant the phase difference between a velocity of 7 km s^{-1} and zero velocity.
(b) Repeat the calculation in (a) using the effective OPD, Δ_{eff}, as given by either Eq. (9.14) or Eq. (9.15) in place of the physical value of 5.0 cm. Use a value of $d\Delta/d\lambda = 5000$.

8.5. An achromatic WAMI is required to observe at 557.7 nm and 630.0 nm. There are three glasses available from which to choose one pair, with characteristics as follows, where the refractive indices are, for 630.0 nm:

$$\text{FK5} \quad n_1 = 1.486094 \quad dn_1/d\lambda = -3.0012 \times 10^{-5}\,\text{nm}^{-1}$$

$$\text{SF4} \quad n_2 = 1.749990 \quad dn_2/d\lambda = -10.9471 \times 10^{-5}\,\text{nm}^{-1}$$

$$\text{BaFS51} \quad n_3 = 1.720052 \quad dn_3/d\lambda = -7.7729 \times 10^{-5}\,\text{nm}^{-1}$$

(a) Identify the two pairs, giving the best achromaticity.
(b) For these two pairs, calculate t_1 and t_2 to achieve field-widening for 630.0 nm, with a total OPD = 6.0 cm. Identify the pair corresponding to the smallest instrument.
(c) Plot $\Delta - \Delta_o$ versus off-axis angle for both the 557.7 and 630.0 emissions for this pair.

8.6. The glasses LaFn21 and LF5 are used to construct a DMI to operate at 630.0 nm; they have the following characteristics:

	n	$dn/dT\,(\text{K}^{-1})$	α
LaFn21	1.784848	3.16×10^{-6}	5.9×10^{-6}
LF5	1.580165	1.73×10^{-6}	8.7×10^{-6}

(a) Calculate t_1 and t_2 for an OPD of 7.5 cm.
(b) Calculate $d\Delta/dT$ for this interferometer.
(c) Express the result of (b) as degrees of phase shift per degree C.

9

OPERATIONAL ATMOSPHERIC SPECTRAL IMAGERS

9.1 INTRODUCTION

Having established the conceptual framework of spectral imaging in the previous eight chapters, the next three demonstrate the application of these principles to the design and operation of actual instruments. After all, it is only in the final results obtained that the success of the various approaches can be judged. Also, it is hoped that, through their application, these chapters will strengthen the understanding and value of the concepts covered. In addition, the descriptions of these selected current and future instruments are also intended to provide a useful body of reference material.

Four of the operational instruments described in this chapter are Doppler Instruments, of which three are Doppler Michelson interferometers (DMI). Of these three, one is a satellite instrument and two are ground-based while the fourth, a satellite-borne instrument is a Fabry–Perot spectrometer. Following that, an infrared etalon spectrometer is described, and then a length-modulated infrared radiometer. Following the description of a Fabry–Perot spectrometer, Chapter 10 is entirely concerned with Michelson interferometers, two of which are proposed DMIs, one for the near infrared (mesosphere) and one for the mid infrared (stratosphere) as well as two infrared Fourier transform interferometers. Selected diffraction grating instruments are described in Chapter 11. For a truly comprehensive description of Earth-observing instruments and missions, the reader may consult the work by Kramer [2002].

9.2 THE WIND IMAGING INTERFEROMETER (WINDII)

9.2.1 Fundamental Spaceflight Considerations

In Chapter 6 the fundamental principles of the Michelson interferometer were outlined, and in Chapter 8 the concept of Doppler Michelson Interferometry was presented. This section describes the first implementation of a DMI instrument in a flight mission. The WIND Imaging Interferometer (WINDII) was launched on NASA's Upper Atmosphere Research

Satellite (UARS) on September 12, 1991, a mission planned for 18 months duration with an optional extension to 30 months [Reber *et al.*, 1993]. WINDII is designed to measure upper atmospheric winds and temperatures in the altitude range 80–300 km using emissions from the naturally occurring airglow. WINDII is one of a complement of ten instruments on the UARS spacecraft and, at the time of writing, it had operated in orbit for more than ten years, with no evident degradation of performance. UARS was the largest scientific satellite then made by NASA, being some 10 metres long with a mass of about 7000 kg, almost filling the cargo bay of the shuttle vehicle from which it was launched. The satellite is earth-oriented; that is, it maintains the same orientation with respect to the horizon, except for monthly yaw-around manoeuvres, when it is rotated by 180° from forward flight to backward flight in order to keep one side of the spacecraft facing the sun. WINDII is a collaboration between the Canadian Space Agency (CSA) and the Centre National d'Etudes Spatiales (CNES) of France in cooperation with NASA. CNES supplied the software for data analysis, the calibration lamp system and the specially designed baffle. The Canadian Space Agency supplied the major portion of the hardware, through a contract to AIT, Ottawa, with CAL Corporation (now EMS Technologies) as principal sub-contractor responsible for the design and fabrication of the flight instrument. The manufacture of the critical Michelson interferometer glass assembly was done at Interoptics of Ottawa (now GSI Lumonics). The author is Principal Investigator, and the WINDII Science Team consists of seven Canadian, seven French and one US co-investigator.

The WINDII design was based on two earlier instrument, WINTERS (Winds and Temperatures by Remote Sensing) which was the instrument originally accepted for flight on UARS with Gerard Thuillier of Service d'Aéronomie du CNRS as Principal Investigator, and the Wide-Angle Michelson Doppler Imaging Interferometer (WAMDII) which was intended to fly on the space shuttle [Shepherd *et al.*, 1985]; WINDII is also a combination of these two names, though for simplicity it also stands for WIND Imaging Interferometer. A developmental model for WAMDII was built by SED Systems in Saskatoon, Canada and tested in the field [Wiens *et al.*, 1988] and it was during this development program that the "solid" Michelson concept described in the previous chapter was developed in collaboration with Jeff Wimperis of GSI Lumonics. The WINDII project as described by Shepherd *et al.* [1993] was begun in 1984 but the WAMDII project was terminated in 1989.

In applying the DMI concept to space measurements, the key elements were: (1) the use of a field-widened MI to allow imaging; (2) the use of a CCD detector for limb-viewing in order to obtain vertical profiles through imaging; (3) two orthogonal look directions in order to measure two horizontal wind components; (4) the use of a filter wheel to allow different airglow emissions to be observed; and (5) the incorporation of baffles to allow observations in the day-time as well as at night. As shown in Chapter 8 the WAMI instrument on which WINDII is based can accept a very large off-axis angle, say 4°, yielding a full field of view of 8° . For ground-based observations of airglow such a large field is acceptable but for limb viewing from a satellite it is not. For a limb-viewing instrument, a vertical distance of 1 km at the limb translates to about 1′ of arc. This angle can be transformed as described in Chapter 3 to the 8° diagonal or about 6° vertical required for WINDII but with a multiplying factor of $6 \times 60 = 360$. This requires a telescope lens having a diameter of 360 times that of the WINDII aperture (8.65 cm), for a diameter of 31 m. This clearly will not work for a satellite measurement, showing that the DMI superiority cannot be employed in this way. The way it can be used is with an imaging detector so that the required elementary field-of-view (FOV) corresponds to one pixel, rather than the size of the entire detector. A detector with 256 vertical pixels can cover the required altitude range with a scale factor of 1.1 km per pixel; this means 280 km which is more than needed to cover the required altitude range, but which allows for the movement of the limb on the image caused by the small ellipticity of the orbit, and by the Earth's oblateness. In conceptual terms the entire limb profile is captured in one image, and in practical terms it means that a scanning mirror is not required to obtain a limb profile. But how does the satellite instrument cope with this much smaller étendue for one pixel compared with the ground-based instrument? The answer is that the long integration path through the limb yields radiances that are much larger than viewed from the ground, by as much as a factor of 50, depending on the thickness of the airglow layer. In addition, all the pixels across one row were combined to yield a single altitude measurement.

The two-dimensional horizontal winds are obtained through limb viewing in two different directions as shown in Figure 9.1, taken from the UARS brochure. Since the measurements are essentially taken at tangent points, the vertical wind components are not sensed; they are small in any case. Two WINDII look directions view to one side of the spacecraft, at 45° and 135° from the velocity vector. The first measurement is taken looking forward, and about eight minutes later the second measurement is taken looking "backward", providing two orthogonal directions. In fact the measurements are made continuously and matched up in space later by the analysis software. Thus the measurements are not made in the plane of the satellite orbit, but to the side, about 1500 km distant from the orbit plane. The UARS satellite has an orbit inclination of 57°, the angle of the orbit plane with the equator, which means that it travels from 57° latitude in one hemisphere to 57° in the other. If WINDII looks "south" when the spacecraft is at 57° S latitude, the observations

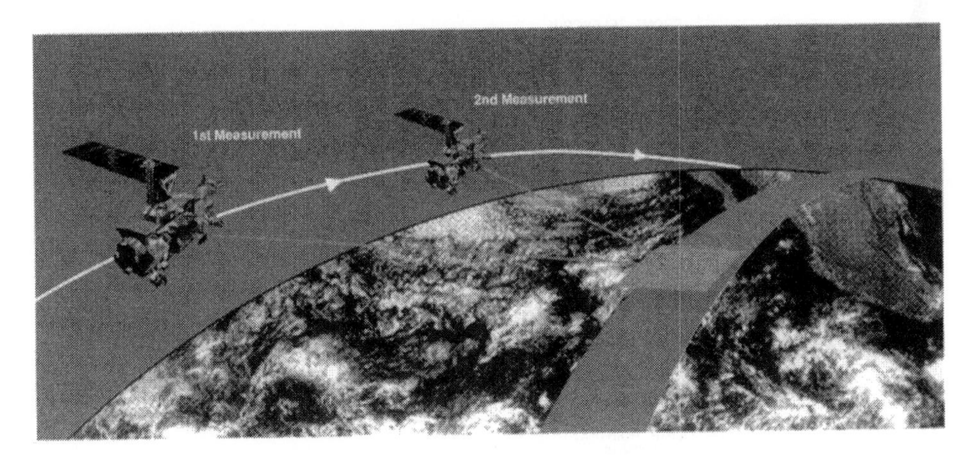

Figure 9.1. Viewing from UARS the same volume of air from two orthogonal directions. Courtesy of NASA Goddard Space Flight Center UARS Project.

are taken at 72° S, but it also looks south when the spacecraft is at 57° N latitude so that the measurements are taken at 42° N, for coverage of 72° S to 42° N.

Owing to torques on the orbiting satellite created by a non-spherical Earth's gravitational field, the orbit precesses relative to the sun, causing the local time to move to earlier times by 20 min each day. After 36 days this amounts to 12 hr, or 180°, which means that the sun has moved from one side of the spacecraft to the other. This is not acceptable as one side of the spacecraft must be designed for sunlit conditions and the other for shadow. WINDII is located on the shadowed side, and views away from the sun. Recalling the discussion of thermal radiation in Chapter 1, the instrument is cooled through its own thermal radiation to space and is warmed by the sun if it is present and by any heat generated by the electronics. All of this must balance at the desired instrument temperature. The CCD is cooled to about −50°C and this is done by coupling it to a radiator plate which provides enhanced cooling by radiating black body energy to the cold *deep space*, which definitely means no-sun conditions. Thus, after 36 days of operation, the UARS spacecraft is "yawed around", so that it flies backwards instead of forwards, keeping the cold side of the spacecraft away from the sun. During this reverse period, WINDII views from 72° N to 42° S so that the full range of ±72 degrees latitude is covered, but not all at the same time.

9.2.2 WINDII Optical System

The optical system, described in more detail by Shepherd *et al.* [1993] is shown in Figure 9.2. The two orthogonal inputs are shown, without the baffle system. The light from each first strikes a *limb pointing mirror*, and then is directed through the first lens element (consisting of several components) of what is called the *front telescope* before being reflected off a *field combiner*. This prism-shaped mirror combines the two fields of view into one optical beam

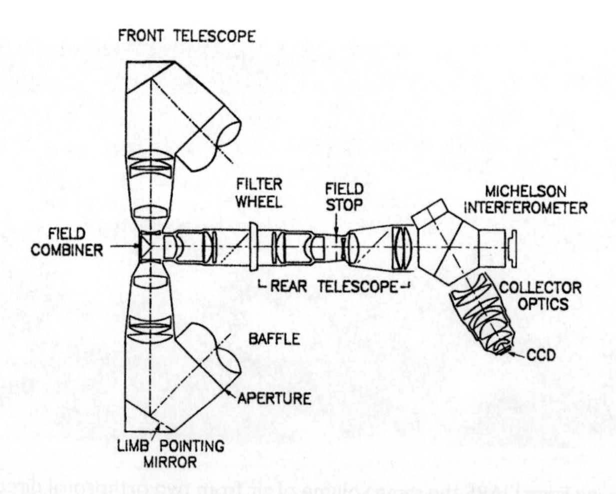

Figure 9.2. Schematic drawing of the WINDII optical system. After Shepherd *et al.* [1993].

that passes through the second element of the front telescope. This light is thus effectively collimated before passing through the filter wheel, which has openings for eight filters. Although the light here is well collimated for a single pixel there is still an angular spread arising from the FOV in front ($8° \times 6°$ with both fields), multiplied by the magnification of the front telescope, which is 1.43. But the collimation is better than would be achieved if the filters were placed elsewhere in a convergent beam and the only way to improve it would have been with larger filters, remembering that the étendue ($A\Omega$) is conserved. The filter wheel is followed by the "rear telescope", of unit magnification, also having two elements within which a field stop is located. On exiting from the rear telescope the light is again in a collimated state as it enters the Michelson interferometer, which is followed by the $f/1$ camera lens with a CCD at its focal plane. The overall result is that the images for the two look directions are located side-by-side on the CCD detector. To meet the demanding requirements of this "wide-field" fast optical system, a custom lens design of the whole system was required; this was done by Ian Powell [1986], then of the National Research Council of Canada.

9.2.3 The Michelson Interferometer

The Michelson interferometer is much like that shown in the photograph of the WAMDII MI in Chapter 8; its beam-splitter consists of two cemented half-hexagons with a low-polarizing semi-reflecting dielectric multilayer [Dobrowolski *et al.*, 1985].

One problem discovered with the WAMDII instrument was multiple reflections between the moveable mirror and the end of the arm. Even though the arm glass had anti-reflection coatings, the weak coherent reflections from these two surfaces, acting as a Fabry–Perot etalon, created additional fringes that are superposed on the Michelson interferometer visibility fringes [Ward *et al.*, 1985]. They do not appear in the phase images because they have the same period as the MI fringes and are subtracted out in the four-point algorithm, but the same is not true for the visibility. To solve this problem for WINDII, wedges were created at the ends of both arms such that the Michelson mirrors were parallel to each other, while the offending surfaces were wedged. By allowing a wedge of about one wavelength, the integrated signal averaged over one wave of the contaminant was effectively zero and the perturbing fringes were eliminated. A related problem arises because the light that exits the Michelson interferometer in the direction of the source is reflected off the interference filter and is then potentially recyled through the interferometer. The solution to this problem is to tilt the interferometer so that only one-half of the available field of view is used. The available field of view of a wide-angle Michelson interferometer is so large that restricting the available portion to one-half of that is not a serious penalty.

The WINDII MI is fully compensated as described in Chapter 8. The variation of OPD with off-axis angle is shown in Figure 9.3 for four wavelengths observed by WINDII. It is seen that the design is highly achromatic over the angular range of use and, at 8° off axis, the difference between extreme wavelengths is only about $\lambda/2$.

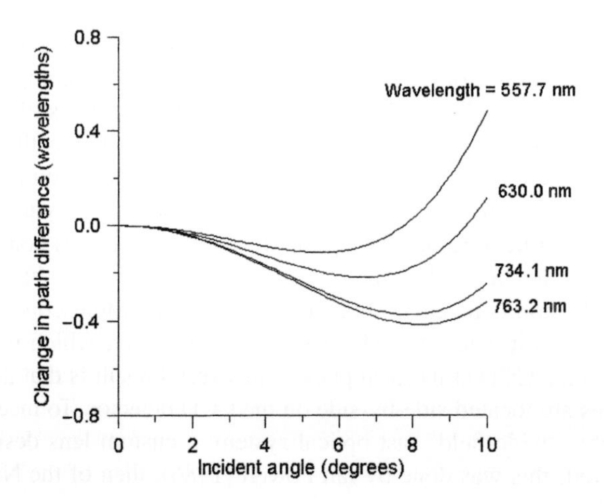

Figure 9.3. The variation in OPD as a function of off-axis angle for WINDII, for four different wavelengths. Courtesy of W. Gault (York University).

Table 9.1. Atmospheric emission lines isolated with the WINDII filters

Filter No.	Centre Wavelength (nm)	FWHM (nm)	Species and Wavelength (nm)	Altitude (day) (km)	Altitude (night) (km)
1	553.1	1.6	Background	70–300	70–300
2	558.4	1.6	$O(^1S)$ – 557.7	90–200	90–110
3	630.7	2.7	$O(^1D)$ – 630.0	180–300	220–300
4	732.9	1.4	$O^+(^2P)$ – 732.0/733.0	250–310	not present
4	732.9	1.4	OH (8,3) $P_1(2)$ 731.6	not observable	80–110
5A	728.8	16.3	OH (8,3) whole band	not observable	80–110
5B	714.8	11.0	Background	not observable	80–110
6	736.9	1.2	OH (8,3) $P_1(3)$ 734.1	not observable	80–110
7	763.2	0.09	$O_2(b^1\Sigma)^PP(7)$, $^PQ(7)$ 763.22, 763.0	80–120	85–110

9.2.4 Interference Filters

The airglow occurs in layers of restricted altitude range for a given emission. Thus several emissions are required to cover the altitude range of about 80–300 km. In addition, some emissions occur only in the day-time, and others only at night. For a species which emits in both the day-time and at night, the altitude ranges are different for the two. The emission species and their altitude ranges are indicated in Table 9.1. Emission lines are isolated using a filter wheel, with eight positions of which one is clear glass to allow for the admission of light from the calibration lamps while the other seven are identified in Table 9.1. All but one of the desired lines is isolated with single interference filter widths of 1.4–2.7 nm, carefully designed so that the filter transmittance for the selected atmospheric line is near the maximum value over the region of the CCD corresponding to the altitude range for this emission. Filter 4 does double duty in that it transmits 732.0 and 733.0 nm emission from O^+ at the top of the window (200–300 km which is where it occurs in the atmosphere), and 731.6 nm OH emission near 85 km at the bottom of the window.

Filter 5 is constructed in two parts, A and B, of which 5B is covered with a prism that deflects the light upward. This technique forms two images on the CCD, one above the other, one corresponding to filter 5A and the other to 5B. For the O_2 Atm (0, 0) band observed with filter 7 the rotational lines are so closely spaced that a 0.09 nm filter was used; the off-axis wavelength shift then limits the region of high transmittance to a narrow range and a single rotational line thus appears as a single Fabry–Perot ring on the CCD. However, there is a series of rotational lines, so that a series of rings is formed. Thus this hybrid Michelson interferometer and Fabry–Perot combination separates different rotational lines onto different portions of the CCD so that each can be analysed separately. Wind and temperature measurements are possible only where rings appear, which complicates the analysis, but each ring (line) gives an independent value of wind velocity, so that is a benefit. Moreover, it allows good background subtraction between the rings, which is important for daytime measurements. This hybrid approach for molecular emissions is discussed in more detail in the next chapter.

9.2.5 Detector

The detector is an RCA SID501 EX thinned back-side illuminated frame-transfer device, with 512×320 pixels of dimension $30 \times 30\,\mu$m. The top 16 rows of pixels are masked off, leaving a $4^\circ \times 6^\circ$ field of view of 160×240 pixels. Half of the CCD is used for imaging and half for storage. To conserve telemetry, a window is formed on the CCD corresponding to the region from which a particular emission originates, and only those pixels within the window are read out. Further, within the window, the pixels are organized into bins, to reduce the amount of data to be transferred, and also to increase the signal-to-noise ratio. Exposures are taken for approximately one second, and then the image is rapidly transferred to the storage area, where it is read out more slowly into the telemetry system. A thermoelectric cooler behind the CCD is connected by straps to a plate that radiates into dark space; achieving a temperature of -50°C. The CCD camera is a specially designed lens with an aperture of $f/1$ to maximize responsivity.

9.2.6 The WINDII Baffle

The baffle in front of the entrance apertures is required to prevent extraneous light scattering from optical surfaces, causing an unwanted background signal. This is extremely important in daylight for observing the faint airglow emissions about only one degree above the bright, sunlit clouds at the Earth's limb. Without a good baffle, day-time measurements would not be possible, because scattered sunlight would completely dominate the signal.

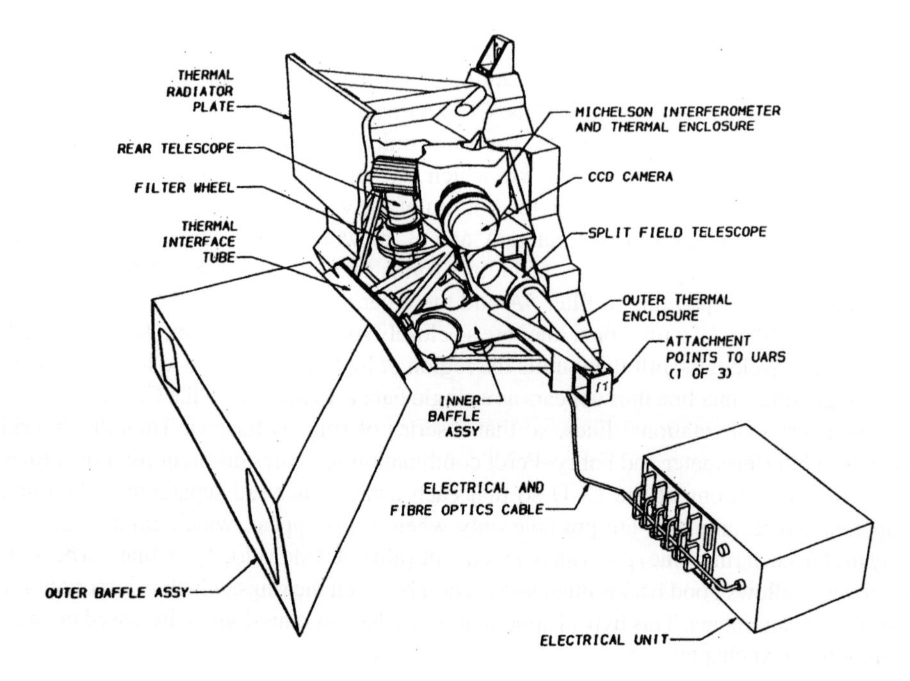

Figure 9.4. Overall view of WINDII. After Shepherd *et al.* [1993].

The principle adopted for the baffle design is that light from the cloud tops, assumed to extend up to 20 km altitude, should not be able to pass directly through the entrance apertures. A sharp vane located at the front lower edge of the baffle is intended to cast a shadow on the input apertures, but in order to do this and keep the baffle section to a reasonable length, it is necessary to reduce the vertical dimension of the aperture during the day-time. A moveable vane is therefore placed over the aperture when the instrument is on the sunlit side of the earth, creating a slot-shaped opening with the long side parallel to the horizon. The daytime aperture is about one-tenth the area of the night-time aperture. This reduces the responsivity but, fortunately, the day-time emissions are generally about ten times brighter than those at night. The WINDII baffle can be seen in the overall view of the instrument shown in Figure 9.4, along with the entrance openings at its front. The beams of the two fields of view cross inside the baffle before reaching the limb-pointing mirrors in the unit identified here as the *split-field telescope*. The cylinder on the side of the *inner baffle assembly* is the calibration source to which the instrument is exposed through the rotation of a mirror, which also acts at the same time as a door to close off input light from the atmosphere. Light from the calibration sources is brought to the calibration unit using fibre optics. After reflection off the field combiner at the centre of the split-field telescope, the light travels to a *fold mirror* under the filter wheel, where it is reflected upwards. After passing through the filter wheel the light enters the rear telescope and then a second fold mirror directs the light into the Michelson interferometer. The CCD camera is arranged so that it is accessible to the radiator plate, to which it is thermally connected, for cooling.

9.2.7 Calibration

A number of quantities must be calibrated in flight in order to obtain the desired results. The most important of these relates to the determination of the phase that corresponds to zero wind velocity. From what has been presented it is evident that it is straightforward to measure the phase of a fringe, but this is a relative, not an absolute measurement. It is not absolute because there is no way to relate the observed phase to zero path difference and, in any case, the wavelengths of the observed lines are not known with sufficient accuracy to predict the exact phase for zero wind. Thus the phase must be referenced against something else. The ideal reference would be a spectral lamp emitting the same emission line as is observed by the atmosphere. Because the observed transitions are forbidden, they are difficult to produce in a small lamp where quenching of the metastable state occurs rapidly on the walls; such lamps cannot be carried on board the spacecraft. However, for most emissions they can be created in the laboratory.

For that reason, conventional noble gas lamps were carried on board, Kr, Ne and Ar, one for each spectral line used for Doppler measurements. Lamps emitting the $O(^1S)$ 557.7 nm atomic oxygen emission, the $O_2(b^1\Sigma)$ emission and the OH emission were fabricated as laboratory sources, and the phase difference of each of these lines with respect to its calibration lamp was measured before flight. It was assumed that, while the WINDII phase would change during the injection into orbit and drift slowly with temperature, after that, the phase difference between these two lines would not change, as the phases of the two move with the instrument. This is strictly true only if the wavelength difference between the two lines is small; for separations more than about 1 nm, a calculated correction must be applied. While WINDII is thermally compensated there is still a residual thermal drift so to accurately track the WINDII phase drift a "phase-calibration" using the relevant spectral lamp (called the frequent calibration) is conducted every twenty minutes. In fact the phase drifts turned out to be rather slow, in part because of the large heat-capacity of the glass interferometer, so the method used in analysis was to collect all the phase-measurements for one day, fit the trend for these measurements and then apply the trended value to all the atmospheric measurements for that day. WINDII also carried a He–Ne laser for visibility calibration, and a tungsten lamp for responsivity calibration. These calibrations were carried out once per week, as part of a larger "infrequent calibration", in which the phases of all of the lamps were measured.

The instrument was fully characterized in the laboratory before flight as described by Hersom and Shepherd [1995]. The characterization database (CDB) was created through an extensive characterization on the ground, using a scanning spectrometer to determine the passbands for all of the elementary bins for all of the filters, an ion laser to measure the MI instrument visibility as a function of wavelength, a calibrated tungsten source for responsivity calibration, and laboratory sources of the atmospheric lines to determine the zero wind phase. This was augmented in orbit through in-flight calibration to track the ground-to-orbit phase shifts and assess the on-orbit trends. During WINDII's first three years in orbit the phase drifted slowly and consistently in exponential fashion through about two fringes, due to some change in the instrument which has not been identified. The CDB is effectively time-dependent, in order to allow for trends. The calibration component for WINDII and its application have been described by Thuillier *et al.* [1998]. Following launch an extensive validation was carried out, using independent wind measurements from the HRDI instrument, the MICADO instrument (described later in this chapter) and ground-based radars, as described by Gault *et al.* [1996a] and Thuillier *et al.* [1996].

9.2.8 Wind Measurement Procedure

The WINDII fields of view are oriented at 45° and 135° from the spacecraft velocity vector, so that each view traces a track off to one side of the orbit, about 1500 km away, as depicted in Figure 9.1. The time delay for viewing the same volume is about 8 minutes, so it is assumed that the wind velocity does not change during that time. In a limb view, the signal obtained corresponds to an integral along the line of sight, as described in Chapter 1, passing through atmospheric layers, each of which has its own emission rate, wind and temperature. The signal received by WINDII is therefore given by Eq. (9.1) where the symbol E is used for volume emission rate rather than V to avoid confusion with the visibility:

$$I_{b,u} = \int_L E(z)[1 + U_b V(z)\cos(\Phi_{b,u} + \phi(z))]\,ds \tag{9.1}$$

Here, u denotes a particular mirror step as before, b denotes a particular bin on the CCD, and z denotes altitude. Quantities like $E(z)$, $V(z)$ and $\phi(z)$ depend only on altitude, because they relate to geophysical quantities, namely volume emission rate, temperature and wind. U is the visibility reduction factor, called "instrument visibility", which is the amount by which the instrument reduces the line visibility V, because of imperfections in the glass interferometer. It depends only on pixel location. By taking a sequence of measurements at different mirror steps u, it is possible to invert the sequence of images obtained in order to obtain profiles of emission rate, wind and temperature.

By expanding the cosine argument the following is obtained:

$$\begin{aligned} I_{b,u} = \int_L E(z) + U_b \cos\Phi_{b,u} \int_L E(z)V(z)\cos(\phi(z))\,dl \\ - U_b \sin\Phi_{b,u} \int_L E(z)V(z)\sin(\phi(z))\,dl \end{aligned} \tag{9.2}$$

$$I_{b,u} = J_1 + U_b \cos\Phi_{b,u} J_2 - U_{b,u} \sin\Phi_b J_3 \tag{9.3}$$

These equations define the J_1, J_2 and J_3 of Chapter 8 in integral terms, which contain the atmospheric information that is to be recovered. In Chapter 8 it was indicated that these quantities could be obtained from Eq. (9.3). through a Fourier-type process, but that is true only when exactly one fringe is sampled with equally spaced steps. Although the step sizes are accurately taken with WINDII, they are not made to the accuracy with which they can be measured. In addition, the step size varies across the field of view so cannot be the same for all bins. Thus the approach is to obtain the step sizes through measurement and calculation and use a more generalized approach which is to write Eq. (9.3) as the following, where the approximate equality is used because of noise:

$$\begin{bmatrix} I_{b,1} \\ I_{b,2} \\ I_{b,3} \\ I_{b,4} \end{bmatrix} \approx \begin{bmatrix} 1 & U_b \cos \Phi_{b,1} & -U_b \sin \Phi_{b,1} \\ 1 & U_b \cos \Phi_{b,2} & -U_b \sin \Phi_{b,2} \\ 1 & U_b \cos \Phi_{b,3} & -U_b \sin \Phi_{b,3} \\ 1 & U_b \cos \Phi_{b,4} & -U_b \sin \Phi_{b,4} \end{bmatrix} \times \begin{bmatrix} J_1 \\ J_2 \\ J_3 \end{bmatrix} \tag{9.4}$$

This equation is solved through a weighted least-squares method; using error covariances as weights provides the minimum attainable error variances. The process is described by Shepherd *et al.* [1993] and in more detail by Rochon [1999]. The J values, which are line-of-sight integrals as noted are inverted to provide height profiles for volume emission rate, temperature and wind, using a linear constrained least squares method as described by Rochon [1999].

9.2.9 Examples of Results Obtained

In this section, a sample of the WINDII results is presented for volume emission rates, winds and temperatures, for both night and day-time observations.

Volume emission rate profiles for the atomic oxygen green line at 558 nm are shown in Figure 9.5 for night-time conditions. The profile (left panel) has been inverted from the data (J_1), as described in Chapter 1. It shows a thin layer of emission centred near 100 km which is caused by the recombination of atomic oxygen formed during the day through the photo-dissociation of O_2 in the Schumann–Runge continuum. Above 250 km there is

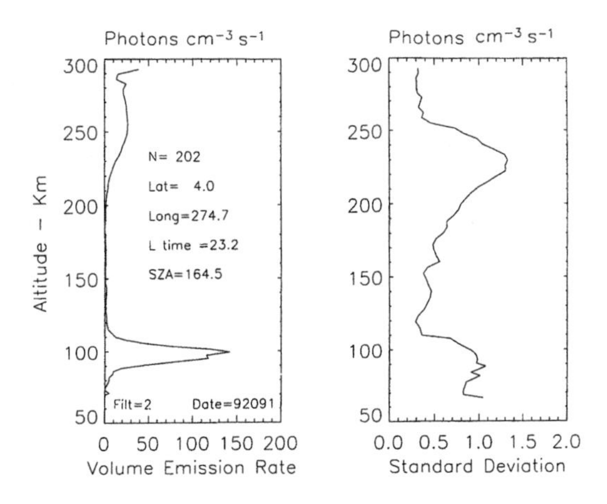

Figure 9.5. Profile of O(^{1}S) 557.7 nm volume emission rate (left) obtained by WINDII at night (solar zenith angle 164.5°) for the conditions shown, and the standard deviation (right). After Shepherd *et al.* [1997].

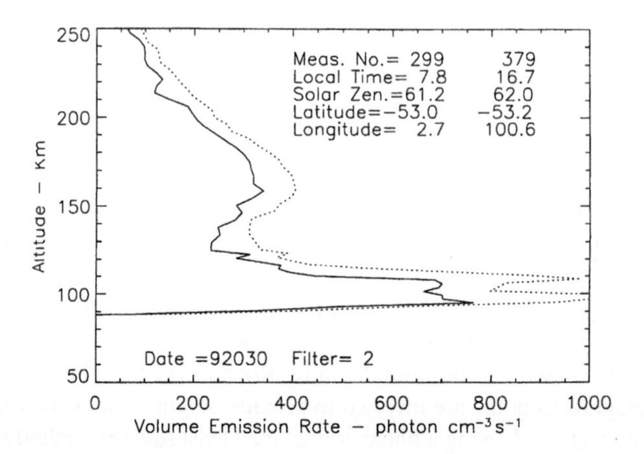

Figure 9.6. Profiles of O(^{1}S) volume emission rate obtained with WINDII during day-time conditions for the same latitude and solar zenith angle, but for two different local times. After Shepherd *et al.* [1997].

weaker emission from the recombination of O_2^+ ions in the ionosphere; this is the non-thermal emission described in Section 8.1. This emission is normally very weak except near the equator as is the case here; the latitude is 4.0°.

The right-hand panel shows the standard deviation of the measurements, which at its worst is about 1 photon $cm^{-3}\,s^{-1}$, and at its best about 0.3. These errors are negligible compared with the observations, but this high signal-to-noise is required for the wind measurements.

In Figure 9.6 is shown the same atomic oxygen 557.7 nm emission but for day-time conditions, for a solar zenith angle of 61°. Two profiles are shown, for the same zenith angle and latitude, but different local times (and thus longitudes). These profiles are obtained from a north-going pass and a south-going pass over the same latitude of −53°. During the daytime the lower layer (E-region) is enhanced and broadened, and another broad layer appears, centred on 165 km altitude (F-region). The enhanced day-time emission is produced through the "prompt" excitation by solar EUV radiation which produces photoelectrons through photoionization. These low-energy electrons are in an energy range with high cross-sections for the excitation of neutral species, many of which produce O(^{1}S) through photochemical reactions. For example, excited $N_2(A^3\Sigma_u^+)$, transfers its energy to O through collisions, producing O(^{1}S). Electron impact on O also produces some O(^{1}S), and the Lyman-β emission can photodissociate O_2, leading to O(^{1}S); this is an important contributor to the E-region emission.

In Figure 1.19 some WINDII data were shown as a function of local time for a latitude of 20 N. From 90 to 110 km data are present for all 24 hours of local time because there is O(^{1}S) emission present both day and night. From 110 to 200 km there are data only between 06 and 18 hr, that is, during day-time hours only. For the lower altitude region there are "tilted" wavefronts evident, and it can be seen that the period is 24 hr. This is the diurnal

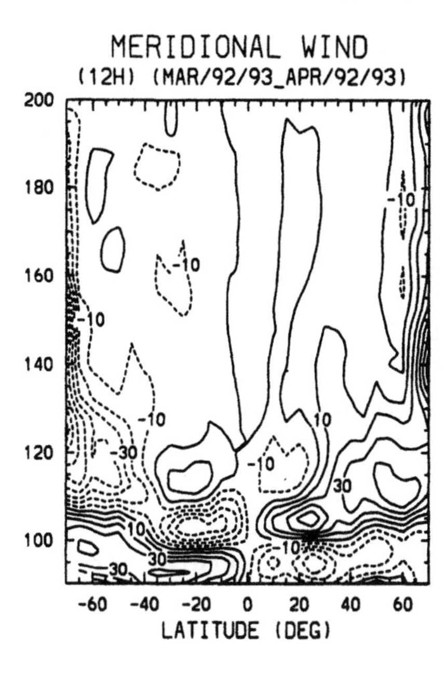

Figure 9.7. The meridional (north–south) wind as a function of latitude for 12 hr local time, as observed by WINDII for the period March/April 1992/93. Solid lines (+) denote northward winds and dotted lines (−) denote southward winds. After McLandress *et al.* [1996a].

tide, produced by heating in the lower atmosphere, through the absorption of solar radiation by water vapour and ozone. The tide propagates upward, growing in altitude as it ascends in order to conserve energy as the atmospheric density decreases. Figure 1.19 shows winds as large as 45 m s^{-1} near 100 km, but these are averaged data and the instantaneous tidal winds can be 70 m s^{-1} or more.

In Figure 9.7, the meridional wind is shown as obtained for the vernal equinox (March and April) for the years 1992 and 1993 by McLandress *et al.* [1996a], plotted as a function of latitude for a local time of 12 hr. Notice the very strong wind "cells" located asymetrically about the equator, tidal winds as shown in Figure 1.21. At 95 km the winds are inwardly directed across the equator and at 105 km the winds are outwardly directed. The vertical structures correspond to a vertical tidal wavelength of about 25 km. The tide begins to dissipate at about 130 km.

A Doppler temperature profile from the $O(^1S)$ 557.7 nm emission is shown in Figure 9.8. The temperature rises monotonically throughout the altitude range 100–180 km, whereas the true temperature levels off at around 160 km. This confirms the existence of non-thermal radiation above 160 km, as was discussed in Chapter 8. The WINDII Doppler temperature measurements have not been fully validated, in part because of the lack of corresponding measurements. In the meantime it has been discovered that excellent mesospheric temperatures can be obtained from the Rayleigh scattering observed by WINDII in its background

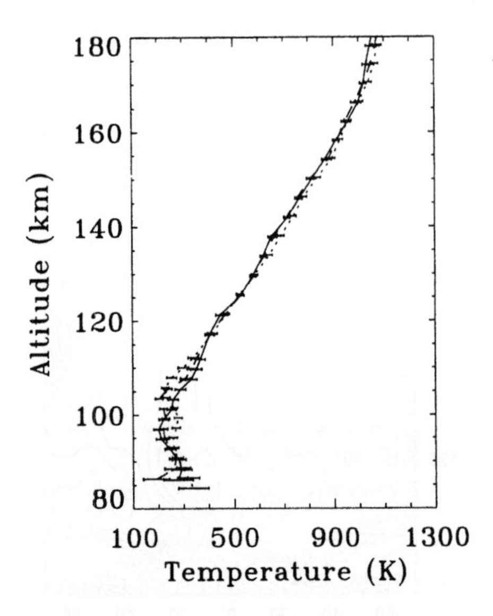

Figure 9.8. WINDII measurement of temperature from the O(^{1}S) day-time emission. After Rochon [1999].

filters. During the day-time, the background filter for the 558 nm green line emission contains only sunlight scattered from the atmosphere through Rayleigh scattering. This is proportional to the atmospheric density and so the signal observed is the integral of that density along the line of sight, which can be inverted to provide relative density as a function of altitude. The atmospheric pressure variation with altitude is linked to the density through the hydrostatic equation, reflecting the fact that the pressure difference dp across an atmospheric parcel of unit volume, is simply that required to support the force of gravity, mNg, where m is the molecular mass and N is the number density so that using the ideal gas law [M. Shepherd *et al.*, 1999]:

$$\frac{dp}{p} = \frac{mg}{kT}\,dz = \frac{dz}{H} \tag{9.5}$$

where g is the acceleration of gravity and H is called the scale height of the atmosphere. This is the hydrostatic equation. Therefore, from the pressure (or equivalently, density) variation with altitude, the scale height and thus the temperature can be found.

An example of the results is shown in the form of temperature maps at an altitude of 85 km for a single day, January 20, 1993, where temperature contours are shown as a function of latitude between −50° and 0°, and longitude as shown in Figure 9.9 [M. Shepherd *et al.*, 1999]. The pattern shows a sequence of valleys and troughs with a north–south alignment. Because, for a given day the satellite orbit is essentially fixed with respect to the sun, all of the observations made at a given latitude have the same local time (the change during the day is actually 20 min) then these patterns are not due to the atmospheric tide, which is dependent on local time. The structures are planetary scale features which are truly longitude

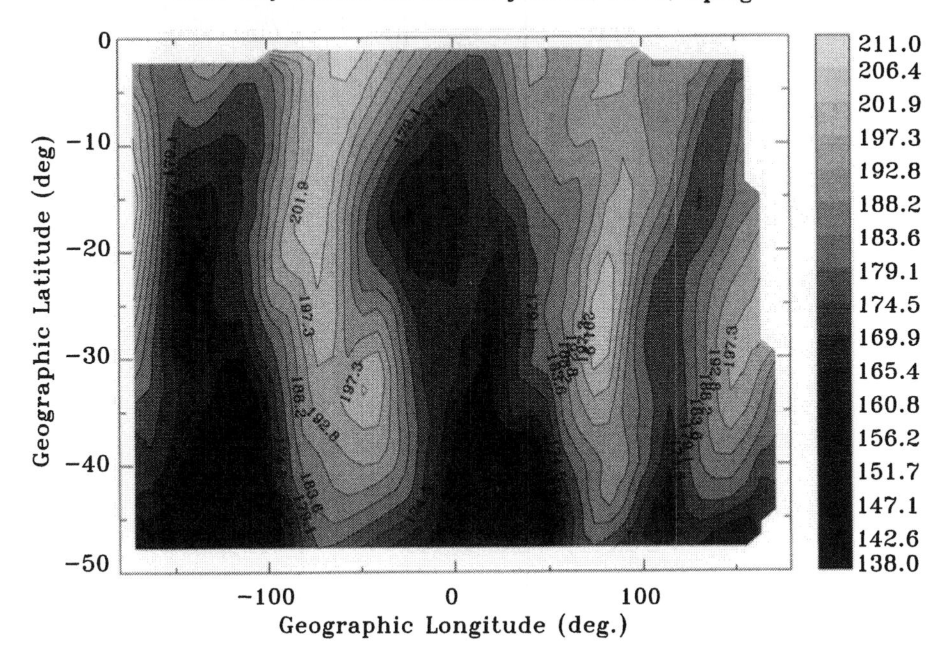

Figure 9.9. Latitude/temperature map of Rayleigh scattering temperatures, observed by WINDII for an altitude of 85 km on January 20, 1993. After M.G. Shepherd *et al.* [1999].

dependent and thus features which are "attached" to the Earth. Over 360° of longitude there are three ridges and three valleys so that there are three complete waves around the Earth. In the valleys the temperature is roughly 170 K and along the ridges about 200 K. This is a very particular type of planetary wave, called the quasi two-day wave because for a ground station the period is observed to be two days. The wave propagates westward at a rate of 60° per day so that the ground station sees the next ridge after a period of two days as shown by M. Shepherd *et al.* [1999]. The quasi two-day wave has also been observed in the WINDII winds [Ward *et al.*, 1996] and the emission rates [Ward *et al.*, 1997].

One of the most important contributions WINDII has made is to simultaneously observe the winds and the atomic oxygen transported by those winds. This has shown that the concentration of atomic oxygen is not at all uniformly distributed in the upper atmosphere, but that it exhibits highly localized features that correlate with the wind field. From the strong tidal signatures near 100 km, as shown in Figure 9.7, one might therefore expect a strong tidal influence on the distribution of atomic oxygen and this is confirmed as shown in Figure 9.10. Here the meridional wind is shown in the upper panel, the $O(^1S)$ emission in the middle panel and the OH emission in the lower panel. In Figure 9.10 (left panels), for the March/April period, for a local time of 02 hr, the diurnal tide is strong and the meridional

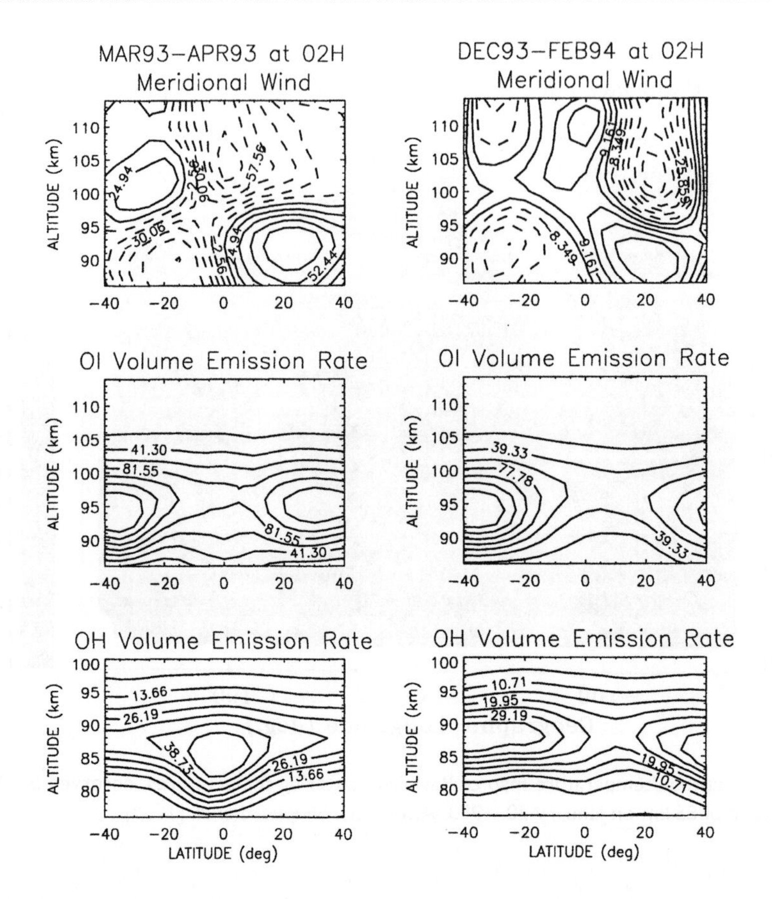

Figure 9.10. The influence of thermospheric winds on the distribution of atomic oxygen as seen in the $O(^1S)$ and OH airglow emissions. Courtesy of S. Zhang (York University).

wind cells are highly symmetrical about the equator. The solid contours indicate northward winds so that near 90 km the winds are blowing away from the equator while higher, near 102 km, they are blowing towards the equator. The influence of these strongly diverging and converging fields on the green line emission near 97 km is seen in the middle panel, where a minimum emission rate is created at the equator. The lower altitude hydroxyl emission near 87 km shows a maximum at the equator which is consistent with the fact that the vertical wavelength of the diurnal tide is about 25 km, so that the two emission layers are about one half-wavelength apart, and therefore are out of phase.

In Figure 9.10(b) which is for the northern winter solstice, the situation is different. The wind field is no longer so symmetrical about the equator, because the diurnal tide is less dominant compared with the semi-diurnal tide. Under these conditions the green line emission still shows a minimum near the equator, but the emission rates are not symmetrical, with significantly higher values in the southern (summer) hemisphere. The hydroxyl emission now shows a minimum at the equator, which is consistent with a longer wavelength of the diurnal tide, bringing the two airglow emissions more closely in phase. This interpretation is confirmed through agreement with TIME-GCM [Roble and Ridley, 1994] model predictions as shown by Zhang *et al.* [2001], providing endorsements of both the instruments and the model. WINDII has also observed greatly enhanced winds in the auroral region produced by geomagnetic storms, up to 700 m s^{-1} [Zhang and Shepherd, 2000].

9.3 ERWIN: AN E-REGION WIND INTERFEROMETER

9.3.1 Introduction

ERWIN is a ground-based wind-measuring instrument that is currently located within the Early Polar Cap Observatory of the National Science Foundation at Resolute Bay (74.3° N, 94.5° W) in Northern Canada. It is based on a concept very similar to that of WINDII, except without the advantage of limb viewing. The challenge was to construct an instrument with adequate sensitivity to measure winds in a relatively short period of time from weak emissions. In order to gain some altitude resolution, three different emissions are used in order to obtain winds from three different heights, the atomic oxygen 557.7 nm green line emission from 97 km, the O_2 Atm (0,1) band emission at 94 km with a wavelength of 866 nm and the hydroxyl emission from 85 km at a wavelength of 843 nm. The O_2 Atm (0,1) band is much weaker than the O_2 Atm (0,0) band observed with WINDII, but since the latter is absorbed in the lower atmosphere the former must be used for ground-based measurements, heightening the demand for good responsivity. Finally, in viewing the E-region $O(^1S)$ emission near 100 km, it is not possible to avoid viewing the higher altitude thermospheric non-thermal component. The objective for ERWIN was to use a sufficiently long OPD that the visibility of the non-thermal component was effectively zero. Since ERWIN was designed following the experience gained with WINDII, and since the requirements were different, some new aspects were involved and it is these that are described in this section. The instrument is operated remotely by modem from York University in Toronto, and it is described by Gault *et al.* [1996b].

9.3.2 Instrument Description

Atmospheric winds are dominantly horizontal, and for an instrument operating alone, the usual method is to direct the field of view at a low angle (e.g., 30°) above the horizon, towards four or eight cardinal compass directions in a circle around the station. These measurements are combined to give a wind vector, assuming a uniform wind or a wind with a uniform gradient in the region sampled. If there is enough redundancy in the measurements, the validity of the assumption can be tested.

The layout of the instrument is shown in Figure 9.11. Light enters via the horizon mirror and passes successively through the insulating window, an interference filter, the telescope, the Michelson interferometer and collecting optics to the photomultiplier detector.

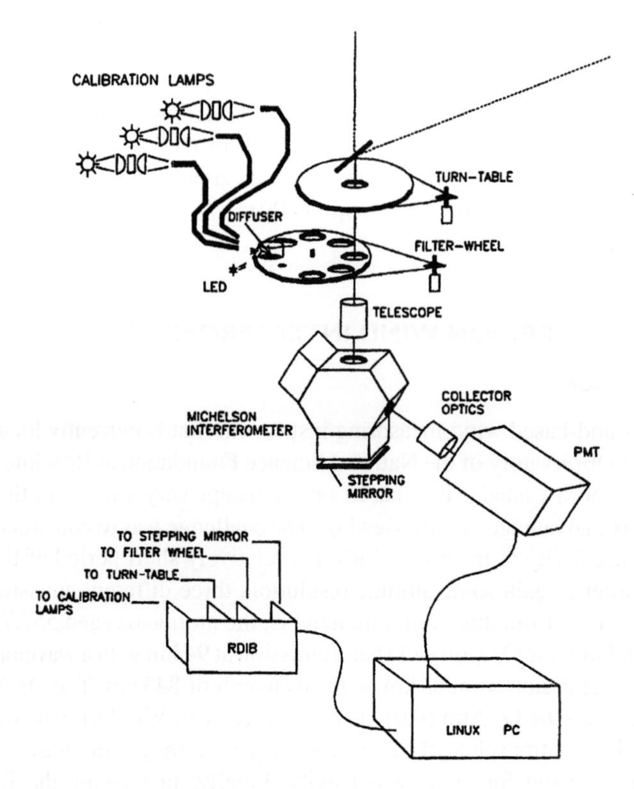

Figure 9.11. General layout of the ERWIN instrument. After Gault *et al.* [1996b].

The horizon mirror is mounted on a turntable which rotates to direct the field of view towards eight possible compass directions at an elevation angle of 30°. A clear view of the zenith is obtained by rotating the turntable past the eighth (NW) position. A mechanism then tips the mirror out of the field of view. Immediately below the turntable is the filter wheel, which has eight positions. One of the filter wheel positions is used for viewing the calibration sources. The detector is a Hamamatsu gallium arsenide photomultiplier (R943-02) which is operated at $-25\,°C$. The dark current is typically $200\,s^{-1}$, well below the level of the weakest signal encountered with the instrument $\approx 10^3\,s^{-1}$ for O_2.

9.3.3 The Michelson Interferometer

The MI uses the technology of WINDII for the beam-splitter and the stepping mechanism, but the design of the arms is unique, consisting of three types of glass and a large air gap. All components of the interferometer are cemented together to form a mechanically stable unit. The glass types and block lengths were chosen to produce the required path difference, field widening and thermal compensation. The stepping mirror is mounted on three piezoelectric cylinders and its position is controlled by feedback from capacitive gap sensors. The interferometer was manufactured by GSI Lumonics and the scanning mirror by Queensgate Systems, U.K., who also built the mirror control system.

Four criteria were important in the selection of the path difference for ERWIN:

- The path difference should be close to optimum for wind measurements.
- The fringes from the broad middle thermospheric O^1 S emission should be suppressed.
- The fringes from blended OH and O_2 lines should be in phase with each other.
- The dimensions of the glass plates and the assembled Michelson should be within practical limits.

The lines of the OH Meinel bands are doublets due to Λ-doubling of the rotational levels and they are much too closely spaced to separate with an interference filter. It is therefore important to choose the path difference so that the component fringes are nearly in phase with each other in order to maintain good visibility. This also reduces the sensitivity of the resultant phase to changes in the relative signal from the two components, which could occur if the filter passband drifts. Similarly in the case of O_2, the lines in the (0, 1) band are too close together to separate with a filter, but they can be fairly well isolated in pairs.

A search was conducted for a path difference between 10 and 12 cm that would produce approximate phase coincidence for an OH doublet and a pair of O_2 lines. This range is located between the minima in the error curves for $O(^1S)$ and OH and includes the minimum in the temperature error curve for $O(^1S)$. With $\Delta \geq 10$ cm, the fringes of the upper $O(^1S)$ emission are suppressed by a factor of 10 or more relative to the lower region.

In addition to the above considerations regarding path difference, the interferometer had to be field widened so it would accept a field of view 10° in diameter. Achromatic compensation is achieved by careful selection of the arm glasses and their lengths. ERWIN also required an air gap in one arm to allow for the movement of the scanning mirror. The glass catalogs contain hundreds of different glasses, but in practice, only a few are readily available at the highest level of quality, as required for an MI. The selection for ERWIN was made from six Schott glasses, BK7, K7, K10, LaKN12, SF11 and SF56. A computer program was first used to match the glasses in pairs, and when two glasses failed to produce a good design, a three-glass design was attempted.

Five-glass combinations produced interesting designs and, of these, the combination BK7/LaKN12/SF11 was selected as most promising. By good fortune, repeated runs of the program using different values of Δ_o produced a value which gave near coincidence for the brightest of both the O_2 and OH pairs of lines. Final trimming of the design was done by adjusting the arm lengths in steps of a few microns, using the refractive indices measured by Schott on the specific melts of the glasses provided. The stepping mirror is made of SF11 to avoid thermal stress in the piezoelectrics and supporting pillars.

9.3.4 Examples of Measurements

ERWIN is designed so it can measure in eight horizon directions rather than four, but so far that capability has not been used, as was used to advantage by Burnside *et al.* [1981].

Observed signal levels range from 10^4 to $10^5\ s^{-1}$ for both OI and OH, and from 10^3 to $10^4\ s^{-1}$ for O_2. Emission rate variations in the source can cause errors because the four exposures must be made sequentially. In order to minimize this effect, the Michelson mirror is scanned in both directions for each measurement and the corresponding signals are co-added. A linear emission rate variation during the measurement would therefore have no effect.

Figure 9.12 shows a sample of the data obtained by ERWIN at Resolute Bay. Winds from the three emissions are shown in the same order they appear in the atmosphere, with $O(^1S)$ winds (about 97 km) at the top, the O_2 Atm band (about 94 km) in the middle and the OH (about 87 km) at the bottom. The data cover a period of almost a day during Jan. 2–3, 1994, and are plotted as a function of time with the vectors showing the direction in which the wind was blowing and the magnitude indicated by the length, according to the scale on the y-axis. The main feature of the observations is a half-day oscillation whose phase evidently propagates downwards. Comparisons with a ground-based FPS at the same site have been shown by Fisher *et al.* [2000].

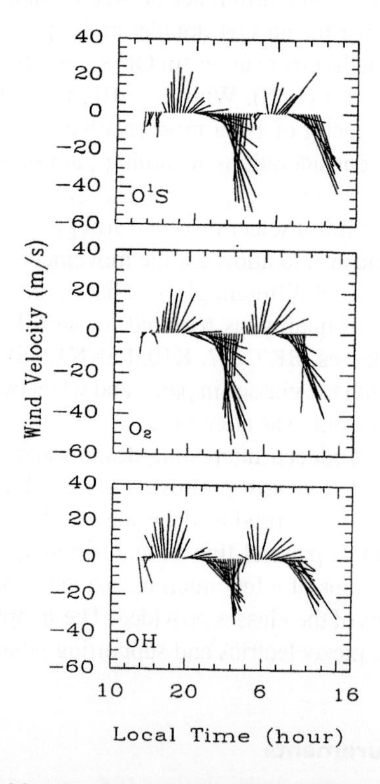

Figure 9.12. Winds observed for the three emission lines at three different altitudes as obtained at Resolute Bay in Northern Canada, 2–3 January, 1994. After Gault *et al.* [1996b].

9.4 MICADO – MICHELSON INTERFEROMETER FOR COORDINATED AURORAL DOPPLER OBSERVATIONS

The MICADO instrument is a ground-based field-widened Doppler Michelson interferometer using an alternative approach to that used in ERWIN which was built several years later. The similarities are to be expected because the French and Canadian scientists involved in MICADO and ERWIN respectively, had been collaborating on the WINDII experiment since 1984. However, there are some interesting differences between MICADO and ERWIN which are outlined in this section. A detailed description of MICADO is given by Thuillier and Hersé [1991]. Instead of using a solid cemented assembly of glasses, MICADO uses an open but rigid design as shown in Figure 9.13. All elements are of fused silica, apart from one compensating mirror of K7 as described below. These elements are fixed to a fused silica baseplate by molecular contact as shown in Figure 9.13, which provides the required stability without the use of cements. The open design allows the interferometer to be scanned through the use of a scanning wedge. This wedge, with a very small prism angle of 2.62 min of arc, is displaced laterally in order to vary its effective thickness. This offers a flexible scanning arrangement which allows small or large steps. Large steps are used to set the wedge to the value of maximum field widening, which is different for different wavelengths. The position of the prism is monitored with a He–Ne laser. The wedge is made of a different material, F5, in order to make the thermal stabilization independent of wavelength as described in Section 8.8.

The interferometer elements are chosen to satisfy the condition of field-widening, and for wavelength independent thermal compensation, taking into account the linear expansion coefficients and refractive index dependence on temperature, as described by Thuillier and

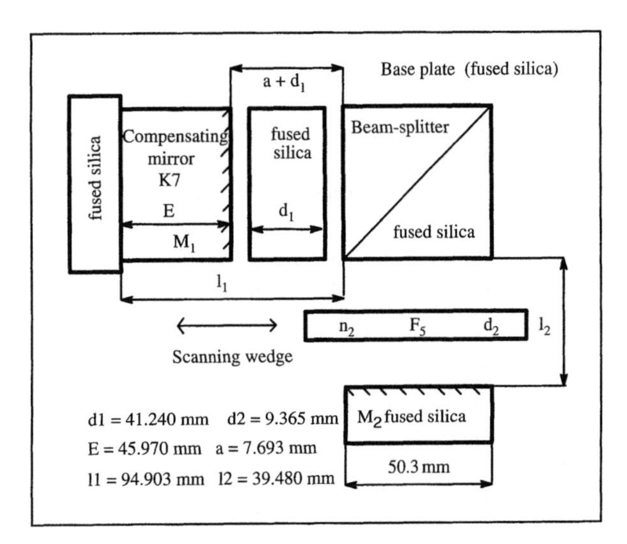

Figure 9.13. The design of the MICADO Michelson interferometer optics. Courtesy of G. Thuillier (Service d'Aéronomie du CNRS).

Shepherd [1985]. Here the compensating mirror of K7 plays an important role in that its linear expansion coefficient provides the appropriate thermal movement of the mirror on its front surface.

Light from the sky is collected with a Cassegrain telescope that can be pointed in different directions. Unlike ERWIN, the background signal is simultaneously measured with a separate photomultiplier (PMT), using light extracted with the beam-splitter shown in Figure 9.14. Also, laser light is injected in a small beam on the axis and detected by a separate photomultiplier. Additional calibration sources (on the left) may be observed by MICADO as well. The system was tested through an induced Doppler shift generated by reflecting laser light off a specially designed wheel. The wheel had teeth of a spiral shape allowing the simulation of a continuously variable velocity between -50 and $50\,\mathrm{m\,s^{-1}}$ and the accurate measurement of velocity was confirmed. MICADO has an accuracy of $4\,\mathrm{m\,s^{-1}}$ for a line-of-sight wind observation of 1 min duration. It was operated during three winter campaigns in Northern Scandinavia in association with the EISCAT radar [Thuillier *et al.*, 1990], and later was installed at the Observatoire de Haute-Provence in Southern France for the validation of WINDII [Thuillier *et al.*, 1996]. During the Scandinavian campaigns it was used to measure vertical winds in the $O(^1D)$ 630 nm emission in the auroral zone by Fauliot *et al.* [1993]. MICADO will have a successor called EPIS (Etudes Polaires par Interféromètrie Svalbard) in which the total emission rate of the atomic line will be

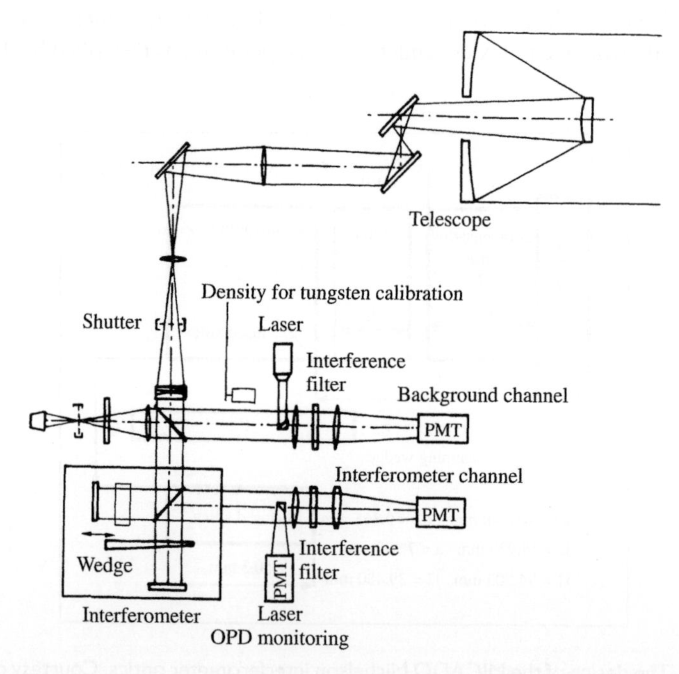

Figure 9.14. Overall configuration of the MICADO system. After Thuillier and Hersé [1991].

measured simultaneously with an interferogram and the background. Svalbard is the Arctic island on which it will be operated (sometimes called Spitzbergen).

9.5 THE HIGH-RESOLUTION DOPPLER IMAGER (HRDI)

9.5.1 Introduction

The high-resolution Doppler imager (HRDI), launched on the Upper Atmosphere Research Satellite on September 12, 1991 is a Fabry–Perot spectrometer intended for the measurement of wind velocities in the stratosphere, mesosphere and lower thermosphere during the day, and the lower thermosphere at night. Winds in the stratosphere are determined by measuring Doppler shifts in absorption of lines in the O_2 Atmospheric band ($b^1\Sigma_g^+ - X^3\Sigma_g^+$) [Hays, 1982]. These features are observed in sunlight scattered into the field of view by aerosols and molecules, as the instrument views the limb of the Earth. Lines from the same band are observed as airglow emission in the daytime in the mesosphere and lower thermosphere, and at about 95 km at night. A much more extended overview is provided in the comprehensive description by Hays *et al.* [1993], with further details given by Grassl *et al.* [1995].

In Chapter 5 the ultimate Fabry–Perot spectrometer was described as one having high resolution in a unique scanned passband for which earlier versions were described by Mack *et al.* [1963] and McNutt [1965]. A unique passband is required for the observation of absorption lines of the stratospheric measurement. HRDI achieved this goal by using three etalons in series in combination with a set of interference filters to provide spectra for a number of regions with a resolution that was compatible with Doppler measurements.

The overall configuration of the optical system is shown in Figure 9.15. Light is collected by a fully gimbaled telescope that allows observations on either side of the spacecraft. This means that observations may be taken at two different local times at the same latitude on a given day, and also that coverage is possible in principle from 72° S to 72° N on any day. Light enters the instrument either from the telescope or from calibration sources according to the position of a scene selection mirror. The light beam is expanded and passes through a dual filter wheel, of which each wheel holds eight filters. The beam is expanded before passing through the etalons, and then is focused by an imaging telescope onto a multi-anode concentric FPS ring detector of 32 anodes, with each ring anode corresponding to a different spectral element according to the fundamental principles of the Fabry–Perot spectrometer as outlined in Chapter 5 and employed by the Dynamics Explorer Fabry–Perot instrument described in that chapter.

9.5.2 Input Optics

The telescope is located on the Earth-facing side of UARS. The primary mirror is a 17.78 cm diameter $f/2$ off-axis parabola, as shown in Figure 9.16. The two-axis gimbaled system allows scanning in both azimuth and zenith directions. The relay optics transfer the light to the light pipe leading to the interferometer independently of the orientation of the telescope. In order to maximize the collected light flux while minimizing the height of the vertical

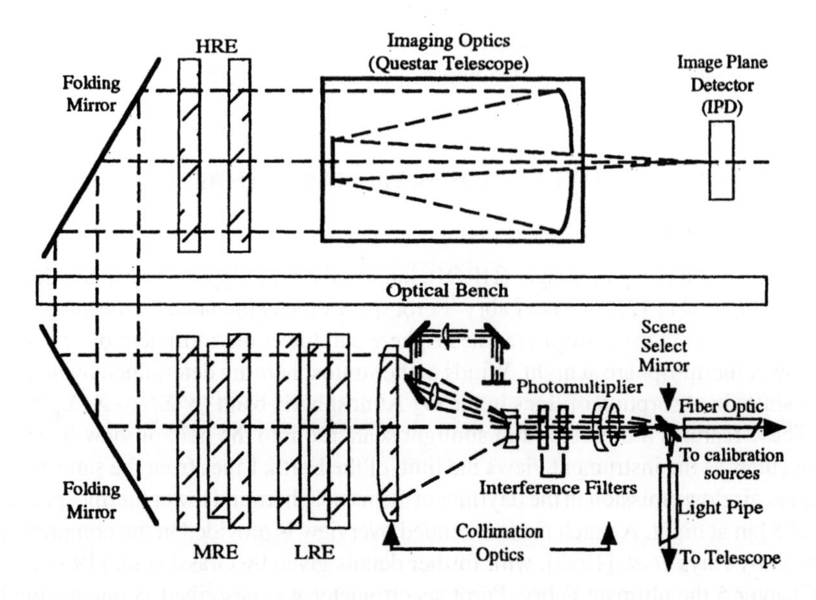

Figure 9.15. The interferometer component of HRDI, including coupling optics. After Hays *et al.* [1993].

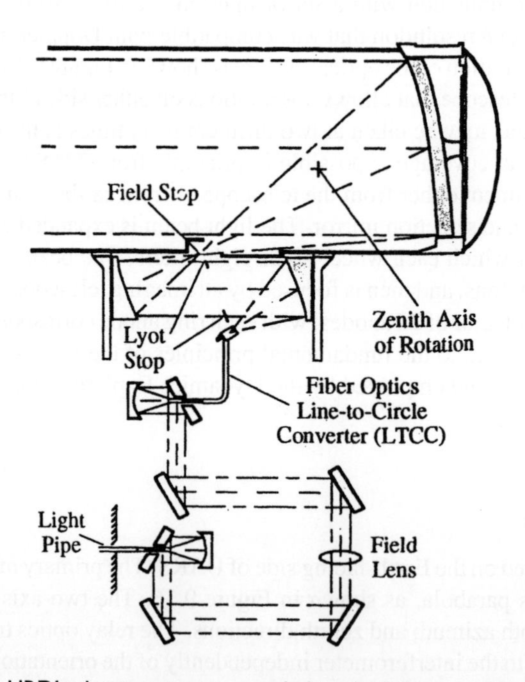

Figure 9.16. The HRDI telescope system, coupled to the light pipe. After Hays *et al.* [1993].

resolution element, a slit field is employed in front. This corresponds to 6 km vertical resolution at the limb and 70 km horizontal for the wide field of view, with an alternative option of 6 km horizontal in the narrow field of view. The slit field is converted to circular for the interferometer, using a fibre optics bundle of 700 optical fibres, each 100 μm in diameter, bundled as a rectangle at one end and a circle at the other. A random arrangement is used in order to scramble variations in emission rate viewed by the telescope. The telescope requires moderately precise (0.05°) pointing with very precise (0.01°) knowledge.

The telescope is coupled to the interferometer with a 3 mm diameter light pipe, which also serves to further scramble the signal. The input optics expand and collimate the beam to 25 mm diameter before it passes though the filters; the divergence at this point is about 2°. The beam is further expanded to 90 mm with a divergence of 0.5° for passage through the etalons. A Questar telescope images the fringes onto the detector; this provides the long focal length of 2160 mm required, in a short physical space. The filters in the filter wheel allow observation of various lines in O_2 atmospheric bands; stratospheric observations are made using lines in the B and γ bands while the mesosphere observations are made with lines in the A band. Other filters are included for observation of H_2O and OH spectral lines and the 557 nm line, $O(^1S)$. The filters are typically 0.8 nm wide. About 1% of the incident light is diverted into a photometer which can make observations through all of the filters available. This provides more sensitive measurements of emission rates than is possible with the interferometer, important for some scientific studies. In addition, the photometer can track stars in order to calibrate the pointing system.

9.5.3 The Etalons

The three 13.2 cm diameter etalons are of high, medium and low resolution (HRE, MRE and LRE respectively). They have surface flatnesses of $\lambda/200$ and similar reflectances to one another of about 0.9 with finesses of approximately 12 (these vary slightly from etalon to etalon and from one wavelength to another). The gap thicknesses, chosen according to an analysis by Skinner *et al.* [1987], are, respectively: 1.0007 cm, 0.1861 cm and 0.0241 cm. While the optical designs of the etalons are similar, the mechanical designs are quite different. The HRE has fixed length spacers of zerodur so that spatial scanning is used to cover its free spectral range with the 32-anode detector described later. The LRE and MRE use piezoelectric posts to change their gap thickness so that all the etalon passbands are made to coincide. These piezoelectric units are servo controlled from capacitors that monitor the etalon spacing, similar in concept to those employed by WINDII, but designed and fabricated by the HRDI investigators. For a linewidth determination, the interference filter is chosen appropriate to a particular line and a spectrum of 32 elements is obtained, covering the free spectral range of the HRE, which is 0.5 cm^{-1}; this is more than enough for a linewidth determination but a full free spectral range needs to be covered since it cannot be predicted where the line will fall within that range. If broader spectral coverage is required, the LRE and MRE are tuned to the next order of the HRE and a further 32 elements are recorded. This can be continued to the limit imposed by the interference filter.

9.5.4 The Detector

The image plane detector (IPD) is a multiple-channel photomultiplier, which is 40 mm in diameter and has 32 equal area anodes. The anode patterns are circular and follow the "equal-area for equal spectral width" formulation of Chapter 5. Three micro-channel plates (MCP) are used to provide gain (electron multiplication) for the system. The 32 elements of the detector simultaneously measure 32 elements of the spectrum at a single location in space with an integration time of 0.096 s. The telescope is stepped upward in order to accomplish an altitude scan.

9.5.5 HRDI Results

Zonal wind measurements from October 1992 to April 1996 covering an altitude range from 10 to 115 km obtained with HRDI are shown in Figure 9.17. Above 50 km the winds are obtained from the O_2 band emission, while below 40 km they are observed in absorption. The stratospheric wind results have been described by Ortland *et al.* [1996]. The tidal fluctuations have been removed so that the longer period variations are evident. In the stratosphere, near 25 km there is a pattern that repeats with a period of 26 months; this is the quasi-biennial oscillation, or QBO. In the mesosphere, near 80 km there is a strong semi-annual oscillation (SAO), but below, around 65 km both the SAO and QBO components

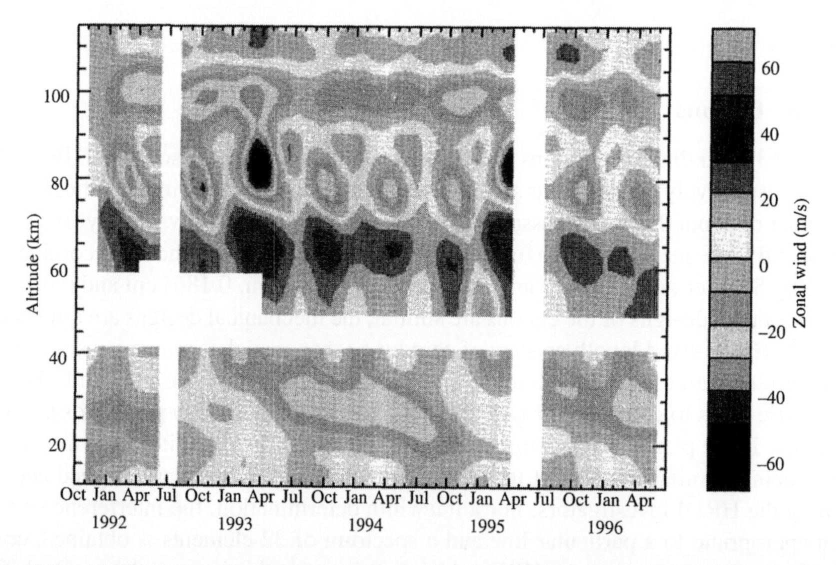

Figure 9.17. HRDI zonal wind measurements for the stratosphere and mesosphere for the period 1992 to 1996. For the stratosphere near 25 km there is a periodic pattern with a period of about 26 months, called the quasi-biennial oscillation or QBO. This can also be seen in the mesosphere near 85 km although it is the semi-annual oscillation that is dominant. Courtesy of W. Skinner (University of Michigan).

appear to be present. At 85 km there also appears to be a QBO component, which indicates the coupling between the stratosphere and mesosphere.

9.5.6 Comparison of HRDI and WINDII

It is of interest to compare HRDI and WINDII because the two instruments are highly complementary in nature, since a number of "opposite" choices were made for the two instruments. For example, the very flexible fully gimbaled telescope was chosen for HRDI while WINDII was designed with fixed view directions. The telescope was advantageous as it allowed viewing on both sides of the spacecraft. This allowed two different local ranges to be observed rather than one for WINDII, and the full latitude range of 72° S to N to be covered on a single day.

The most fundamental difference was that HRDI observed the spectrum thus providing detailed spectral information, while WINDII observed the interferogram which yielded phase shifts based on a smaller amount of information. The HRDI method is illustrated in Figure 9.18(a) in which the spectrum of an observed line is plotted along with the unshifted

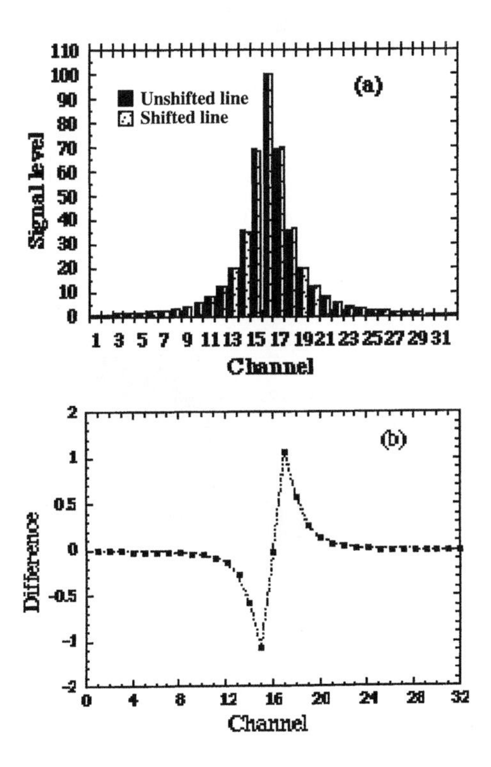

Figure 9.18. Illustrating the difference method of determining line shifts. When the unshifted line is subtracted from the observed line as in (a), the difference signal in (b) is obtained. Courtesy of W. Skinner (University of Michigan).

line. In Figure 9.18(b) the difference signal is shown; it is a very sensitive indicator of the lineshift but it does depend on knowledge of the lineshape. Further, HRDI viewed all spectral elements simultaneously and spatial elements sequentially in time, while WINDII viewed all spatial elements simultaneously but carried out its phase steps sequentially in time. Both instruments were therefore subject to atmospheric variability, but in a different way. It was tremendously satisfying to both investigator teams to find that the winds measured by the two instruments, for the region of space where their coverage overlapped, were in excellent agreement as shown in Figure 9.19. This was true even though the comparisons

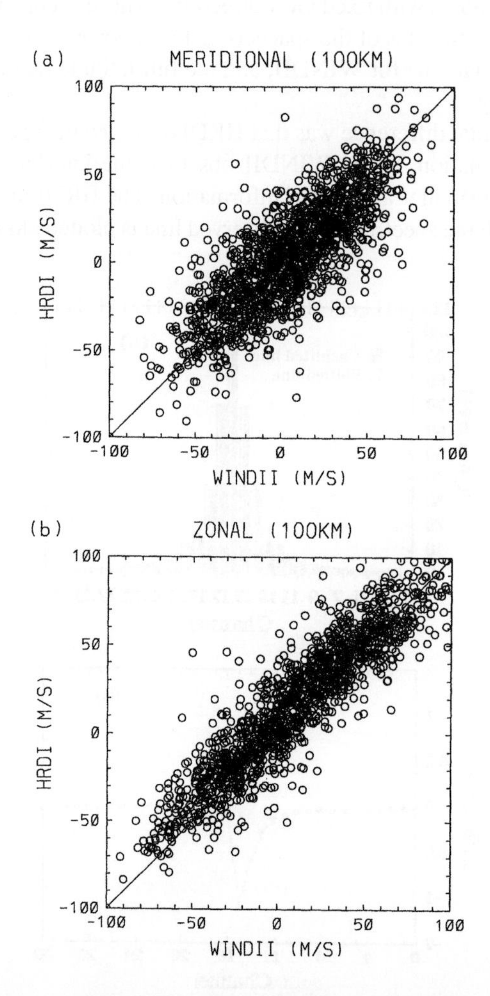

Figure 9.19. Scatterplots of HRDI winds versus WINDII winds at 100 km for: (a) the meridional wind; and (b) the zonal wind, for a two-year comparative dataset. Each point represents a zonal average for a particular day and a particular latitude for either the ascending or descending portion of the orbit. The diagonal solid line has a slope of unity. After McLandress *et al.* [1996b].

Table 9.2. Comparison of responsivity parameters for HRDI and WINDII

Parameter	HRDI	WINDII Night	WINDII Day
Effective area of imaging telescope (cm^2)	55.1	32.7	3.60
Solid angle for one anode or bin (sr)	8.03×10^{-6}	8.9×10^{-6}	8.9×10^{-6}
Photometric sensitivity for one anode or bin (photons s^{-1} R^{-1})	35.2	23.15	2.54
Integration period (s)	0.096	2.0	2.0
Total system transmittance	0.100	0.21	0.21
Detector counting efficiency	0.03	0.5	0.5
Overall responsivity (el R^{-1} (per integration))	1.32×10^{-2}	4.86	0.53

were done for different emissions, O_2 for HRDI and $O(^1S)$ for WINDII, emissions having very different vertical distributions. For example, the measurement of day-time winds with the $O(^1S)$ emission at 100 km require viewing through a great deal of emission at higher altitude as shown in Figure 9.6. This was not true of the O_2 emission which did not extend to high altitude. The agreement was particularly important since the observed winds were much larger than expected from previous experience. If there had not been two instruments achieving such remarkable agreement it would have been difficult to convince the atmospheric dynamics community that these large winds were valid.

A comparison of HRDI and WINDII responsivity parameters is given in Table 9.2. Because of the different approaches, a comparison between the two instruments is difficult to make. The HRDI aperture is larger than that for WINDII, by a factor of 1.7 at night, and 15.3 in the day-time, but the solid angles are very similar where one HRDI anode is compared with one WINDII bin. Because HRDI must scan in altitude to acquire a profile, its exposure time for one altitude is limited to about 0.1 s, which is a factor of 20 less than for WINDII which obtains a full profile in a single image. The overall HRDI system transmittance is smaller by a factor of 2 than for WINDII, probably because of the telescope and its coupling fibre optics. One of the most striking differences is in the quantum efficiency, a factor of 16.6 higher for the CCD detector than for a conventional photocathode. This is a dramatic demonstration of just one advantageous aspect of the CCD.

In general, the fairest comparison is with the overall responsivity in electrons R^{-1}, where that for WINDII is higher by a factor of 30 in the daytime and 300 at night. The combination of system transmittance and quantum efficiency account for all of this difference in the day-time when WINDII has a reduced aperture. At night, when WINDII has its full aperture, it gains a factor of about nine. This difference is consistent with the fact that HRDI did not acquire wind profiles in the mesosphere at night; instead it "stared" at one fixed altitude of 95 km. However, the responsivity does not tell the whole story, since the FPS fringes are sharper than those of the MI and therefore more sensitive to shifts. More importantly, the responsivity advantage of WINDII is possible only because it observes simple spectra, consisting of a single spectral line plus some background. The stratospheric measurements made with HRDI involve more complex spectra which could not have been obtained with WINDII. The most advantageous way to look at these two

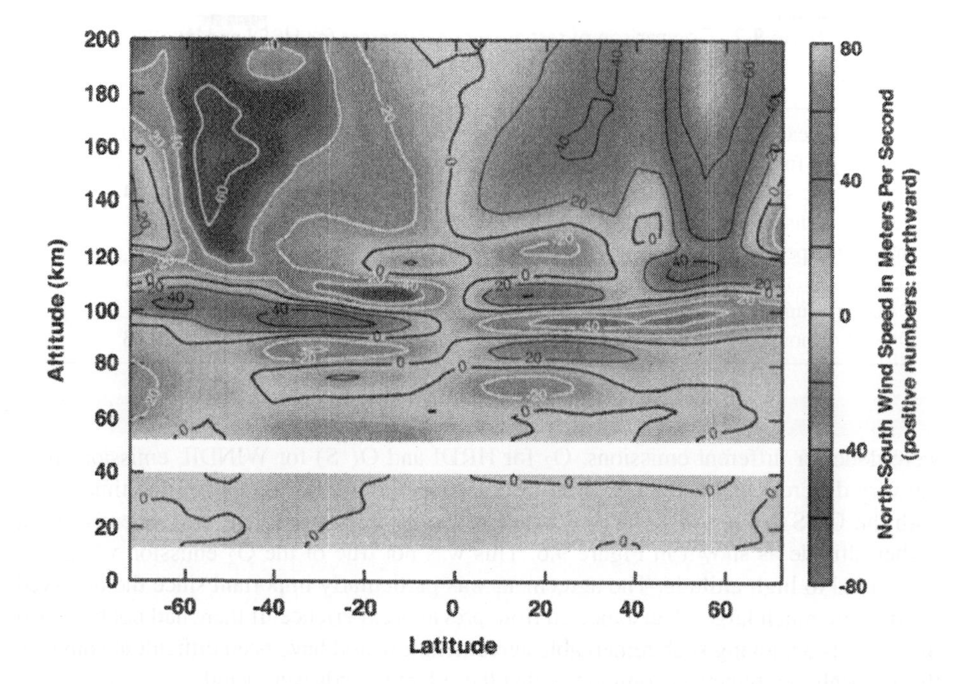

Figure 9.20. The meridional wind as a function of altitude and latitude determined from the combined HRDI WINDII data. Prepared by M.D. Burrage, Courtesy of NASA Goddard Space Flight Center UARS Project.

different approaches is to combine them as is done in Figure 9.20 where the higher altitude WINDII data are combined with the lower altitude HRDI data (with some overlap), to provide wind measurements in the atmosphere from 10 to 300 km. Here the scale factor (-80 to $+80\,\mathrm{m\,s^{-1}}$) obscures the features in the stratosphere. The tides mentioned previously appear strongly in the 70–120 km region but they dissipate above 120 km. Higher still, in the thermosphere, the pattern changes to global scale circulation, away from the equator.

9.6 CLAES: THE CRYOGENIC LIMB ARRAY ETALON SPECTROMETER ON UARS

9.6.1 Introduction

The primary objective of the CLAES experiment is to measure the altitude profiles of temperature and of a series of minor and trace species, important to stratospheric ozone layer photochemistry and radiative structure. CLAES also obtains aerosol absorption coefficients at a variety of infrared wavelengths. In addition to the critical role that the gases CH_4, N_2O, CFC 11 and CFC 12 play in ozone photochemistry, they, along with CO_2, are major greenhouse gases.

As implied by the name, the CLAES instrument involves remote measurement of Earth-limb emission spectra, as was true for WINDII and HRDI except that here the measurements are made in the infrared, like those of ISAMS, between 3.5 and 13 μm. It acquires spectra using Fabry–Perot etalons and the term *cryogenic* refers to the lifetime-limited coolant carried on board to maintain the low instrument temperatures required. This cryogen evaporated as planned after 19 months of the mission, so the instrument then ceased operations. Radiances of vibration-rotation lines are input to retrieval algorithms in order to obtain pressure, temperature and species mixing ratios. CLAES is mounted on the anti-Sun side of the spacecraft (adjacent to WINDII) and views 23.2° below the horizontal. This places a nominal 35 km tangent height in the middle of the field of view. A linear array detector provides instantaneous vertical coverage from 10 to 60 km, with 2.5 km vertical spacing. An adjustment for Earth oblateness and orbit eccentricity is made with a limb-pointing mirror. A detailed description of the instrument is provided by Roche *et al.* [1993], and the following summary is taken from that paper.

9.6.2 Instrument Design

CLAES requires high spectral resolution and high responsivity to isolate and accurately measure weak emissions from trace species such as HCl and NO against intense backgrounds from abundant emitters such as CO_2, H_2O and O_3. There must also be a high degree of out-of-field rejection to ensure that very intense emissions from the hard Earth surface do not contaminate the low-altitude detectors, which are pointed only 0.2° above the surface.

Figure 9.21 shows the major functional elements of the instrument designed to satisfy these requirements. The key features are a tilt-scanned solid Fabry–Perot etalon spectrometer and solid state focal plane detector arrays. A filter wheel is used to select different spectral regions for which spectra are obtained by step-tilting one of four etalons in order to change the incident angle; the approach is similar to HRDI but with a single etalon.

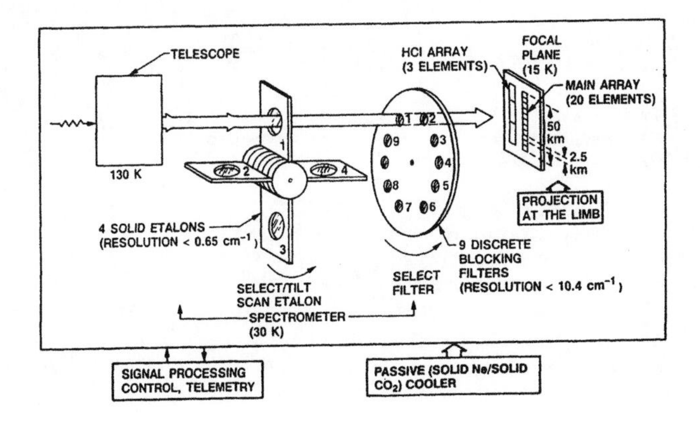

Figure 9.21. CLAES schematic showing etalon paddle wheel, blocking filters and detectors. After Roche *et al.* [1993].

The ingenious paddle wheel combines both etalon selection and tilting in one mechanism. In order to reduce the black body radiation from the instrument which otherwise forms a large background signal a solid-neon-CO_2 cryostat cools optical components and detectors to the temperatures shown in Figure 9.21. The focal plane assembly carries a main array for measurement of spectral radiance between 5.2 and 13 μm which includes all species except HCl. A separate array of three large detectors is used solely for measurement of the extremely weak emissions from HCl at 3.5 μm.

A cutaway drawing of the instrument is shown in Figure 9.22. Light enters the port at the lower right which can be closed by the aperture door shown here in the open position. This deployable telescope aperture door carries a highly stable, instrumented blackbody source for absolute radiometric calibration. In the interests of low scattering and efficient stray-light baffling, an unobscured Gregorian telescope is used in a "z" configuration to accept the entering radiation. This light is collected by the 30 cm diameter primary mirror and brought to a focus where the first field stop provides rejection of the hard Earth-limb image. Light is then collimated by a secondary mirror, as is required for the Fabry–Perot etalons, and a *Lyot stop* is placed at the image of the entrance aperture to suppress diffraction scattering. A Lyot stop is one that is used to block a bright object in the field of view, in this case the Earth, in order to detect nearby fainter objects; they are also used in coronagraphs. A refractive doublet re-images the beam at an intermediate focus, where a tuning fork chopper and the blocking filters are located. The chopper alternates the views between that of the Earth and

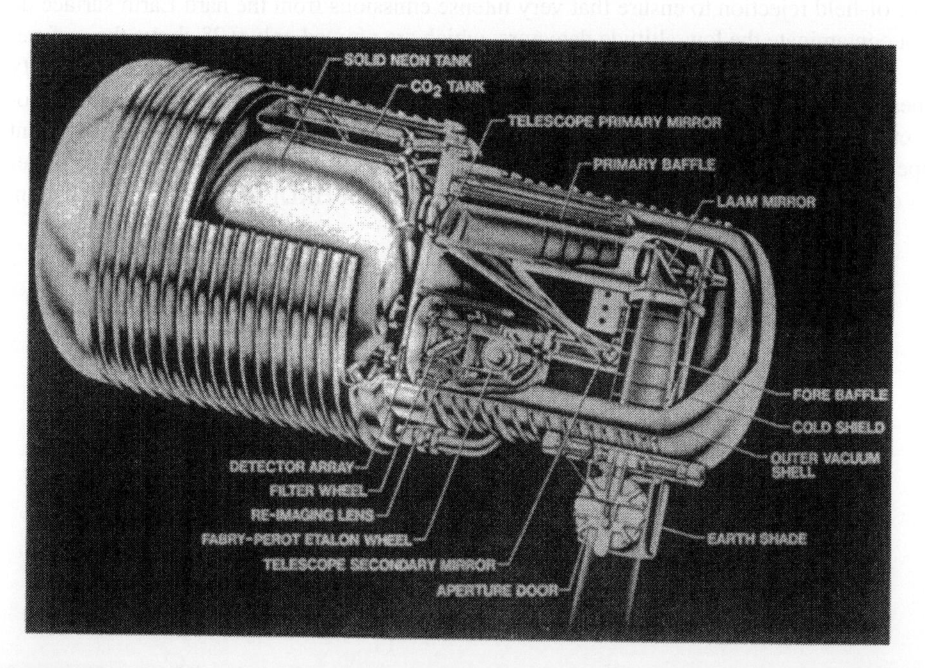

Figure 9.22. Cut-away drawing of CLAES showing the cryostat and optical system. After Roche *et al.* [1993].

a cold surface, so that only the difference between the two is measured. This eliminates any drift in the detectors and removes unwanted background. A second refractive doublet brings the light to near-diffraction limited final focus at the detector arrays.

The detector system consists of two linear arrays of solid state Si:Ga photoconductors. The 20-element main array is used in conjunction with spectral channels 2 through 9 and provides the 2.5 km vertical spacing at the Earth limb. The three-element array is used exclusively with the 3.5 μm HCl channel and sacrifices spatial resolution (to about 14 km) in order to increase the signal. The main array is chopped at 71 Hz, and a synchronous rectification system is used to eliminate dc drift effects and allow for high radiometric stability. The HCl detectors are operated unchopped in conjunction with a dc restore circuit. Measured noise equivalent power (NEP) for the CLAES detectors is 5×10^{-16} W Hz$^{-1/2}$ for the long wavelength channels and 10^{-15} W Hz$^{-1/2}$ for the two shortest wavelength channels. All instrument photon noise backgrounds have been suppressed below the detector noise through cooling and spectral narrow banding. In addition to suppressing photon noise, the cold optics (specifically the 130 K telescope mirrors) also ensure very low and very stable dc background photon emission, an important issue in radiometric accuracy and precision. As described earlier, spectra are obtained by tilt scanning one of the four solid etalons (E1 to E4) between 0° and 23° in conjunction with one or more of the nine selectable discrete interference (blocking) filters. Examples of spectra obtained in flight are shown in Figure 9.23.

9.6.3 Sample CLAES results

Sample results obtained from the CLAES website are shown in Figure 9.24. Concentrations of CF_2C_{12} are shown from 80° S to 80° N latitude, by combining results from just before, and just after, a yaw from forward to backward flight. This is done for two periods near the equinoxes: in the upper panel for the northern spring equinox and in the lower for the fall. The values from all longitudes at a single latitude have been averaged together (i.e. zonally averaged). The figure shows very clearly how CFC-12 rises upward into the stratosphere in a narrowly defined region around the equator, then migrates poleward and downward. In both panels there is a steeper downward slope near the spring poles (north in the upper panel, south in the lower). This is mainly because polar regions are colder in the spring than in the autumn and the colder air descends more steeply. It is particularly noticeable in the lower panel in the spring southern hemisphere, because the Antarctic winter and spring are so much colder than the equivalent seasons in the Arctic. Very similar behaviour is seen in other "long-lived" tropospheric source gases such as methane and nitrous oxide.

9.7 MOPITT – MEASUREMENTS OF POLLUTION IN THE TROPOSPHERE

The MOPITT instrument is derived from principles described in Chapter 7, from the sections on correlation spectroscopy and pressure modulator radiometers. However, in MOPITT these principles are extended through the use of length-modulated radiometry

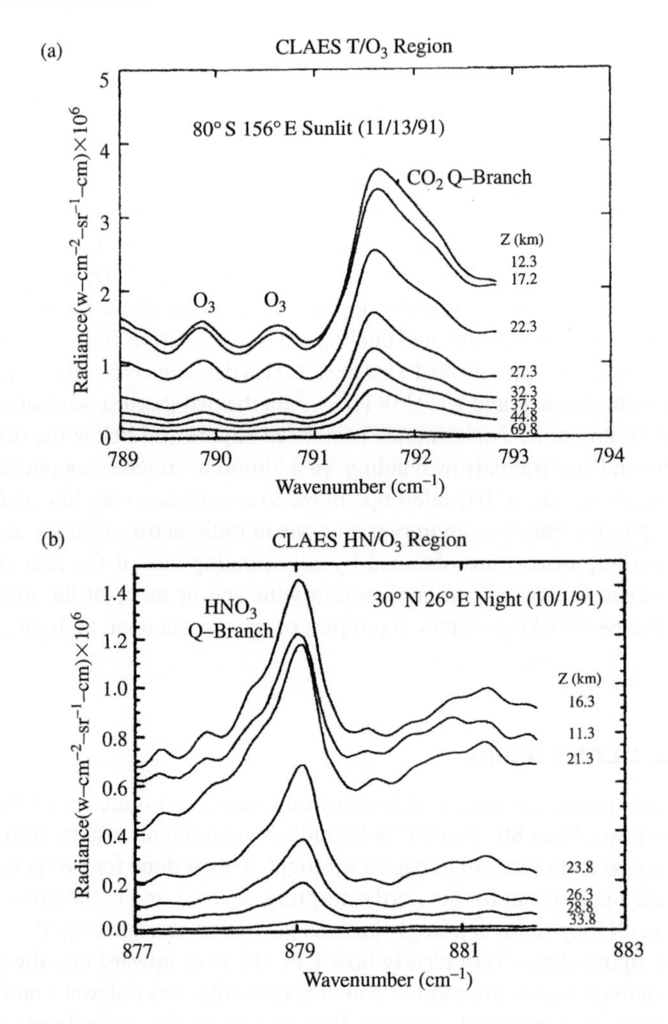

Figure 9.23. CLAES spectra obtained in flight for two spectral regions. After Roche *et al.* (1993).

[Drummond, 1989] in which a rotating disk spinning inside a gas cell changes the length of the cell from one value to another. This changes the absorption in the gas involved, without changing the pressure. MOPITT is the first instrument described in this work that makes measurements in the troposphere, which is difficult because of the influence of clouds and other aerosols. However, the measurements of CO and CH_4 being made by MOPITT are very important in understanding the impact of human activities on the atmosphere. Because of the localized nature of carbon monoxide sources, both high resolution and global coverage are required; this is accomplished through nadir viewing with an elementary field of view of 22 × 22 km. Each detector is a 4 × 1 array, generating a measurement imprint of

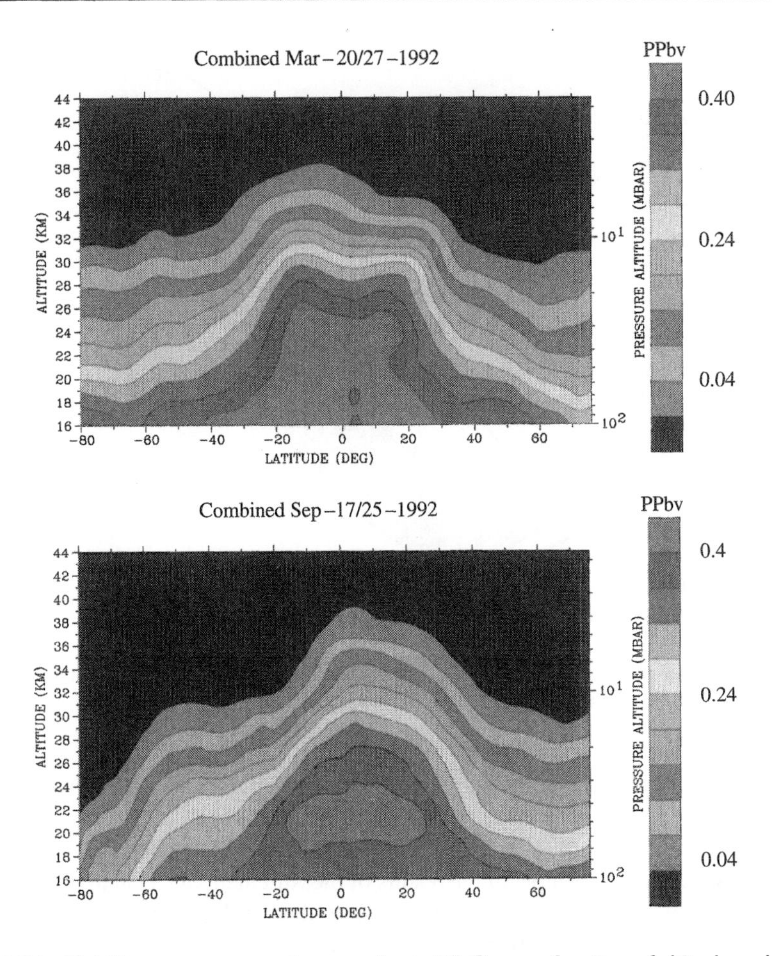

Figure 9.24. CLAES measurements of stratospheric CF_2C_{12} as a function of altitude and latitude for spring and fall equinoxes. Courtesy of A. Roche (Lockheed Martin).

22 × 88 km, but this is extended by cross-track scanning, generating a swath of 88 × 650 km. Successive scans provide a coverage that was illustrated in Figure 1.15.

However, a deep-space view is required to determine the background and calibration is then required. A scan mirror therefore provides nadir, space and calibration views as required. A drawing of the instrument is shown in Figure 9.25, where "up" corresponds to the nadir view. The solid angles of the space views are also shown. The instrument is mounted on a cold plate that is cooled with ammonia, cycled through capillary pumping. The detectors and optics are cooled with two Stirling cycle coolers, operating back-to-back to minimize vibration. The instrument contains eight channels altogether, with a good deal of intended redundancy, using both length-modulated cells for the lower troposphere and pressure-modulated gas cells for the upper troposphere. Profiles of CO are measured from

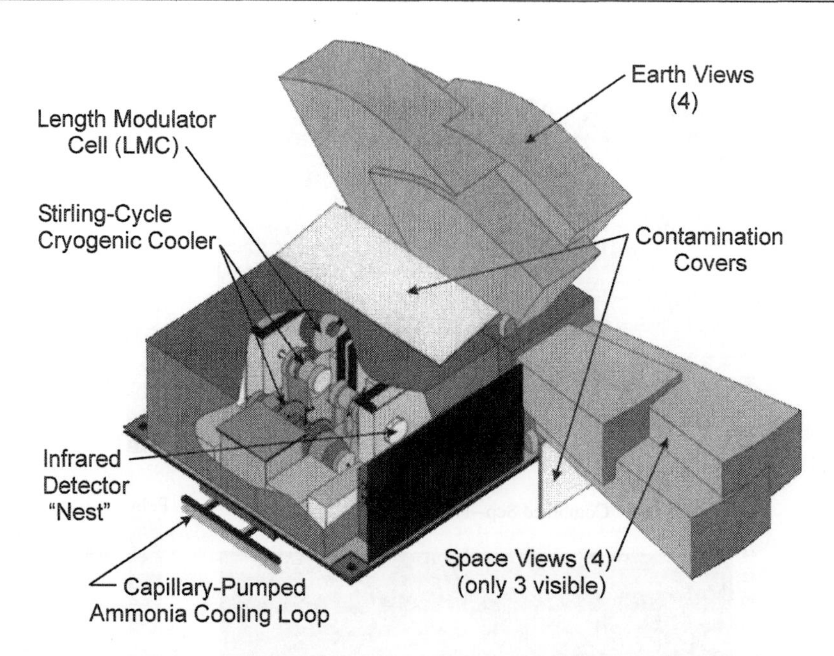

Figure 9.25. Drawing of the MOPITT instrument on the Terra spacecraft. The Earth views are in the nadir direction. Courtesy of J. Drummond (University of Toronto).

upwelling thermal radiation at 4.6 μm wavelength yielding a vertical resolution of 4 km. Column amounts of CO are measured using reflected sunlight at 2.3 μm; this is of course only possible during the sunlit portions of the orbit. MOPITT was launched on Dec. 18, 1999 on the Terra spacecraft. A sample of the results of measurement of CO in the troposphere is shown in Figure 9.26. Early mission plans are described by Drummond *et al.* [1999], and other aspects of the MOPITT project are described in accompanying papers.

Terra is a large spacecraft, similar in size to UARS and is considered the Flagship of the EOS (Earth Observing System) satellites. It is the first of a series of three NASA satellites, Terra, Aqua and Aura that are dedicated roughly according to their names, to atmospheric measurements of the Earth's surface and its troposphere, the oceans and their interactions, and the ozone layer and air quality, respectively. However, there is a considerable degree of overlap of observations of the land surface, oceans and atmosphere in all three missions because of the recognition of the inter-relationship of all three. Terra carries five instruments; ASTER (Advanced Spaceborne Thermal Emission and Reflection radiometer), CERES (Clouds and the Earth's Radiant Energy System), MISR (Multi-angle Imaging Spectro-Radiometer), MODIS (MODerate Imaging Spectroradiometer) and MOPITT. In order to determine the amount of sunlight scattered in different directions for climate studies, MISR views at nine viewing angles, one in the nadir and four in each of the fore and aft directions. This is done in four wavelengths, allowing the determination of aerosols, cloud characteristics and land surface cover. More detailed information about all of these instruments can be found on the Terra website, as well as in the forthcoming published literature.

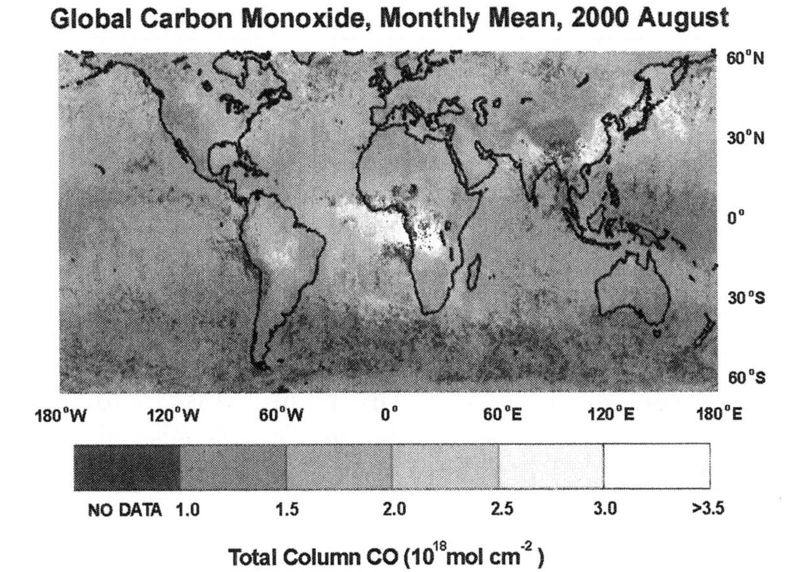

Figure 9.26. Global measurement of carbon monoxide for August, 2000. Note the enhanced concentration over eastern North America, western Brazil, equatorial West Africa, northern India, China and Japan. Courtesy of J. Drummond (University of Toronto) and the MOPITT Science Team.

9.8 PROBLEMS

9.1. (a) Confirm the day and night responsivity values for WINDII as given in Table 9.2 The size of one WINDII pixel is 30 μm × 30 μm and the size of a standard bin is 2 × 25 pixels.
(b) For Figure 1.11 take the E-region integrated emission rate peak as 110 kR for the profile of smallest solar zenith angle. Convert this to an accumulated count in analog digital units (ADU; one ADU = 70 electrons for WINDII).

9.2. Table 9.3 lists WINDII output of inverted J_1, J_2 and J_3 as a function of altitude for a WINDII measurement of the O(^{1}S) 557.7 nm green-line emission at night. These are integrated, i.e. not inverted volume emission rates. They were taken on Jan. 13, 1993 at 01:44:47 UT, at 20.3° latitude and 298.7° longitude. Plot the vertical profiles of volume emission rate, line-of-sight wind and Doppler temperature as a function of altitude. The WINDII OPD is 4.5 cm.

9.3. For ERWIN a Gallium Arsenide photomultiplier was chosen as detector for observations at 870 nm rather than a CCD. This was in large part because for untended operation at a remote site the detector temperature is limited to what is available with thermoelectric cooling, about −25°C, with a comparatively low dark current of 200 el s^{-1} for the photomultiplier and 500 el s^{-1} for the CCD. Justify this conclusion, or otherwise,

Table 9.3. WINDII output needed for Problem 9.2

Altitude	$J1$	$J2$	$J3$
1.1361E + 02	2.7475E + 02	7.9591E + 01	−2.0130E + 01
1.1150E + 02	3.4911E + 02	1.1188E + 02	1.5636E + 00
1.0939E + 02	3.8186E + 02	1.5114E + 02	5.6467E + 00
1.0727E + 02	5.3007E + 02	2.7148E + 02	−3.9073E + 00
1.0514E + 02	9.5271E + 02	6.0758E + 02	−1.6184E + 01
1.0301E + 02	1.8003E + 03	1.3764E + 03	−2.3958E + 01
1.0088E + 02	3.4590E + 03	2.8353E + 03	−8.4089E + 01
9.8740E + 01	5.7105E + 03	4.8040E + 03	−9.2485E + 01
9.6596E + 01	8.5767E + 03	7.1979E + 03	6.0145E + 01
9.4448E + 01	1.1805E + 04	9.8315E + 03	2.9862E + 02
9.2295E + 01	1.2478E + 04	1.0532E + 04	3.3587E + 02
9.0139E + 01	1.1705E + 04	9.7554E + 03	6.7023E + 02
8.7976E + 01	9.7553E + 03	8.1297E + 03	6.6768E + 02
8.5810E + 01	8.1992E + 03	6.7743E + 03	3.6979E + 02
8.3639E + 01	7.1978E + 03	6.0010E + 03	1.7700E + 02
8.1463E + 01	6.5194E + 03	5.4217E + 03	1.0785E + 02
7.9283E + 01	6.0411E + 03	5.0195E + 03	1.7881E + 02
7.7097E + 01	5.7133E + 03	4.7406E + 03	1.9014E + 02
7.4908E + 01	5.5057E + 03	4.4918E + 03	8.4237E + 01
7.2713E + 01	5.2810E + 03	4.3061E + 03	6.4163E + 01
7.0515E + 01	5.0284E + 03	4.1928E + 03	9.4441E + 01

using the values given in Section 9.3, for a signal level of 25 R and quantum efficiencies of 12.1% and 60% for the GaAs photomultiplier and CCD respectively.

9.4. MICADO uses a sliding wedge of 2.62 min arc-wedge angle to scan in OPD and to adjust the field-widened condition from one wavelength to another.
(a) Calculate the lateral prism movement required for an OPD step of 90° in phase for a wavelength of 630 nm where the wedge refractive index is 1.580165 for F5.
(b) At a wavelength of 557.7 nm the refractive index of fused silica is 1.46008 and at 850.0 nm it is 1.4525. Calculate the lateral wedge movement required to shift the wide-angle condition from 557.7 nm to 866.0 nm. The refractive index of F5 glass at 866.0 nm is 1.48137.

9.5. The HRDI etalon spacings are given in Section 9.5.3. Assume that each has a finesse of 12 and plot the relative tandem etalon transmittance over the spectral range of an interference filter characterized by the Lissberger [1968] formula, for a FWHM of 1.45 nm and a centre wavelength of 763.0 nm. Assume the etalon passbands are aligned at 763.0 nm.

9.6. In Figure 1.7, a radiance profile measured by CLAES is shown, with a value of 2.0×10^{-9} W cm^{-2} sr^{-1} (cm^{-1})$^{-1}$ at 50 km. The satellite is at 585 km altitude.

(a) Calculate the SNR for this measurement, using the NEP value given in Section 9.6.2. Use a 15.2 cm diameter collecting mirror, a field-of-view of 2.5×8.0 km at the limb and a system transmittance of 7%. The instrumental passband width is needed; that may be taken from Figure 9.23, assuming that the spectral width is entirely due to the instrument. The electrical passband width is also needed to determine the noise level; it may be taken as the reciprocal of the integration time of 0.128 s. Note that the chopper reduces the effective radiance to 0.45 of its external value.
(b) Compare the atmospheric radiance with that of the thermal radiation from the telescope mirror at 130 K, allowing for an emissivity of the mirror based on 99% reflectance.

10

FUTURE ATMOSPHERIC SPECTRAL IMAGERS

10.1 THE TIMED DOPPLER IMAGER (TIDI)

10.1.1 TIDI Overview

The TIMED (Thermosphere Ionosphere Mesosphere Energetics and Dynamics) mission [Yee *et al.*, 1999] is briefly described at the end of this section. One of its four instruments, the TIMED Doppler Interferometer (TIDI) will investigate the dynamics and energetics of Earth's mesosphere and lower-thermosphere-ionosphere (MLTI) over an altitude range of 60–300 km. This is an ambitious experiment that combines the mesospheric HRDI capability with that of WINDII in the thermosphere. Its design has several novel features that enhance the use of the FPS for the observation of winds from space. TIDI measurements will allow a global description of the vector wind and temperature fields to be obtained, as well as important information on gravity waves, species densities, airglow and auroral emission rates, noctilucent clouds, and ion drifts. TIDI will also contribute to the study of MLTI energetics.

The TIDI interferometer (also called a profiler) primarily measures horizontal vector winds and neutral temperatures from 60 to 300 km, with a vertical resolution ~2 km at the lower altitudes and with accuracies that approach ~3 m s^{-1} and ~2 K, respectively, under optimum viewing conditions. The TIDI design allows for 100% duty cycle instrument operation during day-time, night-time, and in auroral conditions. TIDI views emissions from OI 557.7 nm, OI 630.0 nm, OII 732.0 nm, O_2 At (0-0), O_2 At (0-1), Na D, OI 844.6 nm, and OH Meinel (9-4) and (7-3) to determine Doppler wind and temperature throughout the TIMED altitude range. The temperatures measured are rotational temperatures from spectral line ratios; line ratio measurements also lead to the determination of O_2 concentration.

10.1.2 Instrument Description

In order to measure winds on both sides of the spacecraft but to avoid a complex fully-steerable telescope, TIDI uses four fixed telescopes, viewing at 45° and 135° from the velocity vector on both the warm and cold sides of the spacecraft. It comprises three major subsystems: four identical telescopes, a Fabry–Perot interferometer with a CCD detector, and an electronics box. Light from the selected regions of the atmosphere is collected by the telescopes and is fibre-optically coupled to the interferometer input. The four fields of view are scrambled along with fibres from a calibration input and converted to an array of five concentric circular wedges. This input then passes through a selected filter and a Fabry–Perot etalon, and is imaged onto a CCD via a circle-to-line imaging optic (CLIO) device, as described in Section 5.8.6. The wedge input is required for the CLIO system. Since all five fields are simultaneously imaged onto the CCD, the four viewing directions and the calibration sources are continuously and simultaneously observed. Only one altitude point is observed at a time, and the telescopes execute vertical scans in order to provide the wind and temperature profiles.

The TIDI overall configuration is shown in Figure 10.1, where two of the four telescopes are shown, along with the interferometer assembly. A fibre optic connection is illustrated for one of the telescopes. The four limb-viewing telescopes are of an off-axis Gregorian design, with low scatter optics and baffles and gimbals that allow vertical scans. The telescopes clear aperture is 7.5 cm in diameter, with a focal length of 17.0 cm and an f/number of 2.2. The angular FOV is 2.5° in the horizontal direction and 0.05° in the vertical direction, and is created with the fibre optics configuration described below.

The configuration for the fibre optics is shown in Figure 10.2. The circular bundle (Field 1) accepts light from various calibration sources; called the calibration deck, while Fields 2–5 are slit-like; creating a horizontal field of 2.5° and a vertical field of 0.05° at the limb. The fibres are randomized and brought to the profiler end to form a set of five FP rings, each corresponding to one etalon order, and in a 90° wedge pattern to match the CLIO input. The Doppler measurement package as shown in Figure 10.3 consists of a Fabry–Perot interferometer with a fixed gap etalon, two filter wheels each of eight positions,

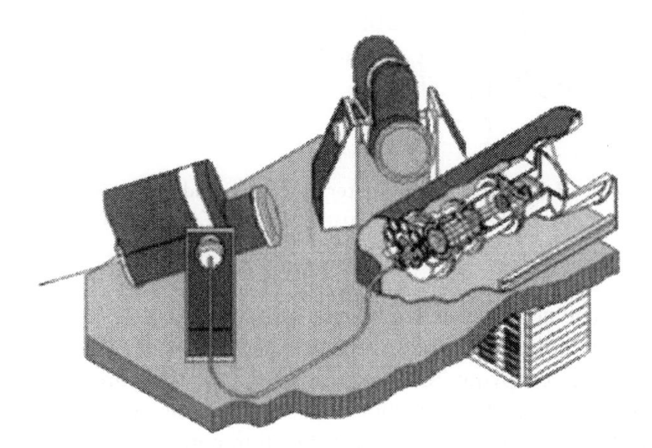

Figure 10.1. Overall configuration for the TIMED TIDI instrument, showing two of the four limb-viewing telescopes along with the interferometer. After Killeen *et al.* [1999].

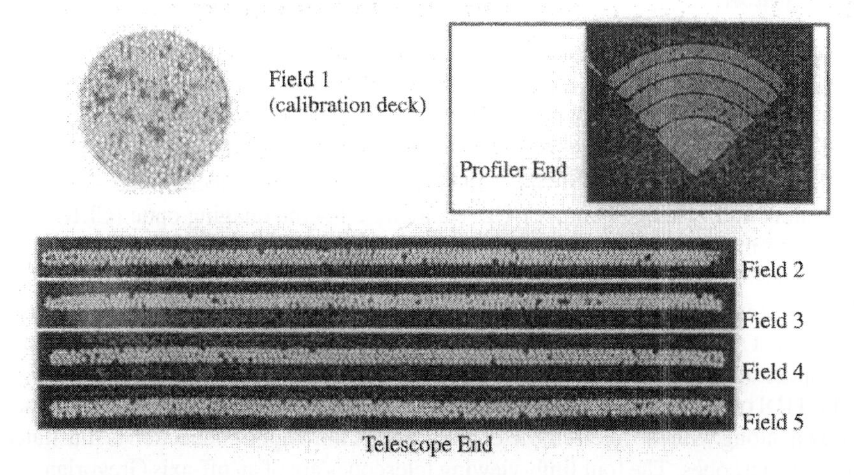

Figure 10.2. Configuration of the fibre optics coupling the input telescopes and the calibration source to the Fabry–Perot etalon system. After Killeen *et al.* [1999].

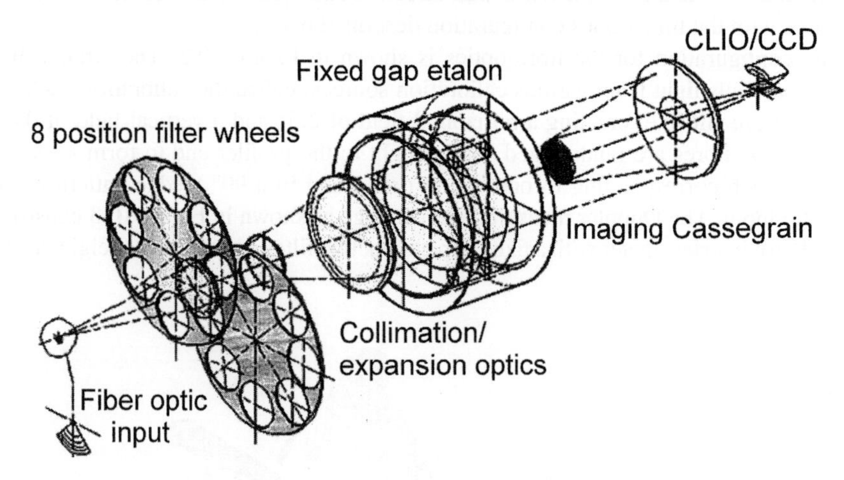

Figure 10.3. Layout of the TIDI spectrometer system, including the fibre optic input, the two filter wheels, the etalon and the CLIO/CCD imaging system. Courtesy of W. Skinner (University of Michigan).

a circle-to-line image converter (CLIO) and a passively cooled CCD detector, providing 32 spectral channels for each of the five fields. The CLIO optics form a slightly wedged pattern on the CCD, occupying 700 pixels in the x (spectral) direction and 140 in the y direction. There is no useful information in the y direction, so all y direction pixels are binned on chip before transmission. The x direction corresponds to the radial direction on the FP fringe and consists of five segments, one for each field. Since each segment corresponds to one order, the segment closest to the axis occupies the most x pixels, and the outermost segment occupies the least. These are binned for each field to provide equal increments in spectral range, namely 32. Thus for each field an FPS spectrum is obtained having 32 spectral elements, corresponding to about 1.2 FSR.

The electronics system contains a hybrid power supply; an 80C51 (UTMC) flight computer; a data acquisition system; a CCD controller; controllers for the filter wheel, shutters and heaters; the telescope servo amplifiers and the calibration lamp power supply.

10.1.3 TIDI Data Coverage

The TIDI telescopes perform limb scans through the terrestrial airglow layers throughout the satellite orbit. Each scan consists of individual views of angular size 2.5° horizontal by 0.05° vertical. The vertical altitude is 2.5 km at the limb, and one complete up/down acquisition cycle requires approximately 105 seconds. TIDI obtains these scans simultaneously in four orthogonal directions: two at 45° forward but on either side of the satellite's velocity vector and two at 45° rearward of the satellite.

With an orbit inclination of 74°, the precession rate of TIMED is such that it takes 60 days to precess 12 hours in local time (3° per day). The advantage of four simultaneous views has its complications in that the viewing conditions can be different for the four telescopes even though the FPS has the same parameters for all four inputs. In particular, the optimum operation of the parameters will be different for day and night conditions. There is an additional problem at the terminator (the day–night boundary) which is shared by all such instruments, in that horizontal gradients are large so that the assumption of horizontal homogeneity for data inversion is least valid. A secondary consideration is the solar scattering angle, the angle between the line of sight to the tangent position and the sun. TIDI will not perform observations if the solar scattering angle is less than 15°.

Consequently, optimum TIDI day-time observations are performed when:

- the solar zenith angle (at the tangent point) is less than 80°,
- the solar zenith angle (at the spacecraft) is less than 90°,
- the solar scattering angle is greater than 15°.

Night-time observations are performed when:

- the solar zenith angle (at the tangent point) is greater than 100°,
- the solar zenith angle (at the spacecraft) is greater than 90°,
- the solar scattering angle is greater than 15°.

10.1.4 TIDI Science Measurement Summary

The filter wheels for TIDI contain a complement of fourteen interference filters carefully chosen to allow full day-time altitudinal coverage of neutral wind and temperature measurements throughout the MLTI. Altitudinal coverage during nocturnal periods is reduced since the terrestrial airglow exhibits a discrete layer behaviour when the sun is not present.

Table 10.1. TIMED Doppler Imager (TIDI) dayside science

Measurement	Feature	Altitude Range
Vector Wind	O_2 Atmospheric (0-1) P11	60–85 km
	O_2 Atmospheric (0-0) P9	85–120 km
	OI (557.7 nm)	90–250 km
	OI (630.0 nm)	200–300 km
	OII (732.0 nm)	170–300 km
Neutral Temperature	O_2 Atmospheric (0-1) P11 and O_2 Atmospheric (0-1) P7	60–85 km
	O_2 Atmospheric (0-0) P9 and O_2 Atmospheric (0–0) P15	85–120 km
	OI (557.7 nm)	100–150 km
	OI (630.0 nm)	200–300 km
O_2 Density	O_2 Atmospheric (0-0) Volume Emission Rate	~100 km
	O_2 Atmospheric (0-1) and O_2 Atmospheric (0-0)	60–90 km
O Density	OII (732.0 nm) and OI (844.6 nm)	150–300 km
O_3 and $O(^1D)$ Density	O_2 Atmospheric Volume Emission Rate	70–95 km

The TIDI dayside measurements are summarized in Table 10.1; the night-side science is similar, but modified according to the available emissions. The full altitude range of 60–300 km is achieved by a careful choice of emission features. At the lowest altitude in the day-time, a line in the (0, 1) O_2 Atm band is used; because the lower vibrational level has quantum number 1 there is no self-absorption along the viewing path. Above 85 km the choice switches to the stronger (0, 0) band for which self-absorption can potentially occur but is not serious at the altitudes of 85–120 km. At night, the O_2 bands are replaced by the OH and sodium (Na D) emissions. For both day and night the atomic oxygen green line $O(^1S)$ emission at 558 nm is used at still higher altitudes, and above that the red line $O(^1D)$ emission at 630 nm. Both Doppler and rotational temperatures are measured, as shown in Table 10.1. The atmospheric concentration of O_2 is measured by comparing measurements of the (0, 1) and (0, 0) bands. Since one is absorbed by molecular oxygen and the other is not, the concentration may be obtained in this way – this works only in the day-time. Atomic oxygen concentrations are obtained during the day-time using solar-excited emissions and during the night from the recombination of atomic oxygen as described in earlier chapters. Ozone concentrations may also be determined from the O_2 Atm band measurements through the linkage in the photochemistry. Although a number of different emissions is required to cover the required altitude range, the wind retrieval will be done through a single inversion involving all emissions.

The TIMED mission carries three instruments in addition to TIDI, dedicated to measurements of the MLTI (Mesosphere, Lower Thermosphere and Ionosphere). GUVI, the Global Ultra Violet Imager is designed to observe the ultraviolet airglow using a scanning imaging spectrograph covering the wavelength range 115–180 nm and using five discrete wavelength channels. It will determine solar energy inputs into the region of the upper atmosphere where ionization of atoms and molecules by ultraviolet radiation is dominant, observing Lyman-α, atomic oxygen and molecular nitrogen emissions. GUVI will globally measure the composition and temperature profiles of the MLTI region, as well as its auroral energy inputs.

SABER (Sounding of the Atmosphere using Broadband Emission Radiometry) is a multi-spectral radiometer covering the wavelength range 1.27–17 μm, intended to measure thermal radiation emitted by the atmosphere over a broad altitude and spectral range. SABER will cover an altitude range from the Earth's surface up to 180 km, measuring constituents including ozone, water vapour, carbon dioxide and nitrogen and hydrogen species, as well as temperature. Its primary science objectives are to explore the MLTI region to determine its energy balance, atmospheric structure, chemistry and dynamics. It will also measure sources of atmospheric cooling, including the airglow.

SEE (Solar Extreme ultraviolet Experiment) is comprised of a spectrometer and a suite of photometers designed to measure solar ultraviolet radiation – the primary energy deposited into the MLTI atmospheric region. Examples of solar ultraviolet radiation on which SEE will focus are solar soft X-rays, extreme-ultraviolet, and far-ultraviolet radiation. The primary objectives of the SEE instrument are to study the solar ultraviolet irradiance to find by how much it varies, how it affects the atmosphere and also how much it heats the atmosphere and changes its composition.

10.2 THE MESOSPHERIC IMAGING MICHELSON INTERFEROMETER (MIMI)

10.2.1 Introduction

The MIMI instrument has the objective of wind measurement throughout the mesosphere by taking advantage of a different O_2 emission, that from the $^1\Delta_g$ band, for which the wavelength for the (0, 0) band is 1.27 μm. Although this band also involves a transition to the ground state, it is highly forbidden, so that self-absorption is a problem only at lower altitudes. In addition, since MIMI uses only individual lines from this band, a combination of strong and weak lines can be used to cover the altitude range from 45 to 85 km.

This emission is produced over this range only in the day-time, through the photolysis of O_3.

$$h\nu + O_3 \rightarrow O_2(^1\Delta_g) + O \tag{10.1}$$

Since the relevant solar flux is known, as is the O_2 concentration to a reasonable approximation, a measurement of the 1.27 μm volume emission rate yields the concentration of ozone. In addition, by measuring the line emission rate ratios of suitably chosen lines in the band, accurate rotational temperatures can be obtained as well. Thus a MIMI mission involving the dynamics, energetics and ozone transport in the mesosphere is feasible. At night, winds are measured from a hydroxyl line at 1.31 μm.

10.2.2 General Description of the Instrument

MIMI consists of two telescopes back-to-back, with the Michelson interferometer between them. The filter wheel is placed after the second telescope, followed by the camera lens and

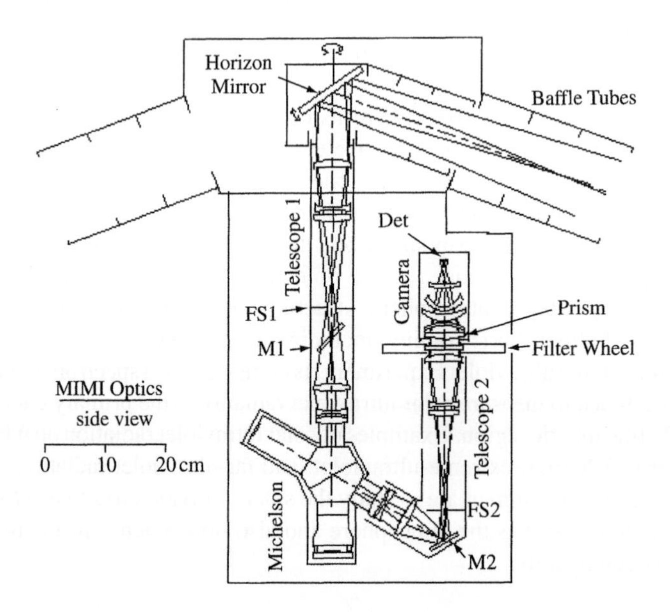

Figure 10.4. Drawing of MIMI optics, showing the Michelson between telescopes 1 and 2, the filter wheel, prism, camera lens and detector. Courtesy of W. Gault (York University).

detector as shown in the optical drawing of Figure 10.4. The entrance aperture is just in front of the first telescope, which looks up into a plane mirror that rotates about a vertical axis and points the field of view (FOV) towards the 45–95 km altitude region just above the Earth's limb, The FOV at the limb is 3° × 3°, defined by the field stop in the first telescope, while at the MI it is 6° × 6°. As was done with WINDII (Chapter 9) the Michelson axis is tilted 3° with respect to the axis of the telescopes in order to avoid recycling light from the interferometer's forward output. An image of the entrance aperture is formed at the Michelson mirrors and again at the filter. At the filter, the FOV is again 3° × 3°. Just behind the filter is a shallow, pyramid-shaped prism that deflects the light and causes the camera lens to form four copies of the field of view at the detector array. This is part of the Michelson scanning system, and is explained below. Two plane mirrors, M1 and M2, are used to fold the optics into a compact shape; the locations are shown in Figure 10.4, but the fold for M1 is not shown. FS1 in Telescope 1 is the true field stop and FS2 is a pseudo-stop in Telescope 2.

Concerning the viewing directions, the approach is similar to that of TIDI but is implemented differently because of the imaging requirement. A pointing mirror rotates fully in azimuth and points the field of view towards the four main observing directions in sequence. These directions are at ±45° and ±135° in azimuth to the direction of motion of the spacecraft. In addition, the mirror can make small adjustments in elevation angle (up to ±2°) to allow for altitude variations of the spacecraft and for the Earth's oblateness.

One MI mirror is divided into four equal quadrants, three of which have coatings of SiO_2 which produce a change in path difference of $\lambda/4$ going from one quadrant to the next. This is a monochromatic version of the Phase Quadrature Interferometer described in Chapter 8. The prism, which is close to the image of the Michelson mirror, is oriented so each prism face corresponds to one mirror quadrant. Each of the four images formed at the detector array therefore corresponds to a different phase step. In this way, four images are recorded simultaneously, each sampling a different point on the interference fringe, and this is one observational mode of MIMI. Because the images are taken simultaneously, emission rate variations during the measurements affect all four images equally, and will not influence the result; this removes a limitation that was present in the WINDII instrument. Such an interferometer has been built and tested in the laboratory; these showed that the Doppler shift produced by a spinning wheel is accurately measured by this method [Gault *et.al.*, 2001].

Although the phase samples are taken simultaneously, it is still necessary to move one of the MI mirrors piezoelectrically in order to perform calibrations. One of the mirrors is therefore mounted on piezoelectrics with capacitive sensors, as in WINDII. This also permits realignment of the interferometer's mirrors during flight, should that become necessary. The stepped MI mirror described above can be used in two ways.

(1) If a single filter is used, four images are obtained at points on the fringe separated by $\lambda/4$ and from these a phase image is derived, without mirror stepping.
(2) In the other method, the interference filter is divided in half, the two halves consisting of different filters aligned so that two images are produced of each of two different emissions.

The Michelson mirror steps are arranged so that the two simultaneous images of each emission are separated by $\lambda/2$ in path difference, so the sum of the two images is proportional to the average emission line intensity during the exposure. Two exposures are required, with the mirror piezoelectrically displaced by $\lambda/4$ in path difference between exposures. The images for each emission are normalized to their sum in pairs, so they are independent of intensity variations. Now the four images for each emission are analysed in the usual way. The advantage of this method is that the two emissions are measured simultaneously. It is believed that this should alleviate the problem of splicing the wind and temperature data from the two regions.

Most optical materials used in the visible can also be used at this wavelength, but a different type of detector is required, as this is outside the spectral range of CCDs. A search found that a HgCdTe (MCT) 256×256 array detector with a cut-off wavelength of 1.4 μm has the best characteristics for the required observing conditions. The instrument is calibrated periodically by interrupting observations and turning the pointing mirror to face the calibration sources. These sources include three microwave-excited lamps for phase and visibility calibrations, and a continuum source for responsivity. The sources are located remotely and communicate with the input device through fibre optic cables.

10.2.3 Michelson Interferometer

The field-widened thermally compensated Michelson interferometer has a 5×6 cm entrance face. This relatively small aperture is possible because of the very high emission rate of the

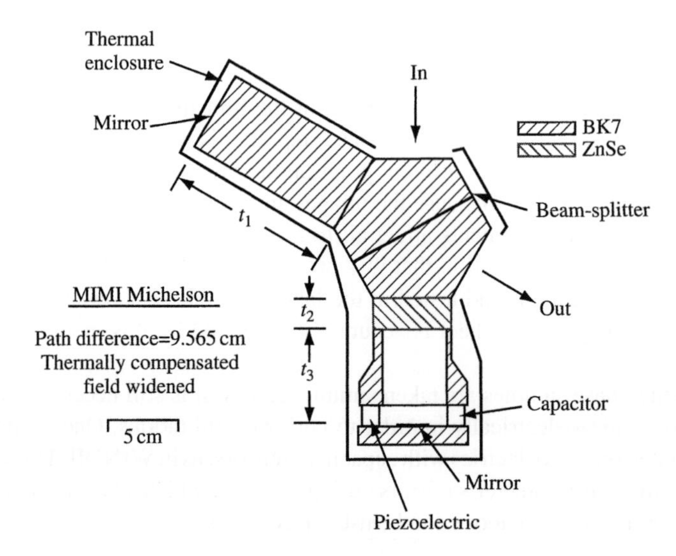

Figure 10.5. The MIMI Michelson interferometer. Courtesy of W. Gault (York University).

$O_2(^1\Delta)$ band. Its design is based on the designs of WINDII and several other Michelson interferometers that have been built by GSI Lumonics. A diagram of the interferometer is shown in Figure 10.5.

The beam-splitter consists of two half-hexagons of BK7 glass with a semi-reflecting coating on one diagonal face. The multi-layer dielectric beam-splitter coating is the same non-polarizing design used for WINDII, adjusted for the longer wavelength of MIMI. The long-path arm of the MI consists of a block of BK7 with a mirror coated directly on the back face. The other arm contains a gap and a plate of ZnSe, which provides thermal compensation for the phase of the fringes. In the short-path arm, the ZnSe plate is cemented to the beam-splitter face and the mirror is coated on a plate of BK7 mounted, via three piezoelectric stacks, to a hollow BK7 spacer attached to the ZnSe. All of these pieces are cemented together, forming a solid, rugged unit. All of the pieces, except for the piezoelectrics and the ZnSe plate, are made of BK7. The linear expansion coefficient of ZnSe closely matches that of BK7 and the piezoelectric has a small footprint on the BK7 to which it is attached. Thermal strain is therefore not expected to be a problem.

A value of $\Delta \approx 10$ cm is used for the OPD, sacrificing some wind accuracy to keep the instrument small. The OH emission imposes another restriction, because the components of its Λ-doubled lines cannot be separated by filters, so Δ must be chosen to make the two sets of fringes approximately in phase with each other. The separation of the line components is 0.0358 nm, and at the wavelength of 1308.5 nm, the separation is equivalent to $d\sigma = 0.2091$ cm^{-1}. The fringes from the two components therefore beat together every $1/d\sigma = 4.783$ cm of path difference, and the second beat occurs at 9.565 cm, close to the desired 10 cm.

10.2.4 Filter Selection

The filter wheel contains six positions, which are allocated according to Table 10.2. The O_2 filters have very narrow bandwidths in order to separate the O_2 lines spatially in the field of view, as each set of O_2 emissions (strong/weak) consists of two individual lines. An additional advantage for day-time measurements is that the narrow bandwidth minimizes the solar continuum background. The narrow bandwidth filters consist of wafer etalons made of fused silica, combined with conventional interference filters to select one of the etalon passbands. Position 5 on the filter wheel is open to allow the instrument to observe the calibration sources and to take star images for correction of the spacecraft's pointing knowledge. Position 6 is closed, and is used for dark current calibrations and for protection of the detector when there is danger of the FOV passing too close to the sun.

10.2.5 Appearance of the O_2 Lines in the Field of View

The combination of the narrow passband with the angular shift of the etalons means that the O_2 lines appear as rings of different radii in the field of view. There is a "strong" emission line pair and a "weak" pair, selected with different filters. The filters are untilted, so the rings are centred in the field of view. The relative emission rates are shown as functions of angle in the field of view in Figure 10.6 for the weak line set. A bandwidth of 0.075 nm was used to give a degree of separation of the lines, and the wavelengths for normal incidence were chosen to give a large spot in the centre of the field for one of the lines. The finesse was assumed to be 20. In order to measure the emission rate ratio of the two lines for the rotational temperature determination, the two lines must be transmitted for the altitude range of interest. For both strong and weak sets, both lines are transmitted over half the height range of the field of view, so coverage is adequate. The spot sizes can be adjusted by fine-tuning of the etalon wavelengths.

10.2.6 MIMI Status

The MIMI instrument was proposed as part of a package for the Canadian SciSat mission, but was not selected. The selected instrument, ACE, is described later in this chapter.

Table 10.2. MIMI filter characteristics

Position	Filter	Central Wavelength (nm)	Bandwidth (nm)	Function
1	O_2 strong / O_2 weak	1264.40 / 1278.42	0.075 / 0.075	Day measurement
2	O_2 strong / OH	1264.40 / 1309.5	0.075 / 1.0	Night measurement
3	O_2 strong	1264.40	0.075	Day/night measurement
4	Background	1040 / 1303.3	3.0/1.0	Rayleigh temperature, Background
5	Open			Calibration, star images
6	Closed			Dark current calibrate

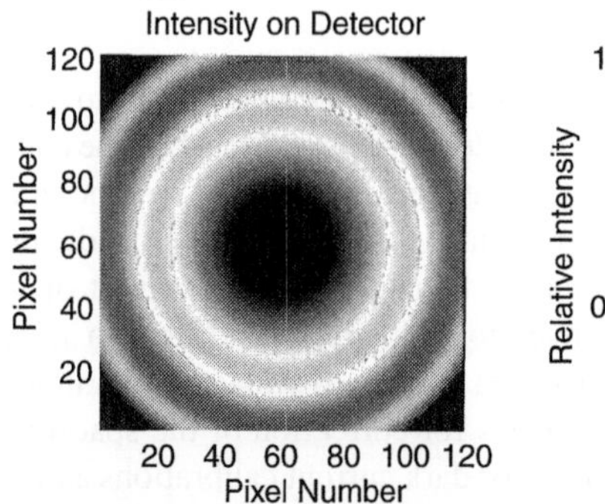

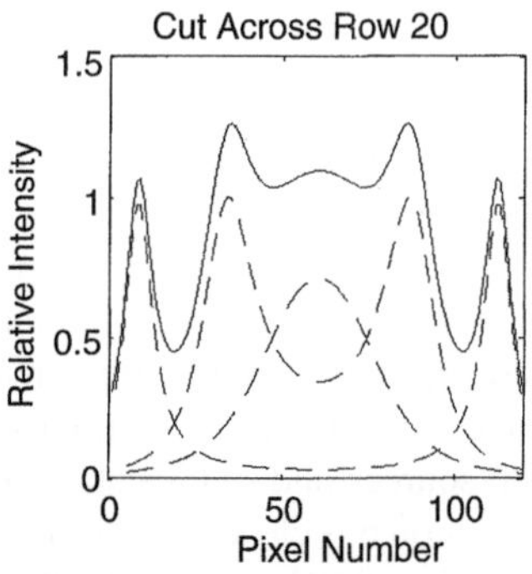

Figure 10.6. Image of O_2 rotational lines as they appear on the array detector (left) and a cut across one row (right), showing the different line components. Courtesy of W.E. Ward (University of New Brunswick).

A variant of MIMI, called WAMI (WAves Michelson Interferometer) which includes a channel for the observation of $O(^1S)$ winds at 558 nm is being proposed for a mission called WAVES- MIDEX, dedicated to the measurement of gravity waves in the middle and upper atmosphere, led by Gary Swenson of the University of Illinois. The Canadian WAMI activity is led by W.E. Ward of the University of New Brunswick with W.A. Gault of York University as Instrument Scientist.

10.3 THE STRATOSPHERIC WIND INTERFEROMETER FOR TRANSPORT STUDIES (SWIFT)

10.3.1 Introduction and Motivation

The Earth's stratosphere extends from approximately 10–50 km and is bounded by the tropopause below and the stratopause above. The stratosphere is so-named because of its strong thermal stratification and owes its existence to the presence of ozone which heats the atmosphere as a result of the absorption of short-wave solar radiation. It is the latitudinal variation of the ozone heating that shapes the general circulation of the stratosphere. Human activity is directly responsible for the destruction of the ozone layer through the anthropogenic release of ozone-destroying chemicals. These pollutants enter the stratosphere in the tropics whereupon they ascend and, at higher levels, spread poleward. This mass circulation is referred to as the Brewer–Dobson circulation and was proposed nearly 50 years ago to explain the observed latitudinal distribution of ozone and the extremely low values of stratospheric water vapour in the tropics.

Long-term records of tropical winds are based solely on observations made at isolated radiosonde stations. Global measurements of stratospheric winds have been made only by HRDI [Ortland *et al.*, 1996] on UARS, which has provided a global climatology of tropical winds. There remains, however, a need for higher resolution measurements made both during the night as well as in day-time to address the problems indicated above. This is the motivation for a DMI approach to this problem, for which the name SWIFT is proposed; Stratospheric Wind Interferometer For Transport studies.

The specific scientific areas of interest are: tropical wind climatologies, the transport of ozone, and data assimilation. SWIFT will have sufficient horizontal resolution to observe equatorial waves and so contribute to studies of tropical dynamics. By the simultaneous measurements of winds and ozone concentrations (as described later), horizontal ozone fluxes can be measured. In data assimilation [Daley, 1991], the observed data are combined with an atmospheric model to provide the most accurate description of the atmosphere consistent with both. An important aspect is that it is the integrated values along the line of sight that are compared with the model; there is no need to invert the data, nor is it necessary to assume horizontal homogeneity. In the future, more and more data will be utilized through data assimilation.

10.3.2 Concept

The SWIFT instrument is intended to take advantage of the demonstrated capability of the WINDII instrument in providing stratospheric wind measurements both by day and at night. There is no suitable stratospheric airglow but, as has been presented earlier, there is a wealth of infrared thermal emission from infrared-active minor constituents. This idea of using such emissions for wind measurement had been proposed earlier by McCleese and Margolis [1983] and by McCleese [1992], not using a Michelson interferometer, but a combination of a gas cell containing the gas from which the emission is observed (as described in Chapter 7) and an electro-optic phase modulator. The atmospheric lines are shifted from the gas cell wavelengths by winds and by the motion of the spacecraft. The electro-optic phase modulator is used to bring the lines back into coincidence, and the applied shift leads to a measurement of the wind. The measurement is made using a suite of lines in the band.

Here, the DMI method is used with a single line. The signal-to-noise ratio for wind velocity measurements using an emission line in the 8–12 μm region is an order of magnitude larger than the signal-to-noise ratio for emission lines in the 4–5 μm region, simply because it is nearer the peak of the Earth's blackbody emission spectrum, as shown in Chapter 1. For this reason the search for a suitable line concentrated on emission lines in the range 8–9 μm. It turned out that several lines of ozone were suitable candidates, offering the capability of measuring ozone concentration along with the winds.

The thermal emission spectrum of the stratosphere as seen from a space-borne limb sounder is complex and contains many closely spaced lines of different emitting species. To obtain a good velocity measurement a strong primary emission line is desired. However, if the emission line is too strong, then self-absorption will prevent the measurement of any useful information from the tangent point of the limb observation. There is also another consideration, unique to an interferometer, in that too narrow a filter will, in the presence of a strong background, shape this continuum such that a modulation of the background by the MI occurs.

In Chapter 8 it was shown that airglow lines have a Gaussian lineshape, reflecting the velocity distribution of the atoms through their Doppler shifts; this type of line broadening is usually called Doppler broadening which under these conditions yields a linewidth of about 0.004 cm^{-1}. Below about 30 km another kind of broadening, called *collisional*, or *pressure broadening* becomes important, creating under these conditions a width of roughly 0.006 cm^{-1}. This broadening arises from the perturbation of the emitted line wavelength during the collision. The theory is complex but the results are summarized by Salby [1996].

The shape of the collisionally broadened line is described by a Lorentzian function as given by:

$$f_c(\sigma - \sigma_o) = \frac{\alpha_c}{\pi[(\sigma - \sigma_o)^2 + \alpha_c^2]} \tag{10.2}$$

where α_c is the half-width at half-maximum of the line. This resembles the Airy function in which $\sin^2(\sigma - \sigma_o)^2$ replaces $(\sigma - \sigma_o)^2$. In either case the function is characterized by a sharp central peak and broad wings. Below 30 km the linewidth is determined by both processes and the combination is called a Voigt profile. Near the peak, the Voigt profile has a Gaussian shape while at the same time it displays the long wings of the Lorentz profile. Since the visibility decreases with increasing optical path difference in a way that depends on the lineshape, an optimum path difference exists for the measurement of wind for each emission line. The optimum path difference is the one that yields the smallest wind error level. Because of collisional broadening, the linewidth varies with altitude which means that an optimum must be sought for the whole altitude range.

For a spherically uniform atmosphere, the Fourier coefficients are given by:

$$\begin{bmatrix} J_1 \\ J_2 \\ J_3 \end{bmatrix} = \frac{1}{f(\sigma_o)} \int_0^\infty \int_0^\infty f(\sigma) B(T, \sigma) \tau(s) \begin{bmatrix} 1 \\ \cos(2\pi\sigma\Delta_o - \phi_o) \\ \sin(2\pi\sigma\Delta_o - \phi_o) \end{bmatrix} \left[\sum_m n_m k_m \right] d\lambda\, ds \tag{10.3}$$

Here B is the Planck function, $f(\sigma)$ is the instrument filter transmittance function, s is the distance from the satellite to the point of observation while n_m and k_m stand respectively for the concentration and total absorption cross-section of species with index m.

Line selection consists of identifying lines that are both "sufficiently" isolated and have "acceptable" strengths. The level of isolation is dictated by the relative positions and strengths of neighbouring lines in consideration of an appropriate line selection filter. The range of acceptable line strengths depends on the signal-to-noise ratio and self-absorption considerations. The spectral regions isolated by the filter for the candidate lines have been subjected to a thorough measurement simulation and inversion process in order to identify those lines that satisfy the 3–5 m s^{-1} accuracy requirement over the 15–40 km altitude range.

The final list of candidate lines over the 1100–1300 cm^{-1} spectral range that meet this requirement in the altitude range of 15–40 km, without knowledge of concentrations of other species, led to a tentative choice of the ozone line at 1133.4335 cm^{-1}. Figure 10.7 shows simulated emission spectra about this line as would be seen by a high-resolution spectrometer looking at the limb. These are shown for tangent heights of 20, 30, 40 and 50 km (with decreasing widths) without (top – linear scale) and with (bottom – logarithmic scale), the filter. The filter transmittance was set to unity at 1133.4335 cm^{-1} for this figure.

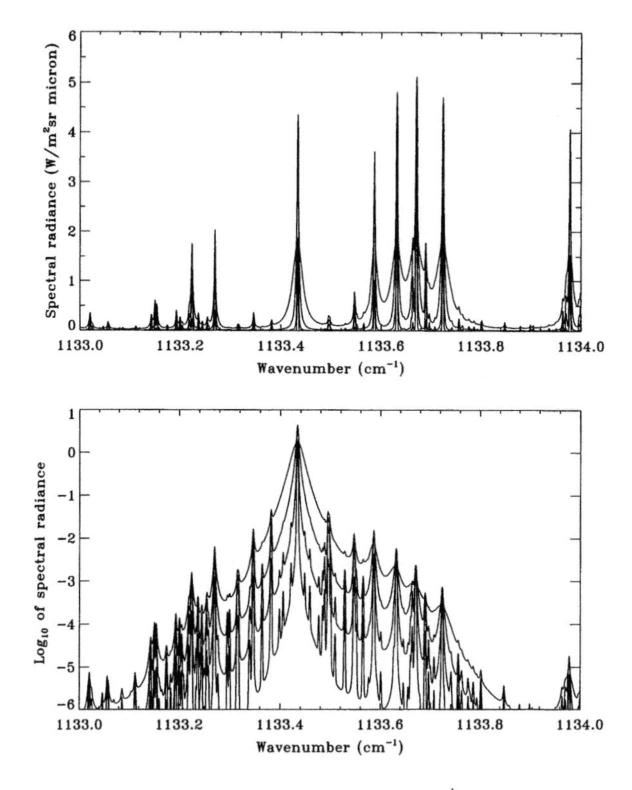

Figure 10.7. Simulated limb emission spectra near 1133 cm^{-1} for 20, 30, 40 and 50 km tangent heights (see text for details). Courtesy of Y.J. Rochon (Meteorological Service of Canada).

The included lines are from O_3 and N_2O. The atmospheric conditions are those of the U.S. 1976 standard atmosphere. All significant lines are from ozone.

The pressure shift of lines near 8 μm is expected to be within the range 0.001–0.004 cm^{-1} atm^{-1} [Toth, 1993]. Once the target line is selected, the related rates will need to be determined in the laboratory, probably using the SWIFT instrument itself. Achievable relative accuracies appear to be on the order of 5–10% [Margottin-Maclou *et al.*, 1996; Regalia *et al.*, 1995]. A value of 0.003 cm^{-1} atm^{-1} with a relative accuracy of 5% has been applied in the simulations.

In addition to winds, SWIFT measurements simultaneously provide the molecular species concentration associated with the target emission line. This requires that the temperature be known to an accuracy of about 2 K, at least above 35 km where self-absorption is weak; as the ozone radiance depends on both the ozone concentration and the temperature. While temperatures of this accuracy are likely to be obtainable from atmospheric models assimilating current temperature data, recent analyses have shown that it is possible to retrieve the temperature for lower altitudes from the SWIFT data itself, increasing the accuracy of the other quantities. Pressure is also needed to allow for the pressure broadening and independent data of 5–10% accuracy for pressure are expected to be available. It may be possible to derive this as well from the derived temperatures, but this had not been demonstrated at the time of writing.

The Doppler wind $w(z)$, the concentration $n_l(z)$ for the species of the candidate line and the temperature can be acquired from the measured $J_1(z)$, $J_2(z)$ and $J_3(z)$. The inversion used for the error analysis is the onion-peeling method of solving for successive layers, beginning at the highest altitude. Given some reference concentration $n_{lo}(z)$, where

$$n_l(z) = \alpha(z) n_{lo}(z) \tag{10.4}$$

the sought concentration parameter becomes $\alpha(z)$. The continuous profiles w, α and T are discretized into homogeneous layers whose boundaries correspond to the tangent height boundaries of the observations.

A run was taken in which the applied inversion approach was the onion-peeling method of solving for successive layers, beginning at the highest altitude. The atmospheric conditions were those of the U.S. 1976 standard atmosphere. The dominant wind error sources for this line/region are random noise and pressure uncertainty at the high and low altitudes, respectively. In addition to satisfying the $5\,\mathrm{m\,s^{-1}}$ threshold, the total wind error level is close to $3\,\mathrm{m\,s^{-1}}$ for most of the altitude region of interest. The uncertainties of the retrieved ozone concentrations are at the level of typically 4–8%. The temperature error is less than 1 K from 17 to 25 km. These error level profiles were obtained by applying inversions to J_1, J_2 and J_3 as described by Eq. (10.3).

Since the above line-search relies on the HITRAN database and on associated simulated spectra, comparison with experimental observations is necessary to affirm the credibility of the simulations. A preliminary comparison of simulation spectra has been conducted with high-resolution transmission spectra from balloon-borne occultation measurements kindly provided by Aaron Goldman. The measurements have a resolution of $0.003\,\mathrm{cm^{-1}}$ and a sampling interval of $0.00125\,\mathrm{cm^{-1}}$. Simulated and measured transmission spectra were compared in the neighbourhood of the various candidate lines for tangent heights of 20.5 and 33.6 km. The balloon altitude was close to 36 km. This preliminary comparison suggests that the relevant lines are accounted for in the simulations and that their relative emission rates and widths appear correctly represented.

10.3.3 Instrument Description

The SWIFT concept is closely modelled on the WINDII instrument on the NASA UARS platform [Shepherd *et al.*, 1993] except that the wavelength used for the measurements is in the infrared as opposed to the visible. The WINDII concept employed three key technologies, phase-stepping interferometry, field-widening and imaging. Despite the restrictions of operating in the infrared, it was found possible to retain these three features.

As described in Chapter 4, infrared instruments are characterized by a large thermal background, in part from the atmosphere, but also from the instrument itself. Most of these instruments, described earlier, used a chopper to detect the difference between a condition of "background + signal" and "background alone". For SWIFT an advantage of the four-step technique is that any background continuum emission which is not modulated by the interferometer is subtracted out in the calculation of the phase. Also, as long as the sequence of mirror steps is more rapid than any temporal variation of the thermal background, then the measurement will not be sensitive to any thermal drift within the instrument. Thus a mechanical chopper is replaced by rapid phase stepping.

Because suitable array detectors are available, the same imaging approach can be used as for WINDII so that, in a single view of the limb, the horizontal wind velocity component along the line of sight will be measured for the entire altitude range. Because of the limited availability of refractive materials in the infrared it is not possible to fully compensate SWIFT in the same way as was done for WINDII, but achromaticity is not required since only one line is observed. Concerning the thermal compensation, experience has shown that the same degree of optical thermal stability as achieved for WINDII is not essential.

The field-widened Michelson interferometer consists of a ZnSe beamsplitter/ compensator, a long solid ZnSe arm and a short air-gap arm which contains the stepping mirror system. The MI is field-widened by an appropriate choice of the element thicknesses. A photograph of the interferometer is shown in Figure 10.8. For the Michelson interferometer, an increase in wavelength by a factor of ten at first sight means an instrument ten times larger. However, the much larger values of refractive index in the infrared alleviate this problem. In addition, because of the uncompensated design in which air in one arm is balanced against high-index ZnSe in the other, the maximum possible OPD is obtained for a given thickness of ZnSe. Because of the success and experience with the solid cemented glass interferometer used for WINDII it was desired to use the same approach with SWIFT,

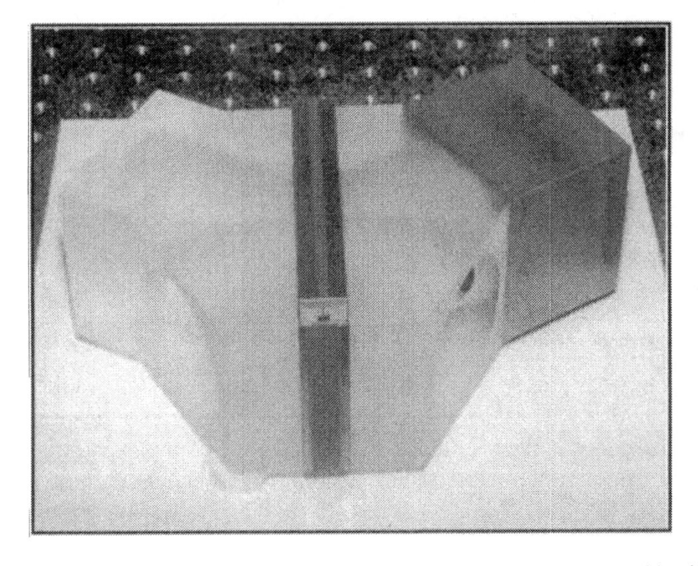

Figure 10.8. Photograph of the SWIFT Michelson interferometer glass assembly. A hexagon of BK-7 is used as the support structure, hollowed out to allow the transmission of the IR radiation. The beam-splitter is between the dark plates of ZnSe in the middle of the hexagon, and the ZnSe plate in the long arm is seen on the right. The darker material is the front surface mirror.

but a significant adaptation was required since only a few materials transmit at 9 μm. Thus an "inside-out" design was developed, in which the hexagonal beam-splitter element, made of BK7, was hollowed out to allow the infrared radiation to pass through to the beam-splitter, which consisted of dielectric layers cemented between two plates of ZnSe, that appear dark in the photograph of Figure 10.8. The ZnSe plate in the long arm is seen as the dark block of glass, with a front-surfaced mirror attached. No scanning mechanism was provided in this prototype; air pressure was used to scan it.

The SWIFT concept is shown in block diagram form in Figure 10.9. Two off-axis telescopes view the limb in the fore and aft directions. Thermal radiation, from the Earth's

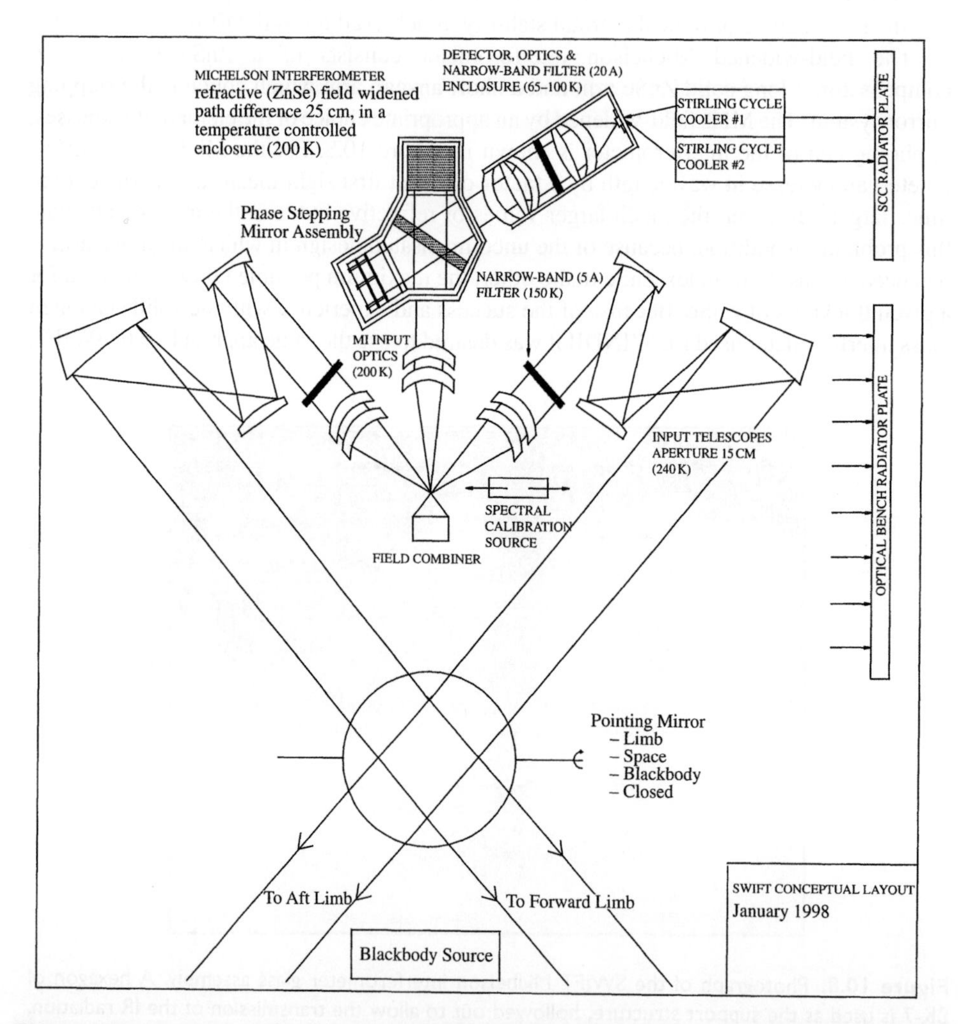

Figure 10.9. Conceptual layout of the SWIFT instrument, showing the two fields of view and the major functional components. Courtesy of G. Buttner, N. Rowlands and A. Scott (EMS Technologies).

surface and warm baffles, is prevented from entering the instrument via intermediate field stops and internal re-imaging of the telescope mirrors. The fore and aft fields of view are combined into one with a reflecting prism. This combined image is re-collimated and presented to the Michelson interferometer input. The final image is formed on the detector by cooled optics. The narrowest bandpass filters are the double etalons (0.5 nm FWHM) which are located in each telescope before the fields are combined. These fore optics and filters can be warmer (~240 K) than the optics and filters closer to the detector (~200 K and ~100 K). The bulk of the instrument cooling can be achieved via passive means with radiators to open space, and this will place constraints on the spacecraft attitude with respect to the sun and the position of the instrument on the spacecraft.

The baseline detector is a 9 μm cut-off HgCdTe array detector with 128 × 128 pixels. The detector and its associated ~0.5 μm wide interference filter will be cooled to ~65 K using a mechanical Stirling cycle cooler. Stirling cycle coolers for space applications are usually paired in order to reduce vibration. Both units will be used together to cool the detector optics and the 20 Å etalon to ~100 K. A thermal link from the coolers also cools the etalons in the fore-optics. Active temperature control (±0.05 K) will be provided for these narrow-band filters. The pointing mirror shown in Figure 10.9 directs the relatively narrow fields of view (1° vertical by 2° horizontal) towards the desired tangent height range (15–65 km) on the limb. Since the spacecraft orbit will not be perfectly circular and because the earth is oblate, active pointing of this mirror will ensure that the region of interest stays within the field of view. This mirror will also periodically (as often as once every 12 seconds) nod the instrument fields of view toward deep space in order to accurately monitor the thermal background flux within the instrument itself. The pointing mirror will also periodically (perhaps once an orbit) redirect the fields of view to a radiometric (blackbody) calibration source in order to accurately derive absolute radiances.

A phase measurement is acquired in ~0.4 s. This fast phase determination mitigates against any effects from the changing thermal background (instrumental or atmospheric) on the detector. Twenty-five of these measurements are co-added to form a measurement image and then the pixels are binned to obtain the required velocity sensitivity. Periodically (perhaps every 15 minutes) the limb mirror points to the phase calibration source, which provides a measurement of the interferometer zero-wind phase, which is essential to the wind measurement. This source will simply be a gas cell at ambient temperature, providing adequate radiance for a spectral line calibration. In order to measure the winds on an absolute scale, a comparison is made with a phase image known to correspond to the zero wind, using an ozone source on the ground. The optics and interferometer will be maintained at ~200 K. Preliminary thermal analysis has shown that a reasonably sized radiator plate (~0.5 m^2) can provide the needed cooling through a thermal link to the optical bench. The actual size of the radiator plates will depend on the orbit and configuration of the host spacecraft. The size of the instrument is approximately 0.5 m^3 and a preliminary estimate of the mass is ~70 kg.

The SWIFT instrument concept also depends on a set of solid etalon filters with clear apertures of 5 cm, bandwidths of 0.5 nm and 2 nm, finesses of at least 20, and which are cooled to ~150 K. During the SWIFT Testbed project, germanium wafers with such large aperture-to-thickness ratios (100–300 respectively) were fabricated [Mani, 2001]. The thin, large diameter (50 mm clear aperture) etalon filters were cooled to 150 K, characterized individually and were also operated in tandem by aligning their passbands under temperature control. The passbands were measured with a Bomem DA-8 system having a spectral resolution of $0.004\,\text{cm}^{-1}$ and were found to meet requirements. However, a tunable filter is currently being investigated as well.

The SWIFT instrument was short-listed by the European Space Agency (ESA) for an Earth Explorer Opportunity Mission which led to discussions with NASDA (National Aeronautics and Space Development Agency), the Japanese space agency. This culminated in the inclusion of SWIFT in a Phase A study for the NASDA GCOM-A1 mission, a study now involving NASDA, the CSA and ESA. GCOM is the Global Change Observing Mission, which is a successor to ADEOS (ADvanced Earth Observing System) and ADEOS-II. GCOM consists of two platforms, GCOM-A, dedicated to atmospheric measurements, and GCOM-B, primarily concerned with Earth surface measurements. The plan calls for a series of two missions, 1 and 2, each of five years duration which, together with ADEOS-II, will provide a continuous fifteen-year set of measurements. The GCOM-A1 mission is planned to be launched in about 2007 [Suzuki *et al.*, 1999].

The SWIFT activity is led by I.C. McDade at York University, with the support of W.A. Gault along with Y.J. Rochon of the Meteorological Service of Canada, and Alan Scott, Neil Rowlands and Gary Buttner of EMS Technologies (as well as a Science Team). The design is evolving and the version flown will incorporate a number of significant improvements over what is described here.

10.4 THE ATMOSPHERIC CHEMISTRY EXPERIMENT (ACE)

The principal goal of the Atmospheric Chemistry Experiment (ACE) mission is to measure and to understand the chemical and dynamical processes that control the distribution of ozone in the upper troposphere and stratosphere. Anthropogenic changes in atmospheric ozone are increasing the amount of ultraviolet radiation received on the ground and may also affect the climate. A comprehensive set of simultaneous measurements of trace gases, thin clouds, aerosols and temperature will be made by solar occultation from a small satellite in a low Earth orbit.

Solar occultation means that the instrument views the sun, but along rays which pass through the Earth's limb as in Figure 1.10. As the satellite moves along the orbit, the sun "rises" or "sets", depending on the direction of satellite motion with respect to the sun, and the viewing ray ascends or descends through the atmosphere, allowing the measurement of absorption spectra at a sequence of tangent point altitudes. Through inversion of these data, vertical profiles of species concentrations are obtained. In order to accurately distinguish one species from another in the spectrum, high resolution is required. Because of the high solar radiance it is possible to obtain high resolution with a comparatively small instrument, and a scanning Michelson interferometer is an excellent choice for this type of measurement. The disadvantage of the solar occultation method is that only two profiles are obtained per orbit, one for sunrise and the other for sunset, about 30 profiles per day. Nevertheless, by accumulating data over a suitable time period, climatological descriptions of the behaviour of the desired species may be obtained. The HALOE (Halogen Occultation Experiment) instrument on the UARS satellite [Russell *et al.*, 1993] offers examples of the types of coverage that have been achieved with solar occultation.

Thus the ACE-FTS instrument is a high-resolution (0.02 cm^{-1}) infrared Fourier transform spectrometer (FTS) operating from 2 to 13 microns (750–4100 cm^{-1}) which will measure the vertical distribution of trace gases and temperature. Aerosols and clouds will be monitored using the extinction of solar radiation at 1.02 and 0.525 microns as measured by two filtered imagers. The vertical resolution will be about 3–4 km from the cloud tops up to about 100 km. The ATMOS instrument made similar measurements from the space shuttle, and produced the spectrum shown in Figure 1.16. However, the ACE-FTS will be much smaller, to be compatible with a small satellite. Because reference spectra of the sun are recorded outside the earth's atmosphere, the ACE-FTS is self-calibrating; that is an advantage of the absorption method, particularly for the determination of long-term trends.

A second instrument called MAESTRO (Measurements of Aerosol Extinction in the Stratosphere and Troposphere Retrieved by Occultation) completes the instrument complement for the ACE mission. MAESTRO is a dual optical spectrograph covering the 285–1030 nm spectral region. It is discussed in the next chapter along with other diffraction grating spectrometers. A high inclination (74°), circular low Earth orbit (650 km) will give ACE coverage of tropical, mid-latitude and polar regions. ACE is a Canadian Space Agency (CSA) mission, the first of a planned series of small scientific satellites called SCISAT; ACE is SCISAT-1, and it is led by Peter Bernath of the University of Waterloo.

The ACE-FTS will be built by ABB-Bomem in Quebec City using a custom design to meet the requirements of the mission, as shown in Figure 10.10. Light enters at the bottom right of the figure and is reflected off the pointing mirror, which is tilted to track the sun as it rises and sets. A small portion of the light is separated out through the 1.5 μm filter and taken to a quadrant detector that controls the pointing of the mirror. The same filter reflects light into the visible and near-infrared imagers, and a dichroic filter (which transmits one wavelength band while reflecting another) extracts light for the MAESTRO instrument. A reflector then forms an image of the sun at a field stop, which determines the size of the field of view, 1.25 mrad for the MI. This must be much smaller than the sun in order to achieve the required vertical resolution of 3–4 km. A collimating reflector then brings near-parallel light to the Michelson interferometer, which uses two cube corners as described earlier. These are frequently called corner cubes, as they are in the figure, even though James Ring always insisted that they are the corners of a cube and should be so named.

The cube corners are mounted on the arms of what is called a *dual-pendulum*, a configuration called the dual pendulum interferometer which rotates as a rigid unit about a centre flex pivot to produce the optical path difference, a technique based on the ABB Bomem MB series of interferometers. Burkert *et al.* [1983] and Fischer and Oelhaf [1996] describe earlier versions of this configuration. Because the same motion moves one corner cube towards the beam-splitter while the other moves away, the extent of motion is one-half of what it would otherwise be for a given OPD. The ray paths are shown in Figure 10.10, where it is clear that the beam returning from each cube corner is displaced from the entering beam. This allows reflection off a fold mirror, placed inside the interferometer, causing the beam to traverse the same path twice, halving again the movement required for a given

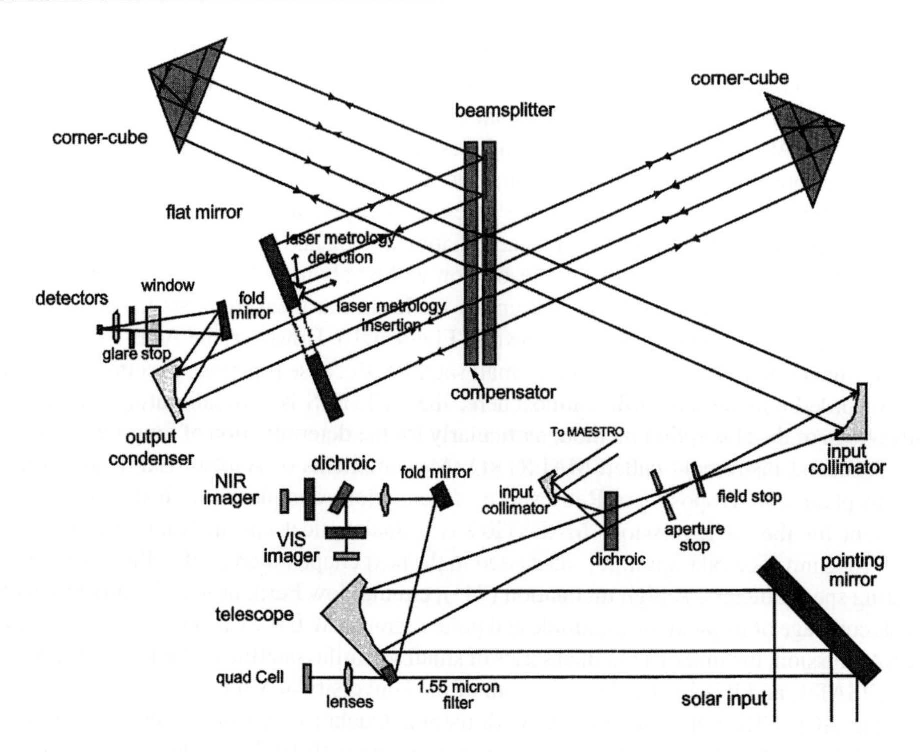

Figure 10.10. Layout of the ACE-FTS Michelson interferometer. After Bernath [2001].

Table 10.3. Main characteristics of the ACE-FTS interferometer

Spectral range	750–4100 cm^{-1}, 2.43–13.3 μm
Spectral resolution	<0.028, 0.056, 0.11, 0.55 cm^{-1}
Sweep duration	2, 1, 0.5, 0.1 s
Spectral stability (relative)	3×10^{-7} (rms) for 180 s
Transmittance uncertainty	<1% (rms)
Dynamic range	0–5800 K
Noise equivalent radiance	0.5% of 5800 K radiance

OPD. This is how the high resolving power is achieved with a small instrument. The ACE design is fully compensated for tilt (rotation) and shear (lateral motion) of both moving and stationary optics inside the interferometer. The OPD is measured with a laser diode operating at 1500 nm. A summary of the main characteristics for the FTS is provided in Table 10.3.

Double-sided interferograms will be Fourier transformed on the ground to obtain the desired atmospheric transmission spectra. The FTS covers the 750–4100 cm^{-1} range using two detectors (InSb and HgCdTe) in a sandwich configuration. The detectors are cooled to less than 110 K (typically 90 K) by a passive radiator pointing towards deep space. The instrument has a mass of 35 kg, a volume of 58 400 cm^3, and a power consumption of less than 40 W. An instrument very similar to ACE-FTS called SOFIS (Solar Occultation FTS for Inclined-orbit Satellite, Suzuki *et al.* [1997]) is part of the Japanese GCOM-A1 satellite mission, already mentioned in connection with the SWIFT instrument, and planned for a launch in 2007.

10.5 THE MICHELSON INTERFEROMETER FOR PASSIVE ATMOSPHERIC SOUNDING (MIPAS)

The MIPAS (Michelson Interferometer for Passive Atmospheric Sounding) instrument is a scanning Michelson interferometer with many similarities to the ACE-FTS instrument just described, but with some very important differences. This is because MIPAS observes atmospheric thermal emission at the Earth's limb, rather than absorption of sunlight. This requires among other things that the instrument be cooled to reach the low signal levels required. The approach is similar to the observations of CLAES, except that since a scanning MI is used, a complete spectrum is obtained over a wide spectral range (4.15–14.6 μm) at a resolving power of about ten times larger. Since it is thermal infrared emission that is observed, vertical profiles are obtained during both day and night portions of the orbit and the horizontal resolution is determined by the time it takes to obtain a vertical profile. An earlier MI that observed thermal atmospheric emission was IMG (Interferometric Measurement of Greenhouse gases), described by Ogawa *et al.* [1994], which was launched on the Japanese ADEOS satellite in 1996. However, IMG used nadir viewing while MIPAS views at the limb. Since MIPAS has a single field of view, a spectrum is obtained for one tangent altitude and then the pointing mirror is stepped vertically to obtain the spectrum for the next altitude point. MIPAS was launched on the ESA Envisat spacecraft on March 1, 2002, so it is no longer a "future" mission; it will measure species concentrations, atmospheric pressure and temperature. The instrument complement for Envisat includes two other atmospheric instruments, GOMOS and SCIAMACHY. Both are diffraction grating instruments which are mentioned in Chapter 11. A simplified optical layout of the MIPAS MI is shown in Figure 10.11 and a detailed description is given in an ESA (2000) document. This is a symmetrical design with two input and two output ports, which are used in a way that is described below. The cube corners are driven simultaneously in opposite directions (horizontally as shown in the Figure 10.12) on independent slideways with linear motors, using mechanical bearings.

The resolution is intended to be sufficient to resolve individual atmospheric lines, or about 0.035 cm^{-1}, for which a maximum OPD of 20 cm is used. The sampling is triggered with a diode laser beam, operating at 1.3 μm. Because the observed signal is thermal radiation at the temperature of the atmosphere, the thermal radiation from the walls of the instrument must be reduced to much less than that produced through cooling to temperatures significantly less than that of the atmosphere. Radiative cooling reduces the temperature of the instrument to about 210 K while Stirling cycle coolers are used to cool the detectors to about 70 K. MIPAS is a large instrument, about 1.36 m long and 1.46 m high and 0.74 m deep, with a mass of about 170 kg. In addition there is an electronics module so that the

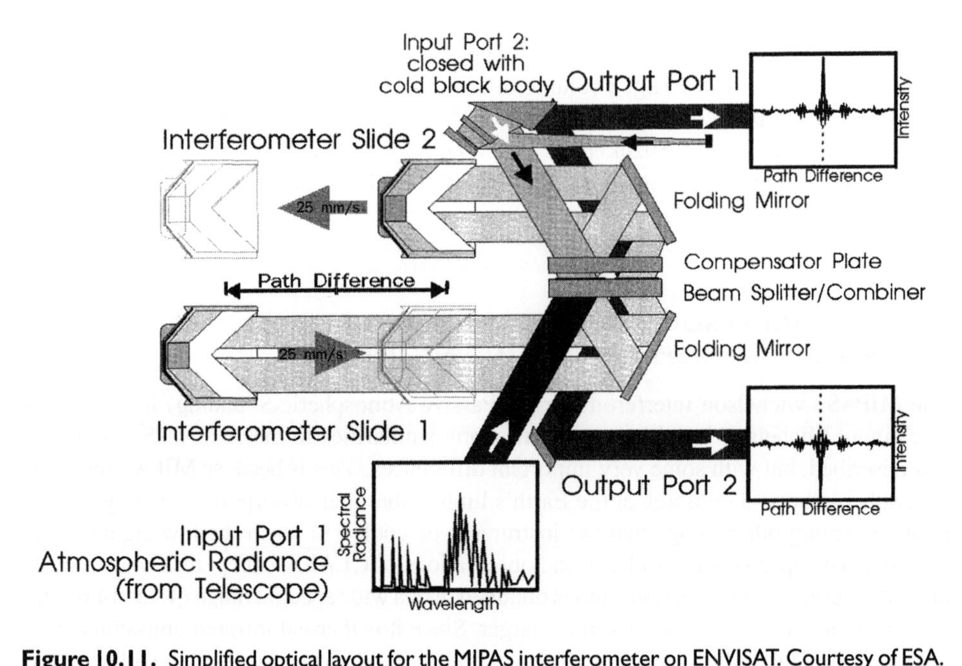

Figure 10.11. Simplified optical layout for the MIPAS interferometer on ENVISAT. Courtesy of ESA.

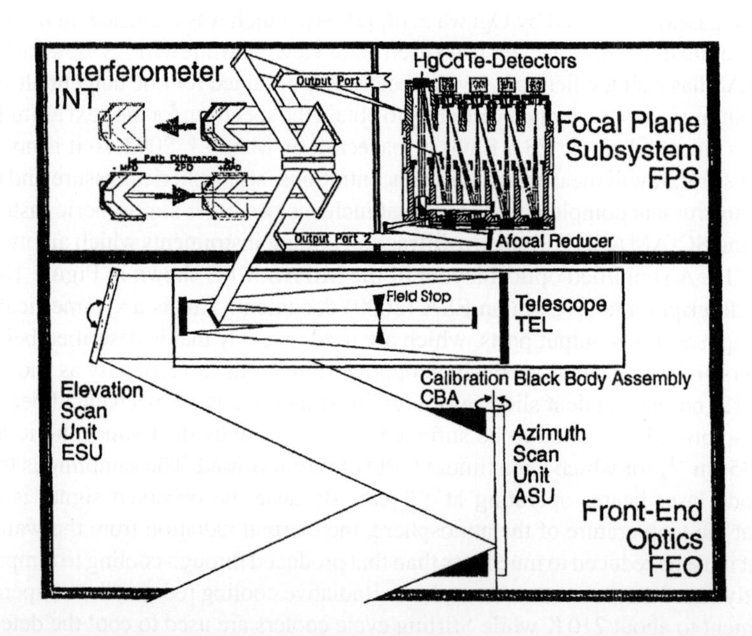

Figure 10.12. Overall schematic of the MIPAS instrument including the input optics and the focal plane detector sub-system. After Fischer and Oelhaf [1996].

total mass of the instrument is 320 kg. Radiation is brought to the instrument through two scan mirrors and an anamorphotic telescope as shown in Figure 10.12. The azimuth scan mirror selects between side-viewing and rearward-viewing baffles while the elevation scan unit allows measurements from 5 km to 250 km; the high altitude views are used only to determine the instrument background radiation. The telescope design is anamorphotic with different magnifications in the vertical and horizontal fields of view. In this way a field of view of 3 km vertical and 30 km horizontal is mapped into a circular field for the interferometer. The second input receives light from the Focal Plane Subsystem which is very cold, about 70 K.

Figure 10.13. The EnviSat spacecraft during integration at ESTEC, ESA's space technology centre, showing the locations of the various instruments. The size of the spacecraft is evident from the workers in the figure. Courtesy of ESA.

Four detectors are required to cover the required spectral range for each output, requiring eight detectors in all, which are assembled in the Focal Plane Subsystem as shown in Figure 10.12. For calibration, the instrument receives radiation from a calibration black body mounted in the Azimuth Scan Unit; it fills the entire aperture of 55.2 mm^2 and has an emissivity above 99.6% which allows accurate calibration of the measured radiances, which is extremely important for an emission instrument.

The Envisat spacecraft is shown in Figure 10.13, during the integration of the instruments into the payload at ESTEC, the ESA Technology Centre located at Noordwijk, near Amsterdam. It is a very large spacecraft, 10.5 m in length, with a mass of 8200 kg and a power capability of 3800 W. Envisat was successfully launched on March 1, 2002.

10.6 PROBLEMS

10.1. The HRDI and TIDI instruments were both developed at the University of Michigan so the design of the latter owes a great deal to the former. Describe the differences between the two instruments as quantitatively as possible with the available information, and identify the two most distinctive differences.

10.2. The WINDII and MIMI instruments were both developed at York University so the design of the latter owes a great deal to the former. Describe the differences between the two instruments as qualitatively as possible with the available information.

10.3. Although the observed wavelength for SWIFT is about ten times as large as that for WINDII, the sizes of the two interferometers are similar. This results from: (i) the higher refractive indices in the infrared, 2.4122 for ZnSe compared to 1.78848 for LaFn21 and 1.580165 for LF5; (ii) the lack of thermal compensation for SWIFT; and (iii) the difference in the linewidths. Discuss these points as quantitatively as possible with the available information. Note that the OPD for WINDII is 4.5 cm while that for SWIFT is likely to be about 15 cm.

10.4. Assume that the pressure shift of the ozone line observed by SWIFT is 0.002 cm^{-1} atm^{-1}. For a wind observation at an altitude of 20 km calculate the percentage pressure error that would cause a wind error of 3 m s^{-1}. Eq. (9.5) may be used to obtain a value of the pressure at 20 km, using a scale height H of 8.0 km.

10.5. The ACE instrument is similar in concept to the earlier ATMOS instrument; both use solar occultation to obtain species concentrations through atmospheric absorption. Describe the differences between the two as quantitatively as possible with the available information.

10.6. The MIPAS instrument has some similarity in concept to the IMG instrument, although the former is intended for limb viewing and the latter was used for nadir viewing. Describe the differences between the two as quantitatively as possible with the available information.

11

GRATING SPECTROMETERS AS SPECTRAL IMAGERS

11.1 INTRODUCTION

In Chapter 3 it was shown that the Fabry–Perot spectrometer and Michelson interferometer both have a superiority, $S = \Omega\mathcal{R}$, that is greater than that for the diffraction grating spectrometer (DGS). The diffraction grating instrument has not been mentioned in the intervening chapters, perhaps leaving the incorrect impression that the instrument has no role to play in atmospheric observations. This chapter is intended to correct that impression, and to describe the past and future role that some selected spectrometers have played in atmospheric observations.

It remains true that the DGS does have lower superiority than the other instruments mentioned; for example, it is impossible to measure atmospheric winds with a DGS. However, there are many low-resolution applications for which this spectrometer is well suited, particularly where radiance levels are high, such as the measurement of scattered light in the day-time atmosphere, or observations by solar occultation. At low resolution, the responsivity for such measurements is more than adequate to achieve the required accuracy and a compact instrument is capable of covering a wide spectral range.

In part because the rectilinear geometry of the DGS makes it highly compatible with the use of rectilinear array detectors, it has found greatest application in the visible region, where CCDs are employed. Linear arrays can be very effectively used to cover a wide spectral range. However, if a two-dimensional CCD is used, the second dimension can be used for a different variable, a spatial dimension, for example. A DGS looking downward from a satellite can have the spatial direction of the CCD aligned across the track so as to measure the spectrum in a swath; a complete spectrum is measured for every spatial element in the swath during one integration time. As the spacecraft moves forward, successive swaths are sequentially measured, building up a two-dimensional image in which the complete spectrum is determined for every pixel; maps can be created for any desired wavelength, or ratio of wavelengths, or combinations of wavelengths according to any desired algorithm. This approach is very widely used in the remote sensing of the Earth's surface, and is described by the term *hyperspectral imaging*.

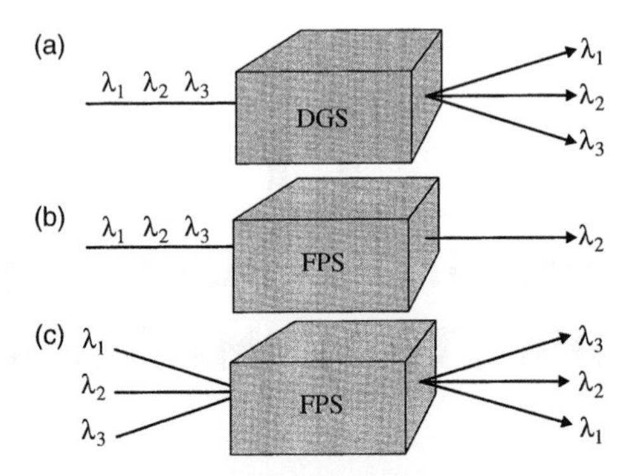

Figure 11.1. Illustrating the ray geometry differences for: (a) a diffraction grating spectrometer; (b) an FPS with a single entering ray; and (c) an FPS with multiple entering rays.

There is a subtle but important difference between the geometries of the FPS and DGS instruments that hinges on the word *diffraction*. For an elementary FPS, an entering ray of white light leaves the device at the same angle at which it enters; the only characteristic involved is the transmittance, which is a function of wavelength. The transmitted ray has an altered spectrum because some component wavelengths are fully (or partially) transmitted while others are reflected. For a DGS an entering ray is diffracted into multiple directions, according to the wavelength. All wavelengths are transmitted but are dispersed in different directions. This is illustrated in Figure 11.1 where in (a) three wavelengths are shown entering the DGS along a single ray; these are dispersed into three different directions. In Figure 11.1(b) the same three wavelengths are shown entering an FPS along a single ray, of which only one is transmitted (the others are reflected but this is not shown), while in (c) the three wavelengths are shown entering in three directions, and all are transmitted in the directions in which they entered. This means that if all wavelengths are to be simultaneously recorded, say on a CCD image, the corresponding field of view must be filled in front, and arranging this can be difficult. The point is that if N spectral elements are to be recorded, the field of view must be N times that corresponding to a single spectral element for an FPS while for the DGS only the field of view corresponding to one spectral element is required. This applies only to the situation where the wavelengths are spatially scanned; if the FPS is stepped in wavelength there is no problem except that the multiplex advantage is lost. Put differently, it means that comparatively speaking, a DGS can be constructed in a very compact configuration.

In the next section, the fundamental aspects of DGS instruments are outlined, and in the sections following, the accomplishments of some previously flown instruments are presented, followed by descriptions of selected instruments for some forthcoming missions. Finally, a hybrid instrument is described, an interferometer that uses diffraction gratings.

11.2 FUNDAMENTAL ASPECTS OF THE DIFFRACTION GRATING SPECTROMETER

It is said that the diffraction grating was invented by the American astronomer David Rittenhouse in 1786 when, inspired by the coloured effects produced by a silk handkerchief, he made a grating by laying hairs across two very finely pitched screws. However, it is Fraunhofer who is given the credit for the diffraction grating as known today; independently in 1821 he made wire gratings and ruled a grating with a diamond point on a mirror surface (for these historical details see Hutley [1982]). But it was not until 1882 when H.A. Rowland, Professor of Physics at Johns Hopkins University, succeeded in building a ruling engine that could fabricate useful gratings, that it became an important spectroscopic device. He also invented the concave grating for which the surface acted as a mirror as well as a grating. Various configurations were then proposed for both plane and concave gratings. One of the early plane-grating configurations was proposed by Ebert, with a single spherical mirror as an autocollimator, a symmetrical system in which the same mirror acted as input collimator and camera. However, it has a problem that is faced by all such instruments, namely aberrations in the optical system associated with the length of the slit. William Fastie [1952], also of Johns Hopkins University, discovered a modified configuration in which the straight slits were replaced by curved ones, as shown in Figure 11.2. The figure is a view from behind the diffraction grating, looking towards the spherical mirror, with the entrance and exit slits on either side of the grating. A point on the entrance slit at a is imaged at the exit slit at a′, but as a line. Similarly a point at b, near the end of the slit is imaged at b′ as a short tilted line. Fastie's solution (now known as the Ebert–Fastie version) was to use curved slits, such that e is imaged at e′, a tilted line but one which is tangent to the exit slit because of the symmetry.

However, when an imaging detector is used, this symmetry is lost. Broadfoot and Sandel [1992] discuss this problem, which is illustrated with their drawing reproduced in Figure 11.3. The solution is to combine the grating and mirror, the latter already a combination of a collimator and camera, by using a curved grating and a curved detector. This offers the designer all the control so as to reduce astigmatism to the required degree. The

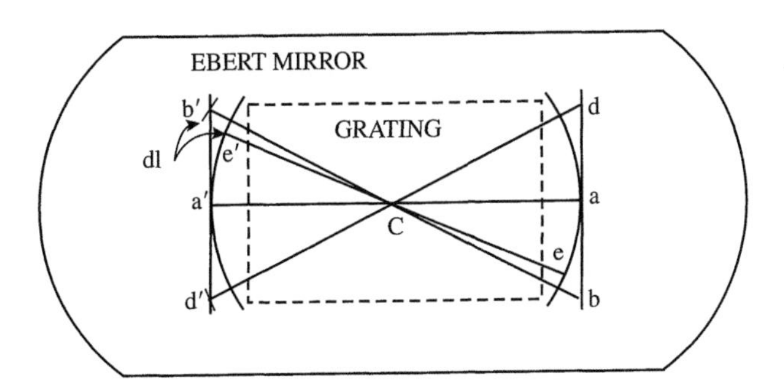

Figure 11.2. Illustrating the aberrations in an Ebert spectrometer with straight and curved slits. Courtesy of G.G. Sivjee (Embry-Riddle Aeronautical University).

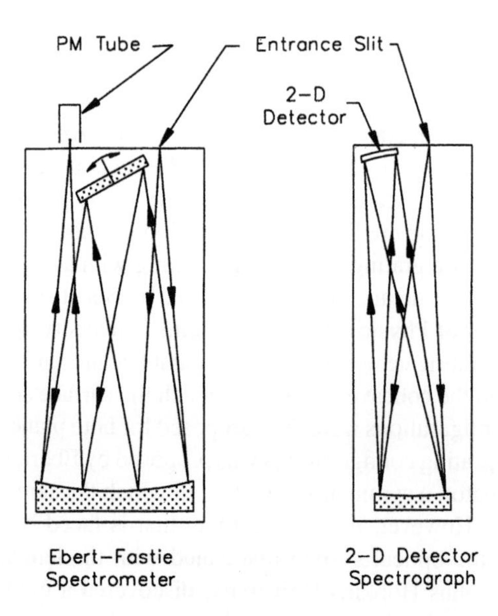

Figure 11.3. Comparison of a conventional Ebert–Fastie spectrometer (left) using a photomultiplier detector and a spectrograph with an imaging detector (right). After Broadfoot and Sandel [1992].

configuration has the advantage of an extremely compact design, and it reduces the number of reflections to one, which is very important for ultraviolet instruments where reflectances are low. This type of instrument has origins in the Wadsworth [1896] and Czerny–Turner (like the Ebert but with separate collimation and camera mirrors) configurations; the design problem is not new as the same considerations apply to spectrographs with photographic film detection. Hutley [1982] describes the theory of the diffraction grating instrument in detail, and describes many of these configurations.

11.3 SELECTED AIRGLOW MISSIONS ACCOMPLISHED

11.3.1 Mariner 10 Ultraviolet Airglow Experiment

An example of the instrument just described is provided by the ultraviolet airglow experiment flown on the Mariner 10 mission by Broadfoot *et al.* [1977a]. The purpose was to measure the abundance and vertical distribution of the most likely constituents in the Mercurian atmosphere. Specific constituents considered were H (121.6 nm), He (58.4 nm), He^+ (30.4 nm), A (86.9 nm), Ne (74.0 nm), C (165.7 nm) and O (130.4 nm). Array detectors was not available then, so 12 discrete detectors were used, then called Bendix Spiraltrons; this capability was later to become Galileo Electro-Optics. These were tubes, about 6 mm long, with a slit at one end acting as the exit slit of the spectrometer, and within which about one-half of the length of the tube acted as a multiplier. Because of the short wavelengths involved, no refractive materials could be used; these were open detectors; that is, they did not have windows. This meant that, following tests in a vacuum chamber, the detectors could not be operated again until they were in space.

Similarly, refractive optics could not be used, and to limit the reflections to the grating only, the Wadsworth-type design was used, in which the grating is illuminated with parallel light. However, this in itself does not solve the whole problem, because fore-optics are required to define a field of view with respect to the slit and grating combination. *Slit-less* spectrographs are sometimes used for taking spectra of stars. The star acts as its own slit, so that each line in the spectrum replicates the star image at that wavelength. But for the Mariner 10 mission more flexibility was needed as to the types of sources, so a mechanical collimator was used instead. This consisted of a series of metal plates out of which a series of slots was cut, like a magnified diffraction grating. By using 16 of these plates at appropriate spacings a field of view of 0.125° half-width was defined. That is, only within this angle could light reach the grating by passing through the slots in all 16 plates. In the other direction the field of view was 3.6°. The instrument layout is shown in Figure 11.4. The instrument was extremely compact, as already indicated, with a height of 14.6 cm, a width of 12.1 cm and a length of 38.1 cm; the mass was 3.82 kg.

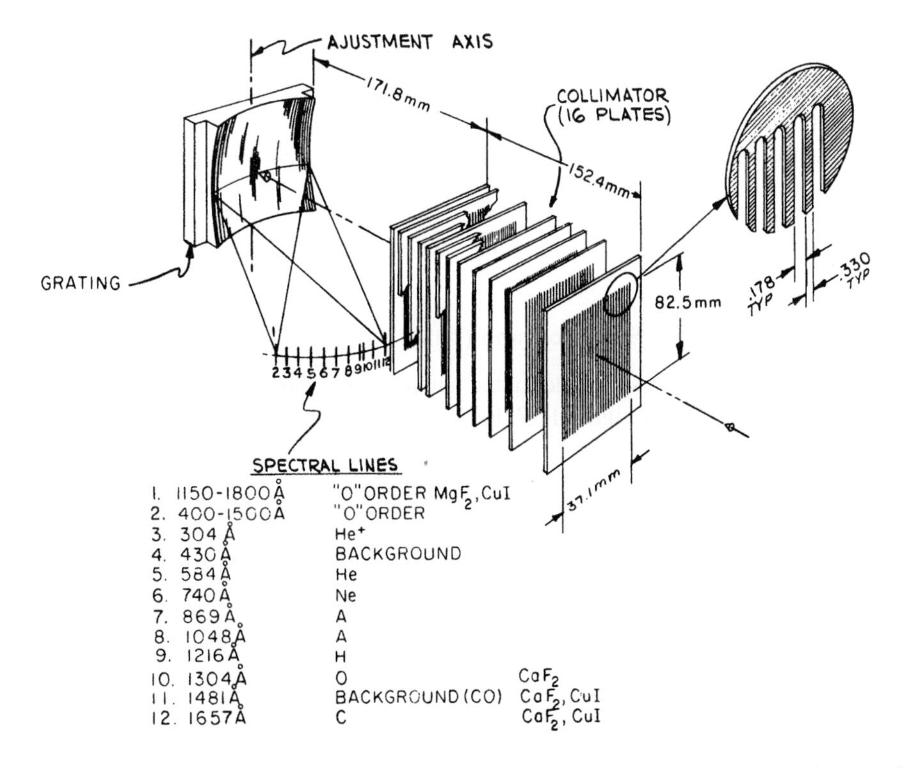

Figure 11.4. The optical layout of the Mariner 10 ultraviolet airglow experiment, showing the mechanical collimator, the concave grating and the twelve individual detectors, along with the species each was intended to observe. After Broadfoot *et al.* [1977a].

11.3.2 The Voyager Mission Ultraviolet Experiment

The Voyager mission was launched in 1977 to explore the atmospheres of Jupiter, Saturn and Uranus, along with some of their satellites. These planets were expected to have similar atmospheres, consisting of H, H_2, He and CH_4. The objectives are described in more detail by Broadfoot *et al.* [1977b]. The ultraviolet instrument is similar to the Mariner 10 instrument, with two important enhancements. First, it used a 128-element linear self-scanned array, which was integrated with a micro-channel plate multiplier.

The instrument was designed to operate in the wavelength range 50–170 nm, in both an extended source mode and occultation mode; this was the second enhancement. In the extended source mode the instrument was very much like Mariner 10, with a mechanical collimator and concave diffraction grating as shown in Figure 11.5. The occultation field is a small aperture field that is offset by 20° from the extended field, and is generated by a special aperture in the collimator and a reflection off the small occultation mirror. The reason for the offset is that the other instruments on the spacecraft view in the same direction as the extended field and this points them 20° away from the sun when the ultraviolet spectrometer views the sun in occultation mode. Broadfoot and Sandel [1992] describe this instrument as having a sensitivity of 0.46 rayleigh s. By this it is meant that a signal of 0.46 R is required to generate a signal-to-noise ratio of unity in one second of integration; the detector noise is given as 4.5×10^{-3} counts s^{-1}.

A sequence of spectra obtained through observing the Jupiter region from a distance is shown in Figure 11.6 from Broadfoot *et al.* [1981]. The spectra are organized according to the distance from the centre of the planet, in units of Jovian radii (R_J). The emissions seen originate in the light from the planet itself, as well as the torus associated with the satellite Io, and the interstellar medium. Both planetary and interstellar Ly-a are seen, as well as auroral H_2 emission, ionized sulphur and oxygen, and interstellar He emission.

11.3.3 Arizona Imager Spectrograph (AIS)

The AIS experiment was responsible for the remarkable airglow spectra obtained from the space shuttle and shown in Figure 1.17. Earlier versions (it has flown on shuttle missions

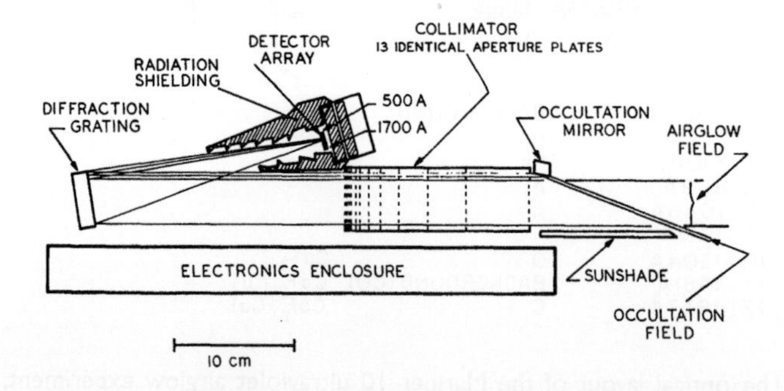

Figure 11.5. Layout of the Voyager UVS showing the two modes of operation, indicated by the airglow field, and the occultation field. After Broadfoot et *al.* [1977b].

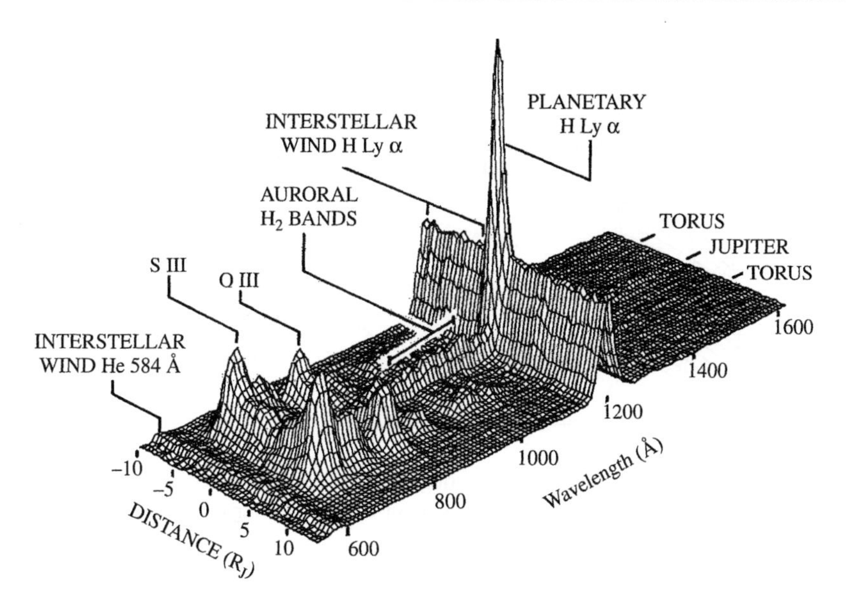

Figure 11.6. Spectra of the Jupiter system, obtained from a distance by Voyager 1. The spectra are organized according to the distance from the centre of the planet, in units of Jovian radii (R_J). The spectra originate from Jupiter, the Io torus and the interplanetary medium. After Broadfoot *et al.* [1981].

many times) are known as the GLO experiment, because the original motivation was to study the spectrum of the shuttle *glow*, caused by the high velocity impact of the vehicle surfaces on the atmosphere. Future possibilities include flight on the International Space Station; but this is as yet uncertain. In this section, the AIS version described by Broadfoot *et al.* [1992] is briefly summarized.

The overall package consists of nine spectrographs and twelve imagers in a remarkably compact package that combines spectra from two imagers onto one CCD, and that combines the image data from the twelve imagers onto two CCDs. This is done using fibre-optic combiners. The nine spectrographs cover the spectral range from 114 to 1090 nm, with a resolution varying from 0.5 to 1.3 nm. The configuration of a dual spectrograph, employing two concave gratings and one CCD is shown in Figure 11.7.

The gratings are aberration-corrected holographic concave gratings. Four dual spectrographs covered the wavelength range 114–930 nm, while a single spectrograph was used for the range 900–1090 nm where the responsivity was lower. Cylindrical optics are used in the fore-optics to simplify the baffling and to minimize the volume of the optical enclosure. They are crossed cylindrical lenses, meaning that the axes of curvature of the two are perpendicular. Both lenses have their focal length at the slit which defines the field of view and the ray paths through the spectrometers.

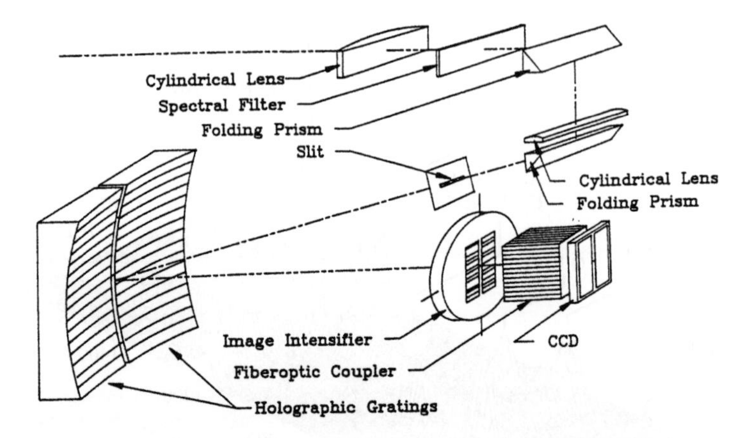

Figure 11.7. Schematic diagram of a dual-grating spectrograph; the spectrum from each grating is dispersed on one-half of the array detector. The input optics involving two cylindrical lenses is also shown. After Broadfoot *et al.* [1992].

11.3.4 Single-Element Imaging Spectrograph (SEIS)

Cotton *et al.* [1994] have shown that imaging may be accomplished with a spectrometer that has only one element, a toroidal diffraction grating. In the dispersion direction, the radius of curvature of the grating is the diameter of the Rowland circle, and the spectrograph acts like a Rowland spectrograph. In this direction the slit, the surface of the curved diffraction grating and the surface of the detector all lie on a common circle, providing minimum astigmatism. In the other direction, called the imaging direction, the radius of curvature is approximately twice that of the Rowland circle diameter so that objects at infinity are imaged on the detector. This is the same arrangement as for a Wadsworth configuration, so the SEIS is said to be a hybrid between the Rowland and Wadsworth spectrometers. Overall, the two-dimensional detector records a normal spectrum in the dispersion direction of the array, but the spectra from the first and second pixel rows from the top correspond to adjacent spatial pixel elements so that an independent spectrum is obtained for every spatial element along a "vertical" line on the target.

This approach was applied in the Tomographic Extreme-ultraviolet SpectrographS (TESS) that were launched on the Tomographic Experiment using Radiative Recombinative Ionospheric EUV and Radio Sources (TERRIERS) satellite on May 18, 1999. Five SEIS spectrographs were employed and the goal was to carry out tomographic imaging of the ionosphere, using emissions produced by the recombination of ionospheric electrons [Cotton *et al.*, 2000]. Two views of this instrument are shown in Figure 11.8. The tomographic data were to be acquired in part from the imaging character of the spectrographs, but also from the different viewing angles obtained through the rotation of the satellite about an axis perpendicular to the orbit plane, the so-called *cartwheel* configuration. Unfortunately, a failure of the attitude control system occurred before any data were obtained.

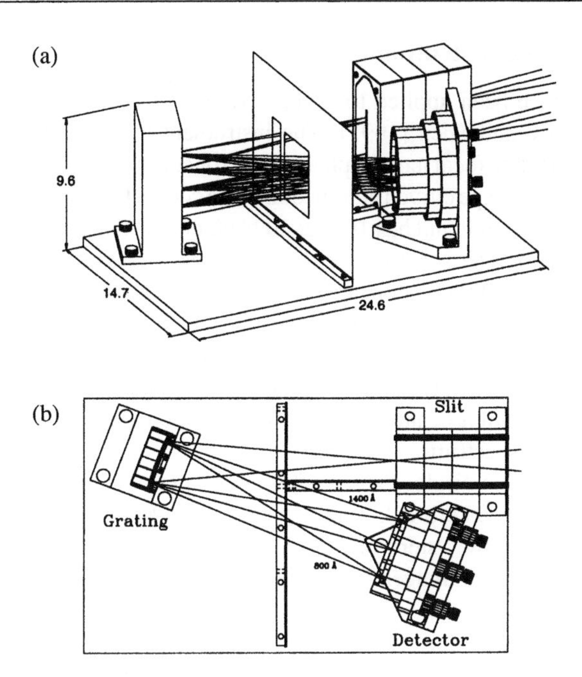

Figure 11.8. Optical layout for the TESS spectrographs as seen in two different views. The dimensions shown are in cm, and it is shown that the spectral range covered is from 80 to 140 nm. After Cotton *et al.* [2000].

Another application of the SEIS instrument has been described by Chakrabarti *et al.* [1999] in what is called the SPINR technique (Spectroscopic and Photometric Imaging with Numeric Reconstruction). Using a SEIS instrument, a two-dimensional image is obtained as well as the spectrum of each pixel in the two-dimensional image, using only a 2-D array detector. Obviously this cannot be done from a single-array readout and in fact is achieved by rotating the instrument, which has a square field of view rather than a slit, and records images at equally spaced angles during the rotation. Suppose that the source consists of a single point, such as a star. Since the SEIS has inherently one imaging and one spectral dimension, the source appears on the image at a position corresponding to its projection on the imaging axis. As the spectrograph rotates, a sinusoidal pattern is traced out by the source, with an amplitude depending on its radial distance from the axis and the phase depending on the azimuth of the source with respect to that axis. With a superposition of sources a superposition of sinusoids appears, from which the image and the spectra can be recovered through image reconstruction.

This technique was demonstrated in a rocket flight with the goal of obtaining a spectral image of the constellation Scorpio. The attitude control system of the rocket was used to rotate the spectrograph around its axis while pointed at the constellation with a spin rate of 0.6 rpm. This provided three complete rotations of the $10° \times 10°$ field of view.

11.3.5 Ground-Based Instruments

Similar principles have been applied to ground-based instruments, an example of which is described by Sivjee and Shen [1997]. The instrument has a modified Czerny–Turner configuration, with a 0.5 m focal length spherical mirror collimator, a 110 mm square grating with a 50 mm long Fastie-type curved entrance slit. The cooled CCD has a dark current of less than 6 electrons s^{-1} per pixel, and has 1024 × 1024 pixels, each 24 μm square. The spectral range was 700–960 nm. The spectrum of the aurora associated with the magnetic cloud event of October 18–20, 1995 is shown in Figure 11.9. Note that the integrated emission rate is given in Rayleigh $Å^{-1}$. This unusual event, which is characterized by solar wind electrons of energies near 500 eV, causes excitation near 160 km which makes the molecular emissions much weaker than in the night-time aurora, though still with significant emission, and creates strong excitation of atomic lines. The atomic oxygen line at 844.6 nm is the strongest feature in the spectrum.

Some truly remarkable airglow spectra have been presented by Slanger *et al.* [1997, 2000] and Slanger and Osterbrock [2000], using the Keck astronomical facility on Mauna Kea, as described by Osterbrock *et al.* [2000]. In Section 3.6.2 it was noted that the performance of a spectral imager cannot be improved by the use of a large telescope in

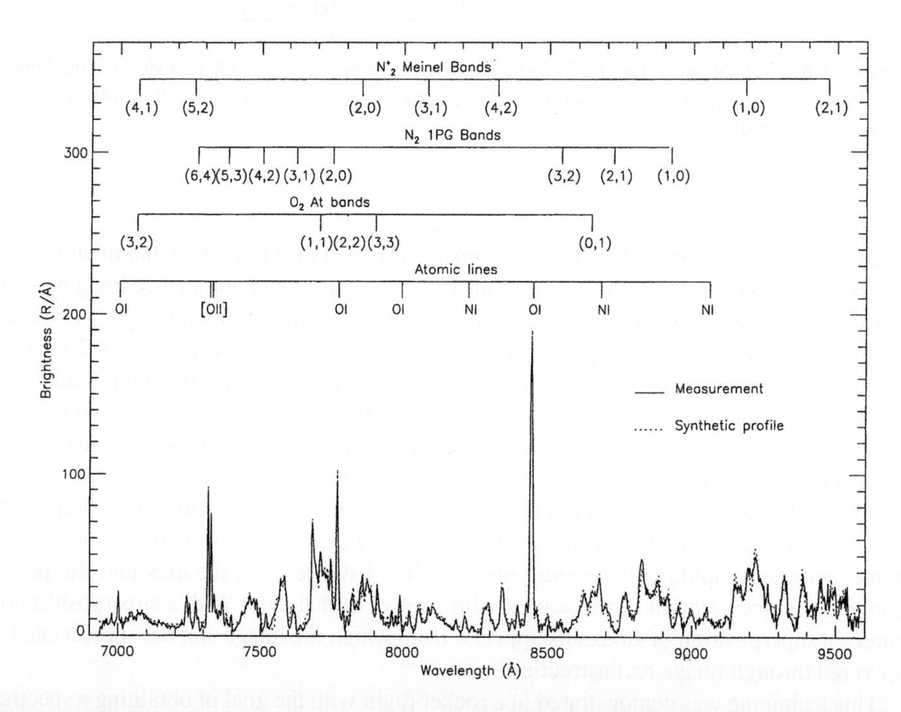

Figure 11.9. Near-infrared spectrum of the aurora associated with the magnetic cloud and corotating stream event of October 18–20, 1995, taken at Sondrestromfjord, Greenland. After Sivjee and Shen [1997].

front, as the responsivity is inherent to the instrument itself; it was concluded that astronomical telescopes are of no benefit to the aeronomer. But that does not prevent a large grating spectrograph from being used with an astronomical telescope, as is done with the Keck telescope on Mauna Kea. Normal astronomical observations yield sky spectra as a product, which may be used for atmospheric investigations. The author is indebted to Dr. Tom Slanger of SRI (Stanford Research International) for providing the following material.

Sky spectra are a by-product of astronomical investigations with large ground-based telescopes. Since the nightglow is always present, it is essential that astronomers record the nightglow spectrum separately, then subtract it away from the spectrum of their object. This is particularly important when the object being viewed is faint. The principal nightglow emission encountered in the visible spectral region is the OH spectrum, but with construction of more powerful and sensitive spectrometers, it is becoming evident that there are many more features capable of contributing to spectral contamination.

The sky spectra are produced simultaneously with the object spectrum, by using an instrumental slit that is long enough so that light can be collected from above and below the stellar image, with no reverse contamination. This light is the sky spectrum, and typically contains no light from other stars, but is limited to the terrestrial nightglow, geocoronal radiation, zodiacal light, and galactic light. The spectrometer is the essential component for nightglow studies; there is little gained by using a 10-metre telescope to view a diffuse source. To date, most of the investigations carried out at SRI have utilized spectra from HIRES (High Resolution Echelle Spectrometer) as described by Vogt [1994]. HIRES operates in the visible and near-IR spectral region (350–1000 nm), and is a relatively conventional in-plane echelle spectrometer with grating cross-dispersion, and is capable of achieving resolutions of $\mathcal{R} = 30\,000$–$80\,000$. It features a backside-illuminated Tektronix 2048×2048 CCD detector, and captures a spectral span of 120–250 nm per exposure. Two cross-dispersers are used – one optimized for the blue, the other for the red. Approximately 10 nm of the spectrum appears in each echelle order. The responsivity of the system is very high, and the camera features $f/1.0$ optics.

Examination of the HIRES spectra has revealed many new nightglow features, over the last several years. These include detection of isotopic emission lines in the O_2 A-band [Slanger *et al.*, 1997], detection of the potassium D1 line [Slanger and Osterbrock, 2000], observation of emission from the first sixteen vibrational levels of the $O_2(b^1\Sigma_g^+)$ state [Slanger *et al.*, 2000], O_2 Chamberlain band emission in the blue spectral region, isotopic OH emission as well as positive detection of the $v = 10$ level [Osterbrock *et al.*, 2000], detection of O-atom Rydberg transitions (to $n = 11$), and detection of $O_2(b^1\Sigma_g^+)$ ionospheric emission.

Figure 11.10 shows a HIRES spectrum in the 708–720 nm region, and contains data from 100 hours of accumulation. The essential features are the two bands of the O_2 Atmospheric Band system, 4–3 and 10–8, with high-J lines of the 3–2 band also appearing. In addition, lines of the OH 7–2 Meinel band are prominent. The lower spectrum is a simulation, which includes the O_2 bands at a temperature of 200 K, with a band intensity ratio of 0.6 : 0.2 : 1.0 for the 3–2, 10–8, and 4–3 bands. The OH rotational lines are not simulated because above $J = 4.5$ the OH rotational levels are not in thermal equilibrium with the $\sim$180 K kinetic temperature of the 87 km emitting region, and thus cannot be simply modelled.

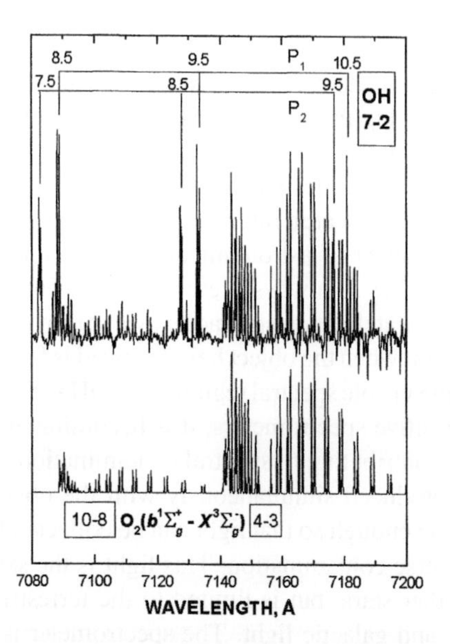

Figure 11.10. Averaged spectrum from Keck/HIRES, showing the 10–8 and 4–3 O_2 Atm. bands. Line pairs from the 3–2 band, with its origin at 705.9 nm, also appear. Note H_2O absorption lines at 715–720 nm, superimposed on the continuum (nightglow+zodiacal+galactic radiation) background. The spectral resolution is 0.02 nm. The lower panel shows the 200 K DIATOM™ simulation of O_2 10–8 and 4–3 Atm. bands. The W.M. Keck Observatory is operated by the California Institute of Technology and the University of California.

11.4 SELECTED ATMOSPHERIC MISSIONS ACCOMPLISHED

11.4.1 Total Ozone Mapping Spectrometer (TOMS)

The TOMS instrument was designed to measure the total column amount of ozone in the atmosphere and it has monitored this quantity almost continuously since November, 1978. Most of the historical images of the Antarctic ozone hole that one sees have been taken with this instrument, which illustrates many of the concepts that have been presented in earlier chapters. TOMS consists of a fixed (i.e. non-scanning) diffraction grating spectrometer with six exit slits located in the near-UV region of the spectrum where absorption bands are present. A rotating chopper exposes the six exit slits, one at a time, to a single detector, but not in the order of monotonically increasing or decreasing wavelength in order to eliminate unwanted trends. The instrument field of view was 3° × 3°, but it viewed into a scanning mirror, scanning perpendicular to the orbital plane in 3° steps from 51° on one side of the spacecraft to 51° on the other, for a total of 35 samples. The scan was in one direction, with a rapid fly-back, completing one scan every eight seconds. Consecutive scans overlapped so that the mapping was continuous. The solar irradiance was measured by viewing a ground aluminium diffuser plate.

The analysis is done through forward calculations in which an ozone amount is assumed and the radiances for the different channels calculated. The difference between predicted and observed values is used to adjust the assumed amount, and the calculation proceeds iteratively. The calculations involve detailed radiative transfer calculations including the geometry, surface pressure, surface reflectance and latitude. The challenge is that, in detecting ozone trends, accurate measurements over a long period of time are required. TOMS has a random error of 2% and the uncertainty in the drift over 14 years is estimated to be 1.5%.

TOMS on Nimbus-7 and Meteor-3 provided nearly continuous measurements from November 1978 to December 1994. After an eighteen-month gap, another TOMS instrument was flown on ADEOS, providing data until June, 1997. Earth Probe TOMS has continued to provide data since July, 1996.

11.4.2 Stratospheric Aerosol and Gas Experiment (SAGE)

The SAGE II instrument was launched on the Earth Radiation Budget Satellite (ERBS) in 1984. This is a holographic diffraction grating instrument which operates in the solar occultation mode, as described in the previous chapter for the ACE instrument. It has seven silicon diodes, measuring all seven channels simultaneously, at 385, 448, 453, 525, 600, 940 and 1020 nm. Solar radiation is reflected off a pitch mirror into a telescope that brings the light to the spectrograph, defining a vertical resolution at the limb tangent point of about 0.5 km. Prior to each sunrise or sunset encounter, the instrument is rotated to the azimuth of the predicted solar position. When the sun appears, the instrument locks onto it and the data are acquired.

SAGE III incorporates a linear CCD array, providing continuous coverage between 280 and 1030 nm at 1 nm resolution, with an additional photodiode providing measurements at 1550 nm which is specifically for aerosol detection. Three instruments are planned to be flown. The first was launched on the Russian METEOR-3M satellite in late 2001, into a sun synchronous orbit, i.e., an orbit with an inclination of 98° for which the orbit plane is fixed with respect to the sun. With this type of orbit the satellite crosses the equator at the same local time throughout the entire mission. This is advantageous when it is desired to present the spacecraft and its instrument always at the same angle with respect to the sun. In this case, it causes the locations of solar occultations to be restricted to high latitudes as shown in Figure 11.11; these locations correspond to the wavy tracks in the polar regions.

The second instrument is scheduled to fly on the International Space Station, beginning in 2004. With a 51° inclination, the orbit moves through all local times in about one month, so that the occultation tracks move rapidly north and south as also shown in Figure 11.11.

11.4.3 Optical Spectrograph and InfraRed Imaging System (OSIRIS)

The OSIRIS (Optical Spectrograph and InfraRed Imaging System) instrument is one of two on the Odin satellite, launched on February 20, 2001 as a joint experiment by Sweden, Canada, France and Finland. The other instrument is a sub-millimetre wave

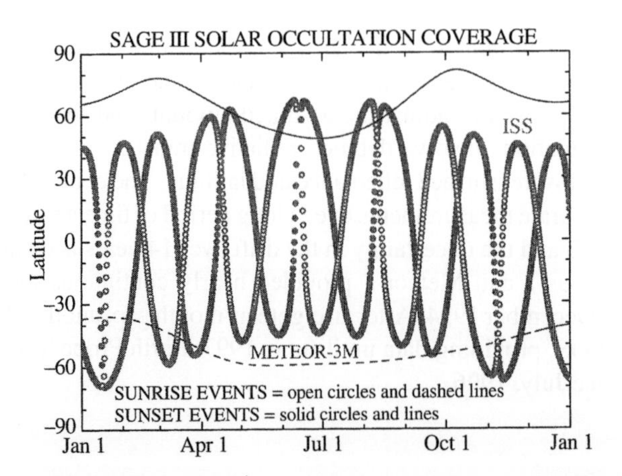

Figure 11.11. Illustrating the SAGE-III coverage obtained with the METEOR-3M sun synchronous orbit as the undulating lines at high latitude, and with the International Space Station at an inclination of 51°, with the oscillating tracks centred roughly on the equator. Courtesy of NASA Langley Research Center, SAGE III Project.

radiometer which uses radio techniques but bridges into the optical domain. It is used both for atmospheric and astronomical measurements. OSIRIS has two components, one a UV/visible limb imaging spectrograph and the other a linear array imager that acquires vertical profiles of the $O_2(^1\Delta)$ emission at 1.27 μm and uses tomographic analysis to derive two-dimensional distributions as described in Chapter 1. OSIRIS is the Canadian contribution to the Odin mission, and is led by Edward Llewellyn of the University of Saskatchewan.

The spectrograph is shown in Figure 11.12, with the infrared imager on the left and the spectrograph on the right. Because the signal in the UV is about 10^{-3} that of the visible region, one of the major design requirements was to minimize the scattering of visible light within the instrument into the UV region. A minimum number of surfaces is employed in an off-axis design, avoiding any central obstruction that could enhance internal scattering, and an aspherized grating avoids the need for a corrector plate. A right-angle prism combined with a field-flattening lens ensures that there is no scatter from the diagonal internal reflection that directs the spectrum into the external CCD package. The slit is imaged horizontally on the limb to enhance the signal, and the limb is scanned through the motion of the spacecraft, for both instruments. A description of the instrument is given by Llewellyn *et al.* [1997].

The sun is the light source for the observations, in which the observed spectrum is determined by Rayleigh and aerosol scattering and by absorption. The column concentrations of the absorbing species are recovered by comparing the spectra measured at different tangent altitudes. This technique, pioneered by Noxon [1975], is called DOAS (Differential Optical Absorption Spectroscopy). It was used for ground-based observations by Solomon *et al.* [1987] but is only now being exploited for satellite observations.

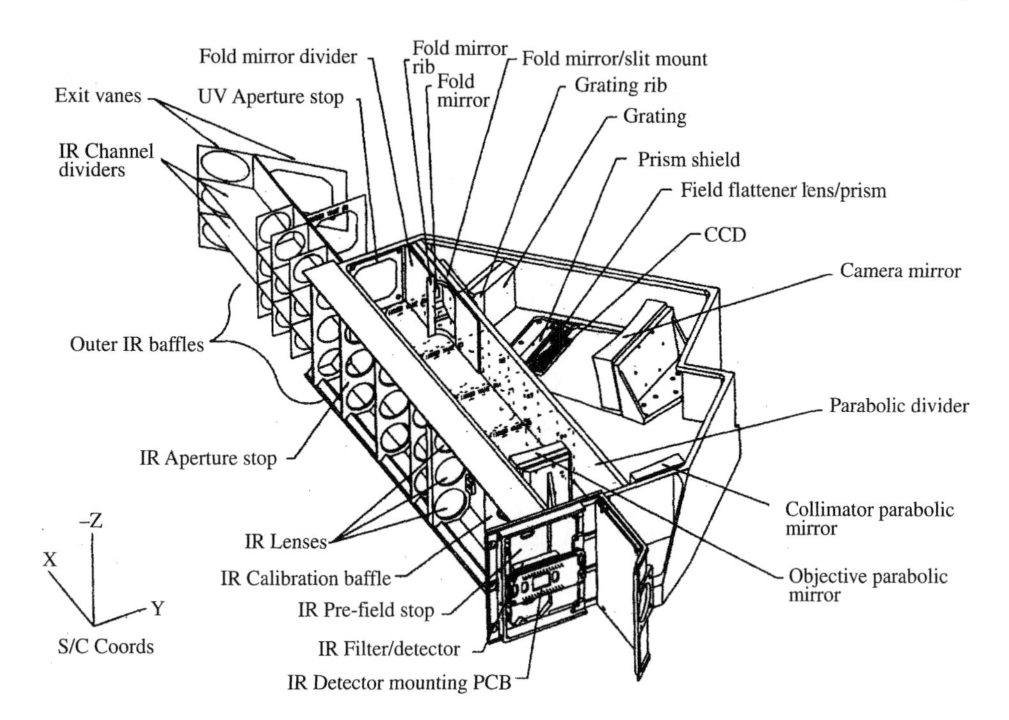

Figure 11.12. The OSIRIS instrument on the Odin spacecraft. The UV/visible spectrograph is on the right, while the infrared imager (note the baffles with the series of holes) is on the left. After Llewellyn *et al.* [1997].

11.4.4 CRyogenic Infrared Spectrometers and Telescopes for the Atmosphere (CRISTA)

The CRISTA experiment was mentioned in Chapter 1 as having achieved better horizontal resolution than similar instruments, through the use of multiple fields of view. It is also unusual in that it is the only infrared instrument mentioned in this chapter, and it is briefly described here because of its major contributions to atmospheric measurement. The increased horizontal resolution partially fills in the gaps between orbits by employing three viewing telescopes separated by horizontal angles of 18°. Each telescope, of 120 mm diameter aperture, feeds a grating spectrometer covering the range 4–14 μm, while the centre telescope accommodates a second spectrometer covering the long-wave infrared, from 14 to 71 μm. To cool the optics adequately, they are placed in a cryostat, and the assembly is mounted in a vacuum tank. The overall configuration is shown in Figure 11.13, where the angled telescopes can be seen, along with the accompanying spectrometers, all fitted neatly into the liquid helium cryostats, and into the vacuum chamber. While compact, the overall size is relatively large, 1.38 m $\times$ 3.0 m long, with a mass of 1350 kg. Its highly successful

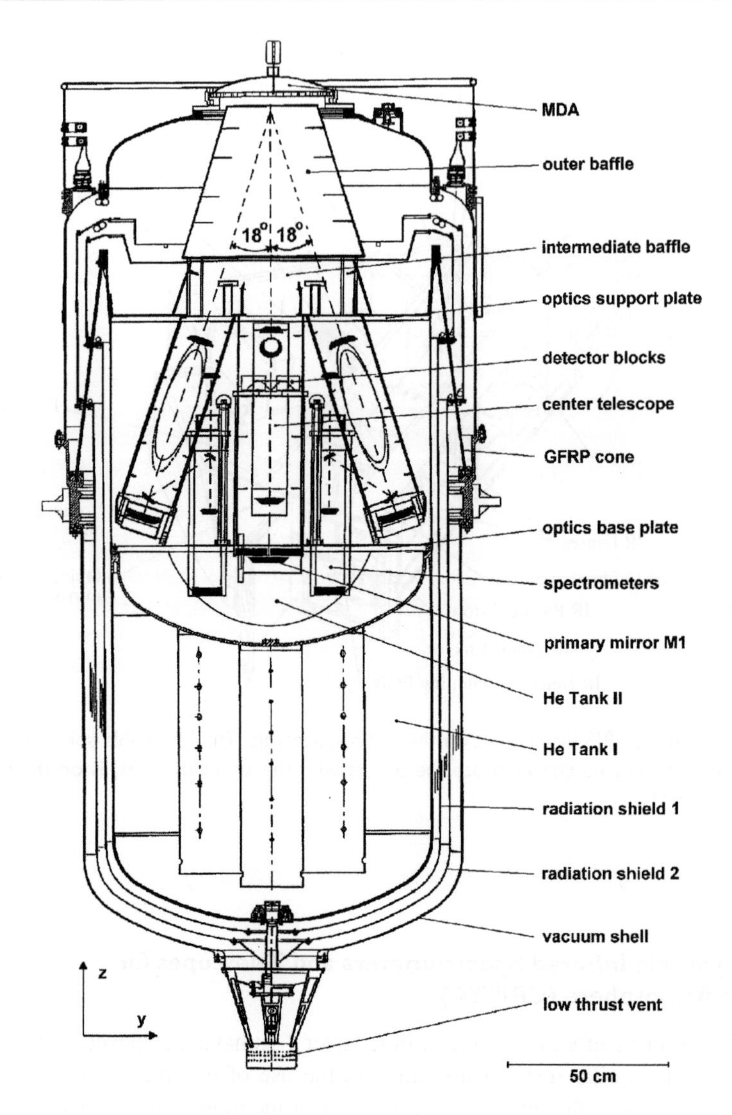

Figure 11.13. The CRISTA instrument comprising three angled telescopes with independent spectrometers, housed in a cryostat within a vacuum chamber. After Offermann *et al.* [1999].

flights were on the space shuttle; the issue of the Journal of Geophysical Research containing the CRISTA description [Offermann *et al.*, 1999] may be consulted for a sequence of papers on the results.

Each spectrometer has up to eight exit slits, each with its own detector. The gratings are scanned by rotation, but because of the multiple slits the scan need be only over 1 μm, which can be executed in 1.2 s. Altitude points are scanned sequentially. The 4–14 μm spectrometers have an étendue of 1.72×10^{-4} cm^2 sr, which is substantial. The corresponding resolving power of 300–600 is adequate for the measurement goals, and contributes to the relatively large étendue.

It may be concluded that imaging spectrographs make their dominant contribution in the visible region, where CCDs are effective, allowing the use of compact instruments covering a wide spectral range. The Michelson interferometers as described in Chapter 10 are more advantageous for the infrared region, so that is where they are applied. The CRISTA experiment demonstrates that excellent performance can be achieved with a diffraction grating instrument in the infrared region with a large instrument.

11.5 FUTURE ATMOSPHERIC MISSIONS USING GRATING SPECTROGRAPHS

11.5.1 GOMOS and SCIAMACHY on Envisat

Like SAGE, GOMOS (Global Ozone Monitoring by Occultation of Stars) employs the occultation method, but is intended to overcome the limitation of two occultations per orbit by using stars rather than the sun. This requires a rather large instrument, with a telescope of 30 cm × 20 cm aperture. In addition the spectral coverage is very broad, covering spectral bands from 250 to 675 nm, 756 to 773 nm and 926 to 952 nm. This is done by using two spectrographs, a UVVIS module covering the first two of the three spectral ranges, and an NIR module which covers the third range. The spectral resolution varies from 0.2 nm in the NIR to 1.2 nm in the UVVIS.

In order to achieve some 600 profiles (occultations) per day it is necessary to be able to make measurements on very faint stars, down to magnitude 4 or 5 (see Problem 11.4 for a definition of stellar magnitudes). A design driver was the need to achieve adequate responsivity in the UV region. In order to use all the signal from the star, a slit-less design is used as described earlier. Each spectral line is an image of the star. This in turn requires a very accurate pointing system in order that the star should not move during the exposure, and thus degrade the spectral resolution. The instrument has a mass of 175 kg, and was launched on the Envisat mission on March 1, 2002, so is no longer a "future" mission.

A single telescope feeds all the modules. Light is first extracted for the star-trackers; only one is required and the second is a redundant back-up. Then the light is split off for the UVVIS and NIR modules. In addition there are two photometers. The detector is a thinned, back-side illuminated frame transfer CCD, specially developed for the mission, with a wave length coverage of 250–950 nm, an unusually large range. The quantum efficiency is 20% at 250 nm, 60% near 600 nm and 20% at 950 nm.

SCIAMACHY (SCanning Imaging Absorption spectroMeter for Atmospheric CHartographY) is also flying on the Envisat spacecraft. SCIAMACHY [Bovensmann *et al.*, 1999] is an advanced version of GOME (Global Ozone Monitoring Experiment [Richter *et al.*, 1998]), which is a nadir-viewing spectrograph operating in the spectral range 240–800 nm; launched on the ERS-2 satellite in 1995. SCIAMACHY has a much more extended spectral range, covering from 240 to 2380 nm, and is much more versatile in its operating modes. It views in the nadir with a cross-track width of 960 km, and a resolution element of 26 × 15 km;

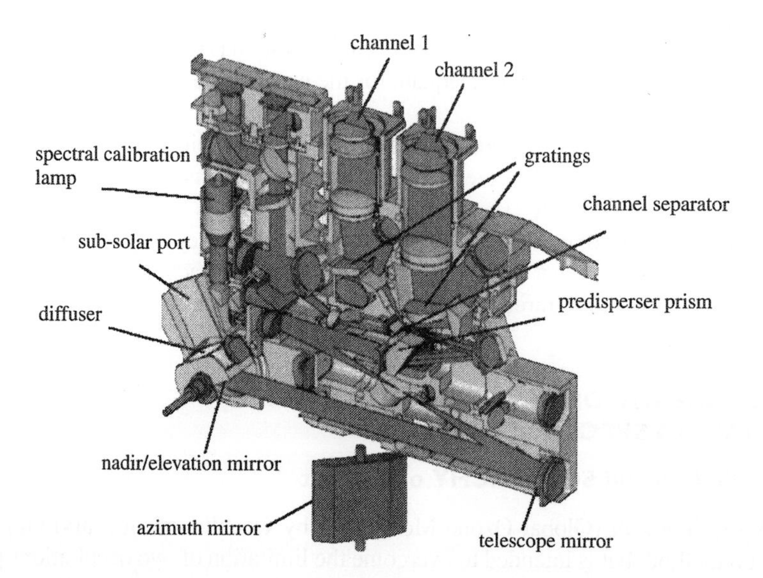

Figure 11.14. Outline drawing of the SCIAMACHY instrument on Envisat. Courtesy of ESA.

or at the limb with a vertical resolution of 2.6 km or in occultation mode, using either the sun or the moon. When the nadir and limb modes are used together, the same volume of atmosphere is viewed, with a time delay of about 7 min. This combination of views allows the determination of column amounts of trace gases in the troposphere. However, this level of versatility and responsivity does require a large instrument, of 198 km mass, and 122 W power. A three-dimensional drawing of the instrument is shown in Figure 11.14.

11.5.2 Ozone Dynamics Ultraviolet Spectrometer (ODUS)

ODUS is, like TOMS, a nadir viewing spectrometer that measures ozone concentrations from the spectrum of back-scattered sunlight from the atmosphere. It is planned for launch on the Japanese GCOM-A1 mission in about 2007, as described by Suzuki *et al.* [1999]. It uses an Ebert–Fastie type of spectrograph design covering the spectral region 306–420 nm with 0.5 nm resolution [Suzuki *et al.*, 1999]. This design, with a plane grating, was chosen in order to obtain the responsivity needed for a horizontal resolution of 20 km; large concave gratings could not be used. A drawing of the instrument is shown in Figure 11.15. It uses a one-dimensional silicon photodiode array detector with 228 elements to obtain the spectrum, and uses a cross-track scan for the mapping. Imaging spectrographs were considered, but it was found that it would be difficult to characterize the instrumental function with sufficient accuracy to achieve the desired precision in ozone amount.

ODUS is designed to map the global distribution of ozone in one day and determine the total ozone amount in a vertical column to an accuracy of 2%, after calibration. The spatial

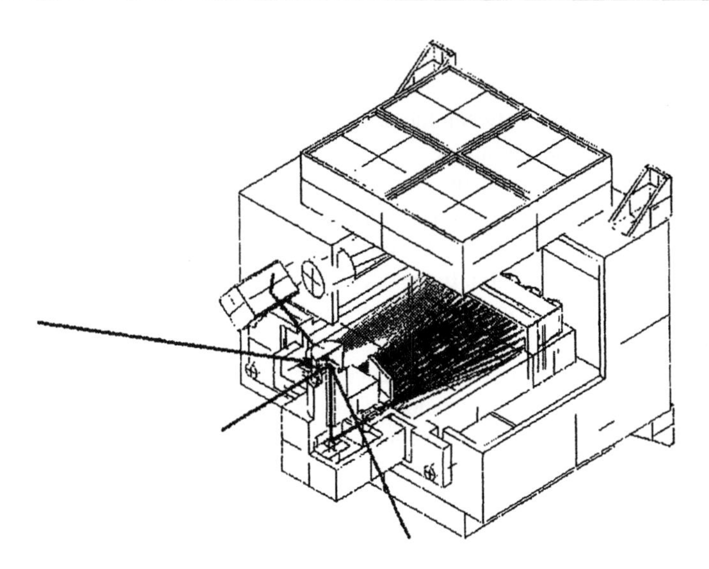

Figure 11.15. Drawing of the Ozone Dynamics Ultraviolet Spectrometer (ODUS), which uses nadir viewing to determine ozone concentrations with a horizontal resolution of 20 km. It will be flown on the NASDA GCOM-A1 mission in 2007. After Suzuki *et al.* [1999].

resolution of 20 km will allow the study of ozone dynamics. The secondary objectives include an observation of volcanic sulphur dioxide, aerosol amount, surface UV-B and surface albedo. It is also expected to detect SO_2 and NO_2 over polluted urban areas, and cloud-top heights from the Ring effect.

The companion instruments to ODUS are currently planned to be SOFIS, a scanning Michelson interferometer operating in solar occultation, and SWIFT, the stratospheric wind interferometer described in Chapter 10.

11.5.3 Measurements of Aerosol Extinction in the Stratosphere and Troposphere Retrieved by Occultation (MAESTRO)

MAESTRO is a remarkably small (5 kg) dual optical spectrograph covering the wavelength range 280–1030 nm in two overlapping units; covering the ranges 280–550 nm and 500–1030 nm. One of the advantages of using two spectrographs is better stray light performance. The detectors are linear EG&G Reticon photodiode arrays with 1024 elements. The design is based on a simple concave holographic grating with no moving parts. The entrance slit is held horizontal to the horizon during sunrise and sunset by controlling the spacecraft roll with a star-tracker and a momentum wheel on the satellite bus. MAESTRO will be flown on the Canadian SCISAT-1 mission with the ACE-FTS instrument described in Chapter 10. The investigation is led by Tom McElroy of the Meterological Service of Canada. Both instruments share a common sun-tracker. MAESTRO will be able to make some near-nadir solar back-scatter measurements like the GOME instrument on the European ERS-2 satellite. MAESTRO will provide data in both the back-scatter and occultation mode and may help to reconcile differences between the data sets collected by distinct instruments using these different observational techniques.

The field of view is $0.1° \times 5°$ in the nadir viewing mode; this corresponds to 1.5×74 km on the ground, with the 74 km direction perpendicular to the spacecraft track. An integration period will be used that smears the 1.5 km direction out to 74 km, making the measurement footprint square, of 74 km on a side. In the occultation mode, the field of view is $0.02° \times 1°$, the additional magnification being provided by the ACE-FTS instrument, yielding a vertical resolution of 1 km.

The most important aspects of the science directly related to having MAESTRO on board the ACE payload, are the extension of the range of measurements to include polar sunrise chemistry (BrO), the measurement of BrO and OClO in the stratosphere, and the ability to detect very small optical depths by virtue of the higher extinction which aerosol and PSC particles have at short wavelengths. These measurements relate to the understanding of the global ozone budget, climate change and tropospheric chemistry. In addition, the acquisition of ozone mapping data in the back-scatter mode will add significantly to the global data set used for mapping ozone and determining long-term ozone change.

11.6 SPATIAL HETERODYNE SPECTROSCOPY (SHS)

It seems appropriate to end this volume with an instrument that integrates features of several different instruments that have been discussed in this and earlier chapters. The SHS instrument employs diffraction gratings, and properly belongs in this chapter. However, it behaves like a Fourier transform spectrometer so relates to Chapter 6, but it also has a configuration like the SISAM, which was described as a modulator in Chapter 7. An elegant description has been given by John Harlander *et al.* [1992]. Harlander was then at the University of Wisconsin-Madison, which is one of the centres of interferometric spectroscopy activity mentioned earlier.

A schematic configuration is shown in Figure 11.16. Light from aperture A_1 is collimated to illuminate a beam-splitter, B.S., as in a Michelson interferometer, except that the mirrors are replaced by diffraction gratings G_1 and G_2, as in the SISAM. The returning rays are united at the beam-splitter and a telescope forms an image of the superposed gratings on an imaging detector. Harlander *et al.* [1992] show that, for wavenumber σ_o (the Littrow wavenumber), for which the returned wavefronts are parallel, the image is uniformly illuminated, as shown in Figure 11.17. However, for a different wavenumber σ_1, for which the tilted wavefronts make an OPD of $\lambda/2$ at the edges, there is a modulation of one wave across the detector, as shown in Figure 11.17, and for an OPD of λ, two waves are formed across the image. For a complex spectrum there is a superposition of spatial frequencies from which the spectra can be obtained through Fourier transformation. There is a highly significant difference from the scanning MI, however, as the whole spectrum is not transformed, only a region around σ_o. This is why it is called a heterodyne system. This system can be very effective where a limited spectral range is required and it is capable of moderately high

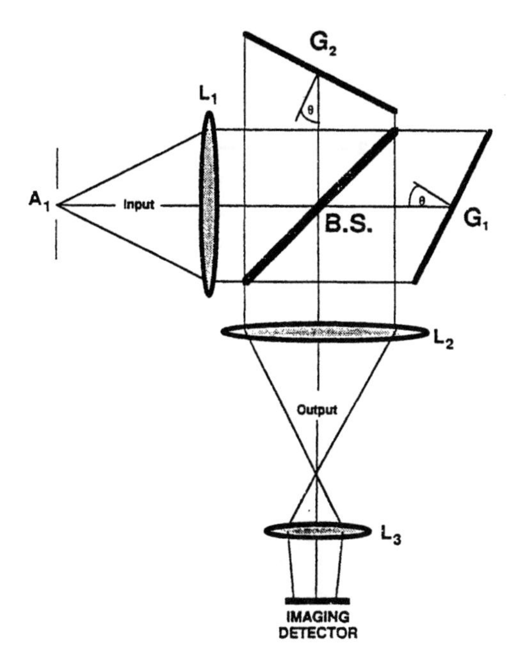

Figure 11.16. Schematic configuration of the Spatial Heterodyne Spectrograph (SHS), which is very similar to the Michelson interferometer, with the mirrors replaced by gratings. After Harlander *et al.* [1992].

resolution. The SHS instrument has a superiority of 2π, the same as the MI or FPS, but may be more effective in the visible region of the spectrum than the scanning MI, and provides a greater spectral range than the FPS without having to be concerned about overlapping orders.

The device becomes highly advantageous at shorter wavelengths, as it can be configured with purely reflective elements as shown by Harlander *et al.* [1992], completely avoiding the need for transmitting materials, including the beam-splitter. Their reflective configuration is shown in Figure 11.18. The beam-splitter is now a diffraction grating, which is possible because the diffraction can take place on both sides of the grating normal. It is usual to make the grating grooves asymmetrical, like a sawtooth, in order to enhance responsivity by reflecting light in the favoured direction. This is called *blazing* the grating. When one purchases a diffraction grating the blaze angle is usually specified. Here the requirement is different: to diffract light efficiently in two directions. The two gratings of the previous design are now replaced by normal MI mirrors, but it does function as an SHS, as before. Because the beams are not superimposed, the aperture is divided into two parts, one of which is used for the source illumination, and the other for the detector. With this arrangement only one side of the interferogram is recorded, but other possibilities exist.

Chakrabarti *et al.* [1999] point out that, since the SHS has a spatial direction and a second dimension that can also be used for imaging, it can be used with the SPINR technique. They

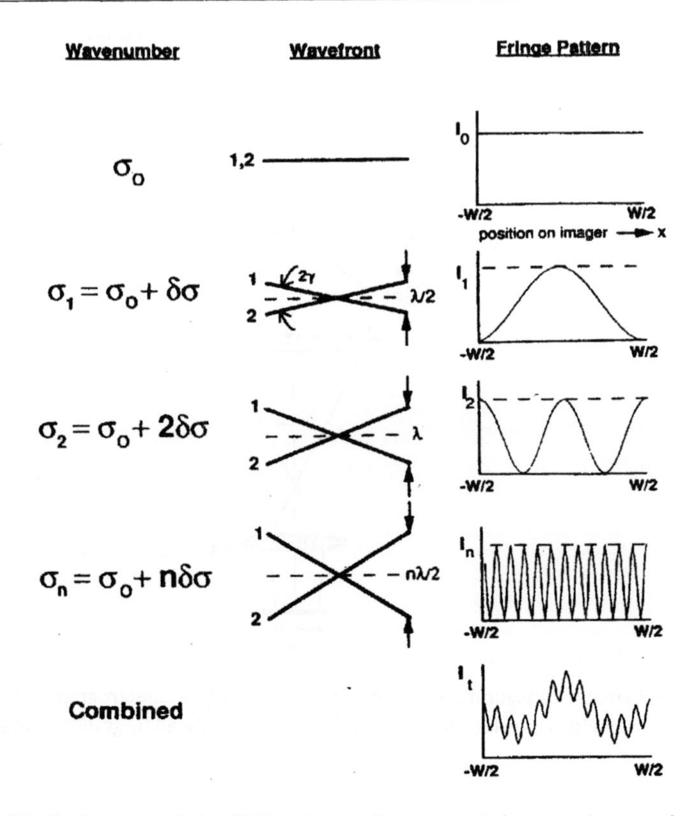

Figure 11.17. Explanation of the SHS concept. For wavenumber σ_0, the wavefronts from the two gratings are parallel, as shown at the top, and the radiance across the image is uniform. For wavenumber σ_1, the wavefronts are tilted as shown, with an OPD of $\lambda/2$ at the edges and the radiance across the image is that of one wave. The combined image pattern for several waves is shown at the bottom. After Harlander *et al.* [1992].

propose a configuration [Chakrabarti *et al.*, 1994] which is different from the above, called a Self-Compensating All-Reflection Interferometric spectrometer (SCARI). In this design, shown in Figure 11.19, the two beams, created in the beam-splitter as before, travel in opposite directions around a common path. This results in a highly stable system, which is described as self-compensating.

The SHS can also be field widened as is described by Harlander *et al.* [1994]. This is done by placing prisms over the gratings in Figure 11.16, such that the virtual images of the gratings are normal to the axis, rather than tilted. The first SHS atmospheric mission is to be a shuttle flight called SHIMMER (Spatial Heterodyne IMager for MEsospheric Radicals), a flight aimed at the measurement of ground state OH through resonance fluorescence observed at the Earth's limb. The observations will be in a spectral width of 2.3 nm centred at 308.9 nm, with a resolution of 5.8 pm. The experiment is described by Englert *et al.* [2000]. The SHS seems destined to make major contributions to atmospheric science.

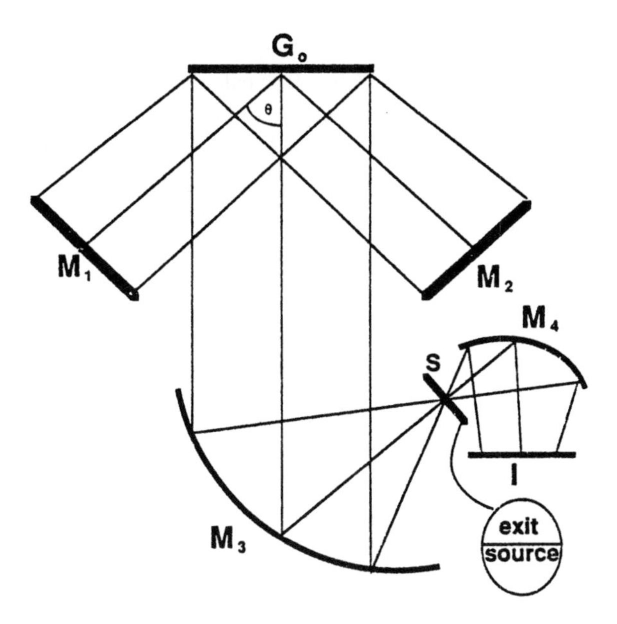

Figure 11.18. An all-reflective SHS instrument. After Harlander *et al.* [1992].

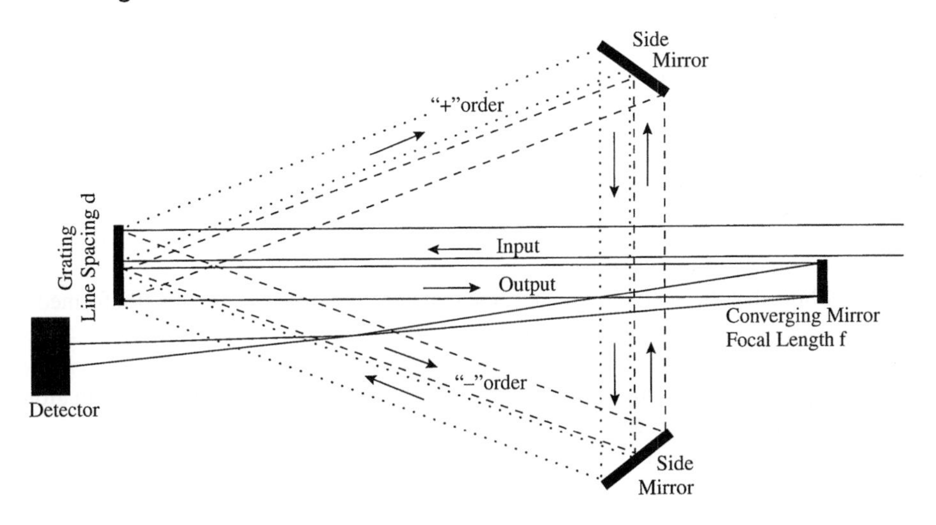

Figure 11.19. Self-Compensating All Reflection Interferometric spectrometer (SCARI) which is compensated through the common paths of the two beams. After Chakrabarti *et al.* [1999].

11.7 PROBLEMS

11.1. Compare an FPS instrument with a DGS instrument, both having 0.1 nm resolution at 500 nm. Both are spatially scanned. For the FPS, use a superiority of 2π, and for the DGS use $S_G = 0.12$.

(a) Thus, using the same $\mathcal{R}$ for both, calculate Ω for each instrument. But the DGS has a slit with an angular length of $b = 0.2$; calculate its width.
(b) Suppose that 30 spectral elements are to be observed by each instrument with spatial scanning. Calculate the angular radius of the input cone of light accepted by the FPS.

11.2. Assess the statement by Broadfoot and Sandel [1992] that the Voyager UVS instrument has a sensitivity of 0.46 R s. The definition of sensitivity may be taken as the emission rate at which the signal-to-noise ratio is equal to unity.

11.3. Compare the TOMS and SAGE instruments which are similar in construction, but for which TOMS is nadir viewing and SAGE is limb viewing in the occultation mode.

11.4. Discuss the requirements on GOMOS in measuring absorption spectra from stars, compared with a solar occultation instrument such as SAGE. The subject of stellar sources was briefly addressed in Section 3.6.2. For occultation, the sun is larger than necessary, with an angular diameter of about $30'$. For vertical resolutions at the limb of 1–2 km, a field of view of about $1'$ is needed. A star similar to the sun has the same radiance but, because of the much greater distance is not a point source, so it does not fill the field of view. Thus astronomers describe the "luminosity" of a star in terms of its measured irradiance at the instrument, and it is specified using a logarithmic scale in *magnitudes*. Magnitude 0 is essentially the brightest source, magnitude 1 is lower by a factor of 2.5, and magnitude 2 is lower by $(2.5)^2$. GOMOS is intended to measure 4th or 5th magnitude stars. On this scale our sun has a magnitude of -26.7.

11.5. Compare the characteristics of ODUS, designed for nadir viewing, with that of MAESTRO, designed for solar occultation measurements. Both are DGS instruments, but differ in configuration and in size.

11.6. As shown in Figure 11.17, the output of the SHS consists of wavefronts tilted by angle ε in opposite directions.
(a) Show that the spatial frequency, the number of fringes per unit length, formed by the intersecting wavefronts is $2\sigma \tan \varepsilon$.
(b) The diffraction grating equation as appplied here is: $\sin\theta + \sin(\theta - \varepsilon) = m/d\sigma$. Noting that $\varepsilon = 0$ corresponds to $\sigma = \sigma_o$, and using small angle approximations, show that the fringe frequency f can be written as: $f = 4(\sigma - \sigma_o)\tan\theta$.
(c) For a wavenumber of 20 000 cm^{-1}, and a grating angle of 45°, calculate the fringe frequency for a wavenumber of 20 100 cm^{-1}.

12

POSTSCRIPT

It was declared at the outset that an objective of this work was to help the reader in finding the most effective spectral imaging instrument for a specific atmospheric measurement. At the end it is clear that doing so is not all that simple. The concept of superiority is very helpful, but does not address all of the issues, for example, the number of spectral elements to be acquired. The introduction of array detectors has significantly changed the dominant influence of superiority. However, if one assumes that the same array detectors are available for all instruments, then the choices are not fundamentally different. For a satellite instrument, the matter of mass, power, volume and telemetry rate may be dominant factors, while for a ground-based instrument it may be the ability to operate in isolation from human assistance. These practical aspects are extremely important.

The scope and ingenuity of interferometric spectroscopists over the last half-century spawned many ideas and many instruments and some sense of that history has hopefully been conveyed here. In the end, the focus comes back to three basic and historical instruments, all invented more than a century ago. The diffraction grating spectrometer, when employed with an array detector, is a very effective instrument for applications where a broad spectral range is to be covered at modest resolution for a source of high radiance, and a simple compact instrument is required. These conditions apply particularly in the visible region. If a broad spectral range is required in the infrared, then the scanning Michelson interferometer is the logical choice; it is available in a wide range of resolutions and sizes. The Doppler Michelson interferometer provides excellent lineshift (wind) measurements where the target spectral line can be suitably isolated, and the target is of low radiance. The Fabry–Perot spectrometer is more effective where the spectrum is complex, linewidths or shapes are required as well, and the source is of higher radiance. In some cases a hybrid instrument may provide the optimum performance. After a century, we have simply returned to these historical giants in spectroscopy: Rowland, Michelson, Fabry and Perot. However, much has been learned from all those who contributed to the field during the past century, particularly the last half-century.

It is possible that a better performance index than superiority can be found, and Jacquinot [1958] did suggest one. He proposed a Factor of Merit, W, given by (using his nomenclature):

$$W = \frac{M\mathfrak{R}}{TB^{\alpha}} \tag{12.1}$$

where M is the number of spectral elements, T is the time of observation, B is the lowest utilizable radiance, and where $\alpha = 1$ for photon noise and 2 for thermal noise. This approach does take factors into account other than the superiority, but perhaps does not incorporate all the subtleties mentioned above. In any case, this approach has not been adopted.

What is done currently is to design a candidate system and calculate, through detailed simulation, the error in the quantity to be derived from the data product, including the data inversion process. If the result does not meet the requirements, then the instrument parameters are modified, and iterated to the final values. While this is clearly the most effective procedure, it is only iterative. It is hoped that the principles outlined in this volume will provide the user with a wise initial choice, so that the iteration will be along the right path. May this volume help you, the reader, in finding that path.

REFERENCES

Abel, P.G., P.J. Ellis, J.T. Houghton, G. Peckham, C.D. Rodgers, S.D. Smith and E.J. Williamson, 1970: The Selective Chopper Radiometer for Nimbus D, *Proc. R. Soc. London*, **A320**, 35–55.

Abreu, V.J. and W.R. Skinner, 1989: Inversion of Fabry–Perot CCD image: use in Doppler shift measurements, *Appl. Opt.*, **28**, 3382–3386.

Airy, G.B., 1877: *On the Undulatory Theory of Optics*, McMillan and Co., London. 159 pp.

Akasofu, S., 1963: The dynamical morphology of the aurora polaris, *J. Geophys. Res.*, **68**, 1667–1673.

Anger, C.D., T. Fancott, J. McNally and H.S. Kerr, 1973: ISIS-II scanning auroral photometer, *Appl. Opt.*, **12**, 1753–1766.

Anger, C.D. and J.S. Murphree, 1976: ISIS-2 Satellite Imagery and Auroral Morphology, *Magnetospheric Particles and Fields*, pp. 223–234, Ed. B.M. McCormac, D. Reidel Publ. Company, Dordrecht, Holland, 331 pp.

Anger, C.D. *et al.*, 1987: An ultraviolet auroral imager for the Viking spacecraft, *Geophys. Res. Lett.*, **14**, 387–390.

Aso T, Å. Steen, U. Brandstrom, B. Gustavsson, A. Urashima and M. Ejiri, 2000: ALIS, a state-of-the-art optical observation network for the exploration of polar atmospheric processes, *Adv. Space Res.*, **26** (6), 917–924.

Babcock, H.D., 1923: A study of the green auroral line by the interference method, *Astrophys. J.*, **57**, 209–221.

Bahsoun-Hamade, F., R.H. Wiens and G.G. Shepherd, 1989: The OI 844.6 nm Emission in Evening Twilight, *Geophys. Res. Lett.*, **16**, 1449–1452.

Bahsoun-Hamade, F., R.H. Wiens, A. Moise and G.G. Shepherd, 1994: Imaging Fabry–Perot spectrometer for twilight observations, *Appl. Opt.*, **33**, 1100–1107.

Baker, D., 1977: *Spectrometric Techniques, Vol. I*, Ed. G. Vanasse, Academic Press, New York, 355 pp.

Baker, D., A. Steed and A.T. Stair, 1981: Development of infrared interferometry for upper atmospheric emission studies, *Appl. Opt.*, **20**, 1734–1746.

Barringer, A.R., 1969: Optical correlator with optimized maximum and minimum correlation masks, US patent 3836254.

Batten, S. and D. Rees, 1990: Thermospheric winds in the auroral oval: Observations of small scale structures and rapid fluctuations by a Doppler imaging system, *Planet. Space Sci.*, **38**, 675–694.

Baumert, L.D., 1964: *Digital communications with space applications*, Ed. S.W. Golomb, Prentice-Hall, Englewood Cliffs, New Jersey, 210 pp.

Beer, R. and D. Marjaniemi, 1966: Wavefronts and construction tolerances for a cat's eye retroreflector, *Appl. Opt.*, **5**, 1191–1197.

Bens, A.R., L.L. Cogger and G.G. Shepherd, 1965: Upper atmospheric temperatures from Doppler line widths – III. Observation of the OI dayglow emission at 6300 Å, *Planet. Space Sci.*, **13**, 551–563.

Bernath, P., 2001: Atmospheric Chemistry Experiment (ACE): An overview, in *Spectroscopy from Space*, Ed. Jean Demaison, Kluwer Academic Publishers, Dordrecht, 356 pp.

Biondi, M.A., D.P. Sipler and M. Weinschenker, 1985: Multiple aperture exit plate for field-widening a Fabry–Perot interferometer, *Appl. Opt.*, **24**, 232–236.

Biondi, M.A., D.P. Sipler, M.E. Zipf and J.L Baumgardner, 1995: All-sky Doppler interferometer for thermospheric dynamics studies, *Appl. Opt.* **34**, 1646–1654.

Bird, J.C., F. Liang, B.H. Solheim and G.G. Shepherd, 1995: A polarizing Michelson interferometer for measuring thermospheric winds, *Meas. Sci. Technol.*, **6**, 1368–1378.

Blamont, J.E. and T.M. Donahue, 1961: The dayglow of the sodium D lines, *J. Geophys. Res.*, **66**, 1407–1423.

Blamont, J. and J.M. Luton, 1972: Geomagnetic effect on the neutral temperature of the F-region during the magnetic storm of September, 1969, *J. Geophys. Res.*, **77**, 3534–3556.

Bluth, G.J.S., S.D. Doiron, C.C. Schnetzler, 1992: Global tracking of the SO_2 clouds from the June, 1991 Mount-Pinatubo eruptions, *Geophys. Res. Lett.*, **19**, 151–154.

Bose, D.N., 1995: Photodetectors, in *Perspectives in Optoelectronics*, Ed. S.S. Jha, World Scientific Publishing Company Pte. Ltd., Singapore, 938 pp.

Bouchareine, P. and P. Connes, 1963: Interféromètre a champ compensé pour spectroscopie par transformation de Fourier, *J. Phys. Rad.*, **24**, 134–138.

Bovensmann, H., J.P. Burrows, M. Buchwitz, J. Frerick, S. Noël, V.V. Rozanov, K.V. Chance and A.P.H. Goede, 1999: SCIAMACHY: Mission objectives and measurement modes, *J. Atm. Sci.*, **56**, 127–150.

Boyle, W.S. and G.E. Smith, 1970: Charge coupled semiconductor devices, *Bell System Technical Journal*, American Telephone and Telegraph, New York, NY, Vol. 49, April.

Bracewell, R.N., 1986: The Fourier transform and its applications, McGraw-Hill, New York, 474 pp.

Bradley, L.C. and H. Khun, 1948: Spectrum of Helium-3, *Nature*, **162**, 412–413.

Brekke, A. and A. Egeland, 1994: *The Northern Lights*, Grøndahl Dreyer, Oslo, 168 pp.

Broadfoot, A.L., S.S. Clapp and F.E. Stuart, 1977a: Mariner 10 ultraviolet spectrometer: airglow experiment, *Space Sci. Instrum.*, **3**, 199–208.

Broadfoot, A.L., B.R. Sandel, D.E. Shemansky, S.K. Atreya, T.M. Donahue, H.W. Moos, J.L. Bertaux, J.E. Blamont, J.M. Ajello, D.F. Strobel, J.C. McConnell, A. Dalgarno, R. Goody, M.B. McElroy and Y.L. Yung, 1977b: Ultraviolet spectrometer experiment for the Voyager mission, *Space Sci. Revs.*, **21**, 183–205.

Broadfoot, A.L., B.R. Sandel, D.E. Shemansky, J.C. McConnell, G.R. Smith, J.B. Holberg, S.K. Atreya, T.M. Donahue, D.F. Strobel and J.L. Bertaux, 1981: Overview of the Voyager ultraviolet spectrometry results through Jupiter encounter, *J. Geophys. Res.*, **86**, 8259–8284.

Broadfoot, A.L. and B.R. Sandel, 1992: Application of the intensified CCD to airglow and auroral measurements, *Appl. Opt.*, **31**, 3097–3108.

Broadfoot, A.L., B.R. Sandel, D. Knecht, R. Viereck and E. Murad, 1992: Panchromatic spectrograph with supporting monochromatic imagers, *Appl. Opt.* **31**, 3083–3096.

Broten, N.W., T.H. Legg, J.L. Locke, C.W. McLeish, R.S. Richards, R.M. Chisholm, H.P. Gush, J.L. Yen, J.A. Galt, 1967: Long Baseline Interferometry: A new technique, *Science*, **156**, 1592–1593.

Burkert, P., F. Fergg and H. Fischer, 1983: A compact high resolution Michelson interferometer for passive atmospheric sounding (MIPAS), *IEEE Trans. Geosci. Remote Sensing* **GE-21**, 345.

Burnett, C.R. and E.B. Burnett, 1983: OH PEPSIOS, *Appl. Opt.*, **22**, 2887–2982.

Burns, G.J., 1906: The light of the sky, *J. Brit. Ast. Assoc.*, **16**, 308.

Burnside, R.G., F.A. Herraro, J.W. Meriwether Jr., and J.C.G. Walker, 1981: Optical observations of thermospheric dynamics at Arecibo, *J. Geophys. Res.*, **86**, 5532–5540.

Chabbal, R., 1953: Recherche des meilleures conditions d'utilisation d'un spectromètre photoélectrique Fabry–Perot, *J. Rech. CNRS*, **24**, 138–186.

Chabbal, R., 1958: Le Spectromètre Fabry–Perot Intègral, *J. Phys. Rad.*, **19**, 246–255.

Chahine, M.T., 1970: Inverse problems in radiative transfer: Determination of atmospheric parameters, *J. Atmos. Sci.*, **27**, 960–967.

Chakrabarti, S., D.M. Cotton, J.S. Vickers and B.C. Bush, 1994: Self-compensating all-reflection interferometer, *Appl. Opt.*, **33**, 2596–2607.

Chakrabarti, S., 1998: Ground based spectroscopic studies of sunlit airglow and aurora, *J. Atmos. Solar-Terr. Phys.*, **60**, 1403–1423.

Chakrabarti, S., T.A. Cook, F. Kamalabadi, D.M. Cotton, V. Taylor, S. Godlin and J.S. Vickers, 1999: *Instrumentation and Techniques for Diffuse/multi-object Ultraviolet Spectroscopy, Ultraviolet-Optical Space Astronomy beyond HST, ASP Conference Series*, Eds. J.A. Morse, J.M. Shull and A.L. Kinney, **164**, 322–332.

Chamberlain, J.W., 1961: *Physics of the Aurora and Airglow*, Academic Press, New York and London, 704 pp.

Chamberlain, J., 1979: *The Principles of Interferometric Spectroscopy*, John Wiley and Sons, New York, 347 pp.

Chamberlain, J.W. and D.M. Hunten, 1987: *Theory of Planetary Atmospheres*, Second Edition, Academic Press, London, 481 pp.

Clairemidi, J., M. Hersé, and G. Moreels, 1985: Bi-dimensional observation of waves near the mesopause at auroral latitudes, *Planet. Space Sci.*, **33**, 1013–1022.

CNRS, 1958: Colloque international sur les progrès récents en spectroscopie interférentielle, *J. Phys. Rad.*, **19**, 185–436.

CNRS, 1967: Colloque sur les mèthodes nouvelles de spectroscopie instrumentale, *J. Physique*, **28**, Supp. C2, 1–341.

Cogger, L.L. and G.G. Shepherd, 1969: Observations of a magnetic conjugate effect in the OI 6300 Å airglow at Saskatoon, *Planet. Space Sci.*, **17**, 1857–1865.

Conde, M. and R.W. Smith, 1998: Spatial structure in the thermospheric horizontal wind above Poker Flat, Alaska, during solar minimum, *J. Geophys. Res.*, **103**, 9449–9471.

Connes, J. and H.P. Gush, 1959: Spectroscopie du ciel nocturne dans l'infrarouge par transformation de Fourier, *J. Phys. Rad.*, **20**, 915–917.

Connes, J. and H.P. Gush, 1960: Étude du ciel nocturne dans le proche infra-rouge, *J. Phys. Rad.*, **21**, 645–654.

Connes, P., 1956: Augmentation du produit luminosité × résolution des interféromètres par l'emploi d'une différence de march indépendante de l'incidence, *Rev. Opt.*, **35**, 37–43.

Connes, P., 1957: Un nouveau type de spectromètre: l'interféromètre à réseaux, *Optica Acta*, **4**, 136–144.

Connes, P., 1958a: L'etalon de Fabry–Perot spherique, *J. Phys. Rad.*, **19**, 262–269.

Connes, P., 1958b: Spectromètre interférentiel a sélection par l'amplitude de modulation, *J. Phys. Rad.*, **19**, 215–222.

Connes, P. and G. Michel, 1975: Astronomical Fourier spectrometer, *Appl. Opt.*, **14**, 2067–2084.

Cormack, A.M., 1963: Representation of a function by its line integrals, with some radiological applications, *J. Appl. Phys.*, **34**, 2722–2727.

Cotton, D.M., T. Cook and S. Chakrabarti, 1994: Single-element imaging spectrograph, *Appl. Opt.*, **33**, 1958–1962.

Cotton, D.M., A. Stephan, T. Cook, J. Vickers, V. Taylor and S. Chakrabarti, 2000: Tomographic extreme-ultraviolet spectrographs: TESS, *Appl. Opt.*, **39**, 3991–3999.

Cuisenier, M. and J. Pinard, 1967: Spectrometre de Fourier à 'oeils de chat' et à balayage rapide, *J. Phys. Coll.* C2, Suppl. 3–4, **28**, C2-97–C2-104.

Curtis, P.D., J.T. Houghton, G.D. Peskett and C.D. Rodgers, 1974: Remote sounding of atmospheric temperature from satellites, V. The pressure modulator radiometer for Nimbus F, *Proc. R. Soc. London*, **A337**, 135–150.

Daley, R., 1991: *Atmospheric Data Analysis*, Cambridge University Press, Cambridge, 457 pp.

Davis, T.N., 1969: Television observation of auroras, in *Atmospheric Emissions*, Ed. Billy M. McCormac, Van Nostrand Reinhold, New York, 563 pp.

Deans, A.J. and G.G. Shepherd, 1978: Rocket measurements of oxygen and nitrogen emissions in the aurora, *Planet. Space Sci.*, **26**, 319–333.

Decker, J.A., 1969: Experimental operation of a Hadamard spectrometer, *Appl. Opt.*, **8**, 2552–2254.

Decker, J.A., 1971a: Experimental realization of the multiplex advantage with a Hadamard-transform spectrometer, *Appl. Opt.*, **10**, 510–514.

Decker, J.A., 1971b: Experimental operation of a 2047-slot Hadamard-transform spectrometer, *Appl. Opt.*, **10**, 1971–1972.

Decker, J.A., 1977: Hadamard-transform spectroscopy, in *Spectrometric Techniques, Volume I* (Ed. G.A. Vanasse), Academic Press, New York and London, 355 pp.

Dobrowolski, J.A., 1981: Versatile computer program for absorbing optical thin film systems, *Appl. Opt.*, **20**, 74–81.

Dobrowolski, J.A., F.C. Ho and A. Waldorf, 1985: Beam-splitter for a wide-angle Michelson Doppler imaging interferometer, *Appl. Opt.*, **24**, 1585–1588.

Drummond, J.R., J.T. Houghton, G.D. Peskett, C.D. Rodgers, M.J. Wale, J. Whitney and E.J. Williamson, 1980: The stratospheric and mesospheric sounder on Nimbus 7, *Phil. Trans. R. Soc. London*, **A296**, 219–241.

Drummond, J.R., 1989: Novel correlation radiometer – the length modulated radiometer, *Appl. Opt.*, **28**, 2451–2452.

Drummond, J.R., P.L. Bailey, G. Brasseur, G.R. Davis, J.C. Gille, G.D. Peskett, H.K. Reichle, N. Roulet, G.S. Mand, J.C. McConnell, 1999: Early mission planning for the MOPITT instrument, *Optical spectroscopic techniques and instrumentation for atmospheric and space research III*, Allen M. Larar; *Ed. Proc. SPIE*, **3756**, 396–402.

Duboin, M.-L., 1974: La raie interdite λ 6300 Å de l'oxygène atomique au cours du renforcement précrépusculaire, Thèse de doctoral d'etat es-sciences physiques l'Université de Paris VI.

Dufour, Ch., 1951: Recherches sur la luminosité, le contraste et la résolution de systèmes interferéntiels a ondes multiples. Utilisation de couches minces complexes. *Ann. Physique*, **6**, 5–107.

Dupoisot, H. and R. Prat, 1979: High resolvance interference spectrometry with a high factor of merit in the visible spectrum, *Appl. Opt.*, **18**, 85–90.

Eather, R.H., 1980: *Majestic Lights: the Aurora in Science, History and the Arts*, American Geophysical Union, Washington, 323 pp.

Elsworth, Y., James, J.F. and Sternberg, R.S., 1974: A field compensated interference spectrometer for the visible region: The optical design, *J. Phys. E: Sci. Instrum.*, **7**, 813–816.

Englert, C.R., J.G. Cardon, M.H. Stevens, R.R. Conway, J.M. Harlander and F.L. Roesler, 2000: SHIMMER, An imaging UV interferometer for the middle atmosphere, EOS, **81**, F943.

ESA, 2000: ESA-Report SP-1229, ENVISAT-MIPAS, An Instrument for Atmospheric Chemistry and Climate Research, March.

Evans, J.W., 1949: The birefringent filter, *J. Opt. Soc. Am.*, **39**, 229–242.

Ezhevskaya, T.B. and S.P. Sinitsa, 1978: Wavefront distortion in cat's-eye reflectors, *Opt. Spectrosc. (USSR)*, **45** (3), 333–336.

Fabry, Ch. and A. Perot, 1896: Mesure de petites épaisseurs en valeur absolue, *Compt. Rend.*, **123**, 802–805.

Fabry, Ch. and A. Perot, 1897: Sur les franges des lames minces argentées et leur application a la mesure de petites épasseurs d'air, *Ann. Chim. Phys.*, **12**, 459–501.

Fastie, W.G., 1952: A small plane grating monochromator, *J. Opt. Soc. Am.*, **42**, 641–647.

Fauliot, V., G. Thuillier and M. Herse, 1993: Observations of the F-region horizontal and vertical winds in the auroral-zone, *Ann. Geophys.*, **11**, 17–28.

Feldstein, Y.I., 1964: Auroral Morphology: I. Location of Auroral Zone, *Tellus*, **16**, 252.

Fellgett, P.B., 1951: Ph.D. thesis, University of Cambridge, Cambridge.

Fischer, H. and H. Oelhaf, 1996: Remote sensing of vertical profiles of atmospheric trace constituents with MIPAS limb-emission spectrometers, *Appl. Opt.*, **35**, 2787–2796.

Fisher, G.M., T.L. Killeen, Q. Wu and P.B. Hays, 1999: Tidal variability of the geomagnetic polar cap mesopause above Resolute Bay, *Geophys. Res. Lett.*, **26**, 573–576.

Fisher, G.M., T.L. Killeen, Q. Wu, J.M. Reeves, P.B. Hays, W.A. Gault, S. Brown and G.G. Shepherd, 2000: Polar cap mesosphere wind observations: Comparisons of simultaneous measurements with a Fabry–Perot interferometer and a field-widened Michelson interferometer, *Appl. Opt.*, **39**, 4283–4291.

Françon, M. and S. Mallick, 1971: *Polarization Interferometers: Applications in Microscopy and Macroscopy*, Wiley Interscience, London, 159 pp.

Frank, L.A., J.D. Craven, K.L. Ackerson, M.R. English, R.H. Eather, and R.L. Carovillano, 1981: Global auroral imaging instrumentation for the Dynamics Explorer mission, *Space Sci. Instrum.*, **5**, 369.

Frank, L.A., J.B. Sigwarth and J.D. Craven, 1986: On the influx of small comets into the Earth's upper atmosphere, I. Observations, *Geophys. Res. Lett.*, **13**, 303–306.

Frank, L. A. and J. D. Craven, 1988: Imaging results from Dynamics Explorer 1, *Rev. Geophys.*, **26**, 249–283.

Frank, L.A., J.B. Sigwarth, J.D. Craven, J.P. Cravens, J.S. Dolan, M.R. Dvorsky, P.K. Hardebeck, J.D. Harvey and D.W. Muller, 1995: The Visible Imaging System (VIS) for the Polar spacecraft, *Space Sci. Revs.*, **71**, 297–328.

Gault, W.A., R.H. Wiens, G.J. Gotshalks, R.A. Koehler and G.G. Shepherd, 1983: PRESTO: PRogrammable Etalon Spectrometer for Twilight Observations, *Appl. Opt.*, **22**, 2551–2555.

Gault, W.A., S.F. Johnston and D.J.W. Kendall, 1985: Optimization of a field-widened Michelson interferometer, *Appl. Opt.*, **24**, 1604–1608.

Gault W.A., G. Thuillier, G.G. Shepherd, S.P. Zhang, R.H. Wiens, W.E. Ward, C. Tai, B.H. Solheim, Y.J. Rochon, C. McLandress, C. Lathuillier, V. Fauliot, M. Herse, C.H. Hersom, R. Gattinger, L. Bourg, M.D. Burrage, S.J. Franke, G. Hernandez, A. Manson, R. Niciejewski, R.S. Vincent, 1996a: Validation of $O(^1S)$ Wind Measurements by WINDII the Wind Imaging Interferometer on UARS, *J. Geophys Res.*, **101**, 10 431–10 440.

Gault, W.A., S. Brown, A. Moise, D. Liang, G. Sellar, G.G. Shepherd and J. Wimperis, 1996b: ERWIN: an E-region wind interferometer, *Appl. Opt.*, **35**, 2913–2922.

Gault, W.A., S. Sargoytchev and S. Brown, 2001: Divided-mirror technique for measuring Doppler shifts with a Michelson interferometer, *SPIE Conference 4306A Proceedings*, San Diego, 21–26 January.

Girard, A., 1963: Spectomètre à grilles, *Appl. Opt.*, **2**, 79–87.

Girard, A., J. Besson, D. Baird, J. Laurent, M.P. Lemaitre, C. Lippins, C. Muller, J. Vercheval and M. Ackermann, 1988: Global results of grille spectrometer onboard Spacelab 1, *Planet. Space Sci.*, **36**, 291–299.

Golay, M.J.E., 1951: Static multislit spectrometry and its application to the panoramic display of infrared spectra, *J. Opt. Soc. Am.*, **41**, 468–472.

Goody, R., 1968: Cross-correlating spectrometer, *J. Opt. Soc. Am.*, **58**, 900–908.

Grainger, J.R. and J. Ring, 1962: Anomalous Fraunhofer line profiles, *Nature*, **193**, 762–762.

Grassl, H.J., W.R. Skinner, P.B. Hays, M.D. Burrage, D.A. Gell, A.R. Marshall, D.A. Ortland, and V.J. Abreu, 1995: Atmospheric wind measurements with the High Resolution Doppler Imager (HRDI) , *J. Spacecraft & Rockets*, **32**, No. 1, 169–176.

Greet, P.A., M. Conde and F. Jacka, 1989: Daytime observation of the sodium layer with a Fabry–Perot spectrometer at Mawson, Antarctica, *Geophys. Res. Lett.*, **16**, 871–874.

Greet, P.A., J. Innis and P.L. Dyson, 1994: High-resolution Fabry–Perot observations of mesospheric OH (6-2) emissions, *Geophys. Res. Lett.*, **21**, 1153–1156.

Gunson, M.R., C.B. Farmer, R.H. Norton, R. Zander, C.P. Rinsland, J.H. Shaw and Bo-Cai Gao, 1990: Measurements of CH_4, N_2O, CO, H_2O, and O_3 in the middle atmosphere by the Atmospheric Trace Molecule Spectroscopy experiment on Spacelab 3, *J. Geophys. Res.*, **95**, 13 867–13 882.

Gush, H.P. and A. Vallance Jones, 1955: Infrared spectrum of the night sky from 1.0 μ to 2.0 μ, *J. Atmos. Terr. Phys.*, **7**, 285–291.

Gush, H.P. and H.L. Buijs, 1964: The near infrared spectrum of the night airglow observed from high altitude, *Can. J. Phys.*, **42**, 1037–1045.

Gush, H.P., 1988: Beginnings of VLBI in Canada, *J. Roy. Astron. Soc. Can.*, **82** (5): 221–232.

Gush, H.P., M. Halpern and E.H. Wishnow, 1990: Rocket measurement of the cosmic-background radiation mm-wave spectrum, *Phys. Rev. Lett.*, **65**, 537–540.

Gush, H.P. and M. Halpern, 1992: Cooled submillimeter Fourier transform spectrometer flown on a rocket, *Rev. Sci. Instrum.*, **63**, 3249–3260.

Hanel, R.A., B. Schlachman, F.D. Clarke, C.H. Prokesh, J.B. Taylor, W.M. Wilson and L. Chaney, 1970: The Nimbus 3 Michelson interferometer, *Appl. Opt.*, **9**, 1767–1774.

Hanel, R.A., 1983: *Spectrometric Techniques, Vol. III*, Ed. G.A. Vanasse, Academic Press, New York and London, 334 pp.

Hansen, G., 1955: Die sichtbarkeit der interferenzen beim Twyman interferometer, *Optik*, **12**, 5.

Hansen, G. and W. Kinder, 1958: Abhängkeit des kontrastes der Fizeau-streifen im Michelson-interferometer vom durchmesser der aperturblende, *Optik*, **15**, 560.

Hariharan, P. and D. Sen, 1961: Double-passed Fabry–Perot interferometer, *J. Opt. Soc. Am.*, **51**, 398–399.

Hariharan, P., 1987: Digital phase-stepping interferometry: effects of multiply reflected beams, *Appl. Opt.*, **26**, 2506–2507.

Hariharan, P., 1989: Phase-stepping interferometry with laser diodes, *Appl. Opt.*, **28**, 27–29.

Hariharan, P., 1991: *Basics of Interferometry*, Academic Press, New York, 213 pp.

Harlander, J., R.J. Reynolds and F.L. Roesler, 1992: Spatial heterodyne spectroscopy for the exploration of diffuse interstellar emission lines at far-ultraviolet wavelengths, *Astrophys. J.*, **396**, 730–740.

Harlander, J., H.T. Tran, F.L. Roesler, K. Jaehnig, S.M. Seo, W. Sanders and R.J. Reynolds, 1994: Field-widened spatial heterodyne spectroscopy: Correcting for optical defects and new vacuum ultraviolet performance tests, *Proc. SPIE*, **2280**, 310–319.

Harvey, J., and the GONG Instrument Development Team: 1988, in E.J. Rolfe (ed.), *Seismology of the Sun and Sun-like Stars*, ESA: The Netherlands, 203 pp.

Hays, P.B. and J.C.G. Walker, 1966: Doppler profiles of the 5577 Å airglow, *Planet. Space Sci.*, **14**, 1331–1337.

Hays, P.B., G. Carignan, B.C. Kennedy, G.G. Shepherd and J.C.G. Walker, 1973: The visible airglow experiment on Atmosphere Explorer, *Radio Sci.*, **8**, 369–377.

Hays, P.B., and C.D. Anger, 1978: Influence of ground scattering on satellite auroral observations, *Appl. Opt.*, **17**, 1898–1904.

Hays, P.B., J.W. Meriwether and R.G. Roble, 1979: Night-time thermospheric winds at high latitudes, *J. Geophys. Res.*, **84**, 1905–1913.

Hays, P.B., T.L. Killeen and B.C. Kennedy, 1981: The Fabry–Perot Interferometer on Dynamics Explorer, *Space Sci. Instrum.*, **5**, 395–416.

Hays, P.B., 1982: High-resolution optical measurement of atmospheric winds from space. 1: Lower atmosphere molecular absorption, *Appl. Opt.*, **21**, 1136–1141.

Hays, P.B. 1990: Circle to line interferometer optical system, *Appl. Opt.*, **29**, 1482–1489.

Hays, P.B., V.J. Abreu, M.E. Dobbs, D.A. Gell, H.J. Grassl and W.R. Skinner, 1993: The high-resolution Doppler Imager on the Upper Atmosphere Research Satellite, *J. Geophys. Res.*, **98**, 10 713–10 723.

Hedin, A.E. and G. Thuillier, 1988: Comparison of OGO 6 measured thermospheric temperatures with the MSIS-86 empirical model, *J. Geophys. Res.*, **93**, 5965–5971.

Hercher, M., 1968: The spherical mirror Fabry–Perot interferometer, *Appl. Opt.*, **7**, 951–966.

Hernandez, G., and O.A. Mills, 1973: Feedback stabilized Fabry–Perot interferometer, *Appl. Opt.*, **12**, 126–130.

Hernandez, G. and R.G. Roble, 1979: Thermospheric dynamics investigations with very high resolution spectrometers, *Appl. Opt.*, **18**, 3376–3385.

Hernandez, G., 1979: Analytical description of a Fabry–Perot spectrometer. V. Optimization for minimum uncertainties in the determination of Doppler widths and shifts, *Appl. Opt.*, **18**, 3826–3834.

Hernandez, G., and Smith, R.W., 1984: Mesospheric wind determinations and the $P_1(2)_{c,d}$ lines of the $X\,^2\Pi$ OH (8-3) band, *Geophys. Res. Lett.*, **11**, 534–537.

Hernandez, G., 1985: Analytical description of a Fabry–Perot spectrometer. IX. Optimum operation with a spherical etalon, *Appl. Opt.*, **24**, 3707–3712.

Hernandez, G., 1986a: Transient response of optical instruments, *Appl. Opt.*, **24**, 928–929.

Hernandez, G., 1986b: *Fabry–Perot Interferometers*, Cambridge University Press, Cambridge, 343 pp.

Hernandez, G., 1987: Afocal coupled etalons. DEM: a high-resolution double-etalon modulator spectrometer, *Appl. Opt.*, **26**, 4857–4869.

Hernandez, G., R.W. Smith, R.G. Roble, J. Gress and K.C. Clark, 1990: Thermospheric dynamics at the South Pole, *Geophys. Res. Lett.*, **17**, 1255–1258.

Hersom, C.H. and G.G. Shepherd, 1995: Characterization of the wind imaging interferometer, *Appl. Opt.*, **34**, 2871–2879.

Hicks, T.R., N.K. Reay and R.J. Scaddan, 1974: A servo-controlled Fabry–Perot interferometer using capacitance micrometers for error detection, *J. Phys. E: Sci. Instrum.*, **7**, 27–30.

Hilliard, R.L. and G.G. Shepherd, 1966a: Wide angle Michelson interferometer for measuring Doppler linewidths, *J. Opt. Soc. Am.*, **56**, 362–369.

Hilliard, R.L. and G.G. Shepherd, 1966b: Upper atmospheric temperatures from Doppler linewidths – IV: A detailed study using the OI 5577 Å auroral and nightglow emission, *Planet. Space Sci.*, **14**, 383–406.

Hines, C.O., 1974: *The Upper Atmosphere in Motion*, Geophysical Monograph 18, American Geophysical Union, Washington, D.C., 1027 pp.

Hines, C.O. and D.W. Tarasick, 1987: On the detection and utilization of gravity waves from airglow, *Planet. Space Sci.*, **35**, 851–866.

Hirschberg, J.G. and P. Platz, 1965: A multichannel Fabry–Perot spectrometer, *Appl. Opt.*, **4**, 1375–1381.

Hirschberg, J.G., W.I. Fried, L. Hazelton and A. Wouters, 1971: Multiplex Fabry–Perot interferometer, *Appl. Opt.*, **10**, 1979–1980.

Hirschberg, J.G., 1974: Field-widened Michelson interferometer with no moving parts, *Appl. Opt.*, **13**, 233–234.

Hirschberg, J.G. and D.F. Cornwall, 1979: Field-widened Michelson spectrometer with no moving parts. Part 2, **18**, 2726–2727.

Houghton, J.T., F.W. Taylor and C.D. Rodgers, 1984: *Remote Sensing of Atmospheres*, Cambridge University Press, Cambridge, 343 pp.

Houston, W.V., 1927: A compound interferometer for fine structure work, *Phys. Rev.*, **29**, 478–484.

Huffman, Robert E., 1992: *Atmospheric Ultraviolet Remote Sensing*, Academic Press Inc., San Diego, 320 pp.

Hunten, D.M. 1953: A rapid-scanning auroral spectrometer, *Can. J. Phys.*, **31**, 681–690.

Hunten, D.M., F.E. Roach and J.W. Chamberlain, 1956: A photometric unit for the airglow and aurora. *J. Atmos. Terr. Phys.*, **8**, 345–346.

Hunten, D.M., H.N. Rundle, G.G. Shepherd and A.Vallance Jones, 1967: Optical upper atmospheric investigations at the University of Saskatchewan, *Appl. Opt.*, **6**, 1609–1623.

Hutley, M.C., 1982: *Diffraction Gratings*, Academic Press, London, 330 pp.

Jacka, F, 1984: Application of Fabry–Perot spectrometers for measurement of upper atmosphere temperature and winds, Chap. 2 in *Middle Atmosphere Handbook for MAP*, Ed. R.A. Vincent.

Jacquinot, P. and Ch. Dufour, 1948: Conditions optiques d'emploi des cellules photo-électriques dans les spectrographes et les interféromètres, *J. Res. CNRS*, **6**, 91–103.

Jacquinot, P., 1950: Quelques recherches sur les raies faible dans les spectres optiques, *Proc. Phys. Soc. B.*, **63**, 969–979.

Jacquinot, P., 1954: The luminosity of spectrometers with prisms, gratings or Fabry–Perot etalons, *J. Opt. Soc. Am.*, **44**, 761–765.

Jacquinot, P., 1958: Caractères communs aux nouvelles méthodes de spectroscopie interférentielle: facteur de mérite, *J. Phys. Rad.*, **19**, 223–229.

Jacquinot, P., 1960: New developments in interference spectroscopy, *Rep. Progr. Phys.*, **23**, 267–312.

Kanasewich, E., 1981: *Time Sequence Analysis in Geophysics*, The University of Alberta Press, Edmonton, 480 pp.

Kaneda, E., T. Mukai and K. Hirao, 1981: Synoptic features of auroral system and corresponding electron precipitation observed by KYOKKO, in *Physics of Auroral Arc Formation*, Eds. S.-I. Akasofu and J.R. Kan, *AGU Geophysical Monograph*, **25**, 24–30.

Kaneda, E. and T. Yamamoto, 1991: Auroral substorms observed by UV-imager on Akebono, in *Magnetospheric Substorms*, Eds. J.R. Kan, T.A. Potemra, S. Kokubun and T. Iijima, *AGU Geophysical Monograph*, **64**, 235–240.

Katzenstein, J., 1965: The axicon-scanned Fabry–Perot spectrometer, *Appl. Opt.*, **4**, 263–266.

Kerr, R.B., J. Noto, R.S. Lancaster, M. Franco, R.J. Rudy, R. Williams and J.H. Hecht, 1996: Fabry–Perot observations of helium 10830 Å emission at Millstone Hill, *Geophys. Res. Lett.*, **23**, 3239–3242.

Khun, H., 1950: The use of two Fabry–Perot interferometers in series for the detection of weak satellites, *J. Phys. Rad.*, **11**, 425–426.

Killeen, T.L., B.C. Kennedy, P.B. Hays, D.A. Symanow and D.H. Ceckowski, 1983: Image plane detector for the Dynamics Explorer Fabry–Perot interferometer, *Appl. Opt.*, **22**, 3503–3513.

Killeen, T.L., W.R. Skinner, R.M. Johnson, C.J. Edmonson, Q. Wu, R. Niciejewski, H. Grassl, D. Gell, P. Hansen, J. Harvey and J. Kafkalidis, 1999: TIMED Doppler Interferometer (TIDI), *Optical spectroscopic techniques and instrumentation for atmospheric and space research III*, Allen M. Larar; *Ed. Proc. SPIE*, **3756**, 289–301.

Kramer, H.J., 2002: *Observation of the Earth and its Environment*, 4th Edition, Springer Verlag, Heidelberg, 1514 pp.

Laws, D.A., 1962: Primary distortion of an extended image by a plane grating, *Optica Acta*, **9**, 69–71.

Lissberger, P.H., 1968: Effective refractive index as a criterion of performance of interference filters, *J. Opt. Soc. Am.*, **58**, 1586–1590.

Llewellyn, E.J., D.A. Degenstein, I.C. McDade, R.L. Gattinger, R. King, R. Buckingham, E.H. Richardson, D.P. Murtagh, W.F.J. Evans, B.H. Solheim, K. Strong and J.C. McConnell, 1997: OSIRIS – An application of tomography for absorbed emissions in remote sensing, in *Applications of Photonic Technology*, **2**, pp. 627–632, Eds. G.A. Lampropoulos and R.A. Lessard, Plenum Press, New York.

Lu, Z.J., W.A. Gault and R.A. Koehler, 1988a: A new scanning method for field-compensated Michelson interferometers, *J. Phys. E: Sci. Instrum.*, **21**, 68–71.

Lu, Z.J., R.A. Koehler, W.A. Gault and F.C. Liang, 1988b: A dynamic alignment system for scanning Michelson interferometers, *J. Phys. E: Sci. Instrum.*, **21**, 71–74.

Lyot, B., 1944: Le filtre monochromatique polarisant et ses applications en physique solaire, *Ann. Astrophys.*, **7**, 31.

Mack, J.E., D.P. McNutt, F.L. Roesler and R. Chabbal, 1963: The PEPSIOS purely interferometric high-resolution scanning spectrometer. 1. The pilot model, *Appl. Opt.*, **2**, 873–885.

Mani, R., 2001: *SWIFT instrument testbed for stratospheric wind measurements*, Ph.D. thesis, York University.

Margottin-Maclou, M., F. Rachet, A. Henry and A. Valentin, 1996: Pressure-induced line shifts in the O_3 band of the nitrous oxide perturbed by N_2, O_2, He, Ar and Xe, *J. Quant. Spectrosc. Radiat. Transfer*, **56**, 1–16.

McCleese, D.J. and J. Margolis, 1983: Remote sensing of stratospheric and mesospheric winds by gas correlation electrooptic phase-modulation spectroscopy, *Appl. Opt.* **22**, 2528–2534.

McCleese, D.J., 1992: The stratospheric wind infrared limb sounder: Investigation of amospheric dynamics and transport from EOS, *Proceedings of the International School of Physics "Enrico Fermi": The use of EOS for studies of atmospheric physics*, J.C. Gille and G. Visconti (Eds.), North Holland, Amsterdam.

McClintock, W.E., C.A. Barth, R.E. Steele, G.M. Lawrence and J.G. Timothy, 1982: Rocket-borne instrument with a high-resolution microchannel plate detector for planetary UV spectroscopy, *Appl. Opt.*, **21**, 3071–3079.

McDade, I.C., D.P. Murtagh, R.G.H. Greer, P.H.G. Dickinson, G. Witt, J. Stegman, E.J. Llewellyn, L. Thomas and D.B. Jenkins, 1986: ETON 2: Quenching parameters for the proposed precursors to $O_2(b^1\Sigma_g^+)$ and $O(^1S)$ in the terrestrial nightglow. *Planet. Space Sci.*, **34**, 789–800.

McDade, I.C. and E.J. Llewellyn, 1991: Inversion techniques for recovering the two-dimensional distributions of auroral emission rates from tomographic rocket photometer measurements, *Can. J. Phys.*, **69**, 1059–1068.

McDade, I.C., N.D. Lloyd and E.J. Llewellyn, 1991: A rocket tomography measurement of the N_2^+ 3914 Å emission rates within an auroral arc, *Planet. Space Sci.*, **39**, 895–906.

McLandress, C., G.G. Shepherd and B.H. Solheim, 1996a: Satellite observations of thermospheric tides: Results from the Wind Imaging Interferometer on UARS, *J. Geophys. Res.*, **101**, 4093–4114.

McLandress, C., G.G. Shepherd, B.H. Solheim, M.D. Burrage, P.B. Hays and W.R. Skinner, 1996b: Combined mesosphere/thermosphere winds using WINDII and HRDI data from the Upper Atmosphere Research Satellite, *J. Geophys. Res.*, **101**, 10 441–10 453.

McLennan, J.C. and J.H. McLeod, 1927: On the wavelength of the green auroral line in the oxygen spectrum, *Proc. Roy. Soc.* (London), **A 115**, 515.

McLeod, J.H., 1954: The axicon: A new type of optical element, *J. Opt. Soc. Am.*, **44**, 592–597.

McNutt, D.P., 1965: Pepsios purely interferometric high-resolution scanning spectrometer. II Theory of spacer ratios, *J. Opt. Soc. Am.*, **55**, 288–292.

McWhirter, I., D. Rees and A.H. Greenaway, 1982: Miniature imaging photon detectors 3. An assessment of the performance of the resistive anode IPD, *J. Phys. E.: Sci. Instrum.* **15**(1), 145–150.

Meaburn, J., 1976: *Detection and Spectrometry of Faint Light*, Reidel, Dordrecht, 270 pp.

Meinel, A. and M. Meinel, 1991: *Sunsets, Twilights, and Evening Skies*, Cambridge University Press, Cambridge, 163 pp.

Mende, S.B., R.H. Eather, and E.K. Aamodt, 1977: Instrument for the monochromatic observation of all sky auroral images, *Appl. Opt.*, **16**, 1691–1700.

Menke, W., 1989: *Geophysical Data Analysis: Discrete Inversion Theory*, Academic Press, Toronto, 289 pp.

Menzel, W.P. and J.F.W. Purdom, 1994: Introducing GOES-I: The first of a new generation of geostationary operational environmental satellites, *Bull. Am. Meteor. Soc.*, **75**, 757–781.

Mertz, L., 1959: *Heterodyne Interference Spectroscopy*, Stockholm Congress.

Mertz, L., 1965: *Transformations in Optics*, Wiley, New York, 116 pp.

Michelson, A.A., 1927: *Studies in Optics*, University of Chicago Press, Chicago and London, reprinted in the *Phoenix Science Series*, 1962, 176 pp.

Miller, J.R. and G.G. Shepherd, 1968: Auroral measurements using rocket-borne photometers, *Ann. Géophys*, **24**, 305–312.

Murcray, D.G., F.H. Murcray, W.J. Williams, T.G. Kyle and A. Goldman, 1969: Variation of the infrared solar spectrum between 700 cm^{-1} and 2240 cm^{-1} with altitude, *Appl. Opt.*, **8**, 2519–2536.

Nagaoka, H., 1917: On the regularity in the distribution of the satellites of spectrum lines, *Proc. Phys. Soc. London*, **29**, 92–119.

Narayanan, R., J.N. Desai, N.K. Modi, R. Raghavarao and R. Sridharan, 1989: Dayglow photometry: a new approach, *Appl. Opt.*, **28**, 2138–2142.

Naylor, D.A., T.A. Clark and R.T. Boreiko, 1981: Balloon-borne stratospheric far-IR spectral absorption measurements: the design of a solar telescope and high-resolution Michelson interferometer combination, *Appl. Opt.*, **20**, 1132–1144.

Nelson, E.D. and M.L. Fredman, 1970: Hadamard spectroscopy, *J. Opt. Soc. Am.*, **60**, 1664–1669.

Neo, Y.P and G.G. Shepherd, 1972: Airglow observations with a Hadamard photometer, *Planet. Space Sci.*, **20**, 1351–1355.

Newcomb, S., 1901: A rude attempt to determine the total light of all the stars, *Astrophys. J.*, **14**, 297–312.

Nilson, J.A. and G.G. Shepherd, 1961: Upper atmospheric temperatures from Doppler line widths – I. Some preliminary observations on OI 5577 Å in aurora, *Planet. Space Sci.*, **5**, 299–306.

Nossal, S., F.L. Roesler, M.M. Coakley and R.J. Reynolds, 1997: Geocoronal hydrogen Balmer-α line profiles obtained using Fabry–Perot annular summing spectroscopy: Effective temperature results, *J. Geophys. Res.*, **102**, 14 541–14 553.

Noxon, J.F., 1968: Day airglow, *Space Sci. Revs.* **8**, 92–134.

Noxon, J.F., 1975: Nitrogen dioxide in the stratosphere and troposphere as measured by ground based absorption spectroscopy, *Science*, **189**, 547–549.

Offermann, D., K.-U. Grossmann, P. Barthol, P. Knieling, M. Riese and R. Trant, 1999: The CRyogenic Infrared Spectrometers and Telescopes for the Atmosphere (CRISTA)–Experiment and middle atmosphere variability, *J. Geophys. Res.*, **104**, 16 311–16 325.

Ogawa, T., H. Shimoda, M. Hayashi, T. Imasu, A. Ono, S. Nishiyama and H. Kobayashi, 1994: IMG, Interferometric Measurement of Greenhouse gases from space, *Adv. Space Res.*, **14**, (1), 25–28.

Oguti, T., E. Kaneda, M. Ejiri, S. Sasaki, A. Kadokura, T. Yamamoto, K. Hayashi, R. Fujii and K. Makita, 1990: Studies of aurora dynamics by aurora-TV on the Akebono (EXOS-D) satellite, *J. Geomag. Geolectr.*, **42**, 555–564.

Okano, S., J.S. Kim and T. Ichikawa, 1980: Design of a multiple-zone aperture and application to a Fabry–Perot interferometer, *Appl. Opt.*, **19**, 1622–1629.

Ortland, D.A., W.R. Skinner, P.B. Hays, M.D. Burrage, R.S. Lieberman, A.R. Marshall and D.A. Gell, 1996: Measurements of stratospheric winds by the High Resolution Doppler Imager, *J. Geophys. Res.*, **101**, 10351–10363.

Osterbrock, D.E., R.T. Waters, T.A. Barlow *et al.*, 2000: Faint emission lines in the blue and red spectral regions of the night airglow, *Publ. Astron. Soc. Pacific*, **112**, 733–741.

Papoulis, A., 1962: *The Fourier Integral and its Applications*, McGraw-Hill Book Co., New York, 318 pp.

Pease, C.B., 1991: *Satellite Imaging Instruments*, Ellis Horwood Ltd., Chichester, England, 336 pp.

Peck, E.R., 1948: Theory of the corner-cube interferometer, *J. Opt. Soc. Am.* **38**, 1015–1024.

Peteherych, S., G.G. Shepherd and J.K. Walker, 1985: Observation of vertical winds in two intense auroral arcs, *Planet. Space Sci.*, **33**, 869–873.

Peterson, A.W. and L.M. Kieffaber, 1973: Infrared photography of OH airglow structures, *Nature*, **242**, 321–322.

Philip, A.G.D., K.A. Janes and A.R. Upgren, 1995: *New Developments in Array Technology and Applications*, Kluwer Academic Publishers, Dordrecht, 397 pp.

Phillips, K.J.H., 1992: *Guide to the Sun*, Cambridge University Press, Cambridge, 386 pp.

Piotrowski McCall, S.H.C., J.A. Dobrowolski and G.G. Shepherd, 1989: Phase shifting thin film multilayers for Michelson interferometers, *Appl. Opt.*, **28**, 2854–2859.

Powell, I., 1986: Optical system design for the WAMDII instrument, *J. Soc. Photoopt. Instrum. Eng.*, **655**, 198.

Rao, P.K, S.J. Holmes, R.K. Anderson, J.S. Winston and P.E. Lehr (Eds.), 1990: *Weather satellites: systems, data, and environmental applications, Am. Meteor. Soc., Boston*, 504 pp.

Rayleigh, Lord (Strutt, R.J.), 1930: Absolute intensity of the aurora line in the night sky and the number of atomic transitions required to maintain it, *Proc. Roy. Soc. (London)*, **A 129**, 458–467.

Reber, C.A., C.E. Trevathan, R.J. McNeal and M.R. Luther, 1993: The Upper Atmosphere Research Satellite (UARS) mission, *J. Geophys. Res.*, **98**, 10 643–10 647.

Rees, D., T.J. Fuller-Rowell, A. Lyons, T.L. Killeen and P.B. Hays, 1982: Stable and rugged etalon for the Dynamics Explorer Fabry–Perot interferometer: 1. Design and construction, *Appl. Opt.* **21**, 3896–3902.

Rees, D., A.H. Greenway, R. Gordon, I. McWhirter, P.J. Charleton and Å. Steen, 1984: The Doppler Imaging System: Initial observations of the auroral thermosphere, *Planet. Space Sci.*, **32**, 273–285.

Rees, M.H., 1989: *Physics and Chemistry of the Upper Atmosphere*, Cambridge University Press, Cambridge, 289 pp.

Regalia, L., A. Barbe, J.-J. Plateaux, V. Dana, J.-Y. Mandin, and M.-Y. Allout, 1995: Nitrogen-broadening and nitrogen-shifting coefficients in the ν_3 band of N_2O, *J. Molec. Spectrosc.*, **172**, 563–566.

Richter, A., F. Wittrock, M. Eisinger and J.P. Burrows, 1998: GOME observations of tropospheric BrO in northern spring and summer 1997, *Geophys. Res. Lett.*, **25**, 2683–2686.

Ring, J. and J.W. Schofield, 1972: Field compensated Michelson spectrometers, *Appl. Opt.*, **11**, 507–516.

Roach, F.E., L.R. Megill, M.H. Rees and E. Marovich, 1958: The height of nightglow 5577, *J. Atmos. Terr. Phys.*, **12**, 171–186.

Roach, F.E. and J.L. Gordon, 1973: *The Light of the Night Sky*, D. Reidel Publishing Company, Dordrecht, Holland, 125 pp.

Roble, R.G., and E.C. Ridley, 1994: A Thermosphere-Ionosphere-Mesosphere-Electrodynamics General-Circulation Model (TIME-GCM) – Equinox solar-cycle minimum simulations (30–500 km), *Geophys. Res. Lett.*, **21**, 417–420.

Roche, A.E., J.B. Kumer, J.L. Mergenthaler, G.A. Ely, W.G. Uplinger, J.F. Potter, T.C. James, L.W. Sterritt, 1993: The Cryogenic Limb Array Etalon Spectrometer (CLAES) on UARS: Experimental description and performance, *J. Geophys. Res.* **98**, 10 763–10 775.

Rochon, Y.J., 1999: *The Retrieval of Doppler Winds, Temperatures and Emission Rates for the WINDII experiment*, Ph.D. Thesis, York University.

Rodgers, C.D., 1976: Retrieval of atmospheric temperature and composition from remote measurements of thermal radiation, *Rev. Geophys. Space Phys.*, **14**, 609–624.

Rodgers, C.D., 1990: Characterization and error analysis of profiles retrieved from remote sounding measurements, *J. Geophys. Res.*, **95**, 5587–5595.

Roesler, F.L. and J.E. Mack, 1967: Pepsios purely interferometric high-resolution scanning interferometer. IV. Performance of the Pepsios spectrometer, *J. Physique*, **28**, Suppl. C2, 313–320.

Ronchi, V., 1964: Forty years of history of a grating interferometer, *Appl. Opt.*, **3**, 437–451.

Russell, J.M. III, L.L. Gordley, J.H. Park, S.R. Drayson, W.D. Hesketh, R.J. Cicerone, A.F. Tuck, J.E. Frederick, J.E. Harries, P.J. Crutzen, 1993: The halogen occultation experiment, *J. Geophys. Res.*, **98**, 10 777–10 797.

Salby, M.L., 1996: *Fundamentals of Atmospheric Physics*, Academic Press, 627 pp.

Sandercock, J.R., 1971: The design and use of a stabilised multipassed interferometer of high contrast ratio, *Light Scattering in Solids*. (M. Balkanski, ed.) pp. 9–12, Flammarion Press, Paris.

Seyrafi, K. and S.A. Hovanessian, 1993: *Introduction to Electro-optical Imaging and Tracking Systems*, Artech House, Boston, 260 pp.

Scherrer P.H., R.S. Bogart, R.I. Bush, J.T. Hoeksema, A.G. Kosovichev, J. Schou, W. Rosenberg, L. Springer, T.D. Tarbell, A. Title, C.J. Wolfson and I. Zayer, 1995a: The solar oscillations investigation – Michelson Doppler Imager, *Solar Physics*, **162**, 129–188.

Scherrer, P.H. *et al.* (as above), 1995b: *The solar oscillations investigation – Michelson Doppler Imager*, *The SOHO Mission*, B. Fleck, V. Domingo and A. Poland (Eds.), Kluwer Academic Publishers, Dordrecht, The Netherlands.

Scrimger, J.A. and D.M. Hunten, 1957: Absorption of sunlight by atmospheric sodium, *Can. J. Phys.*, **35**, 918–927.

Shepherd, G.G., 1960: A Fabry–Perot spectrometer for auroral and airglow observations, *Can. J. Phys.*, **38**, 1560–1569.

Shepherd, G.G., C.W. Lake, J.R. Miller and L.L. Cogger, 1965: A spatial spectral scanning technique for the Fabry–Perot spectrometer, *Appl. Opt.*, **4**, 267–272.

Shepherd, G.G., T. Fancott, J. McNally and H.S. Kerr, 1973: ISIS-II atomic oxygen red line photometer, *Appl. Opt.*, **12**, 1767–1774.

Shepherd, G.G., 1974: How about radiance response?, *Appl. Opt.*, **13**, 1734–1734.

Shepherd, G.G., A.J. Deans and Y.P. Neo, 1978: SCIMP – A scanning interferometric multiplex spectrometer, *Can. J. Phys.*, **56**, 681–686.

Shepherd, G.G., 1979: Dayside cleft aurora and its ionospheric effects, *Revs. Geophys. Space Phys.* **17**, 2017–2033.

Shepherd, G.G., W.A. Gault, D.W. Miller, Z. Pasturczyk, S.F. Johnston, P.R. Kosteniuk, J.W. Haslett, D.J.W. Kendall and J.R. Wimperis, 1985: WAMDII: Wide-Angle Michelson Doppler Imaging Interferometer for Spacelab, *Appl. Opt.*, **24**, 1571–1584.

Shepherd, G.G., W.A. Gault and R.A. Koehler, 1991: The development of wide-angle Michelson interferometers in Canada, *Can. J. Phys.*, **69**, 1175–1183.

Shepherd G.G., G. Thuillier, W.A. Gault, B.H. Solheim, C. Hersom, J.M. Alunni, J.-F. Brun, S. Brune, P. Charlot, L.L. Cogger, D.-L. Desaulniers, W.F.J. Evans, R.L. Gattinger, R. Girod, D. Harvie, M. Hersé, R.H. Hum, D.J.W. Kendall, E.J. Llewellyn, R.P. Lowe, J. Ohrt, F. Pasternak, O. Peillet, I. Powell, Y. Rochon, W.E. Ward, R.H. Wiens and J. Wimperis, 1993: WINDII: The wind imaging interferometer on the upper atmosphere research satellite, *J. Geophys. Res.*, **98**, 10 725–10 750.

Shepherd, G.G., R.G. Roble, C. McLandress and W.E. Ward, 1997: WINDII observations of the 558 nm emission in the lower thermosphere; the influence of dynamics on composition, *J. Atmos. Solar-Terr. Phys.*, **59**, 655–667.

Shepherd, M.G., W.E. Ward, B. Prawirosoehardjo, R.G. Roble, S.-P. Zhang and D.Y. Wang, 1999: Planetary scale and tidal perturbations in mesospheric temperature observed by WINDII, *Earth, Planets and Space*, **51**, 593–610.

Siegmund, O.H.W., R.F. Malina, K. Coburn and D. Werthimer, 1984: Microchannel plate EUV detectors for the Extreme Ultraviolet Explorer, *IEEE Trans. Nucl. Sci.* **NS-31**, 776–779.

Siegmund, O.H.W., M. Lampton, J. Bixler, S. Chakrabarti, J. Vallarga, S. Boyer and R.F. Malina, 1986: Wedge and strip readout systems for photon-counting detectors in space astronomy, *J. Opt. Soc. Am.*, **A3**, 2139–2145.

Sipler, D.P., M.A. Biondi and R.G. Roble, 1983: F-Region neutral winds and temperatures at equatorial latitudes: Measured and predicted behaviour during geomagnetically quiet conditions, *Planet. Space Sci.*, **31**, 53–66.

Sivjee, G.G. and D. Shen, 1997: Auroral optical emissions during the solar magnetic cloud event of October 1995, *J. Geophys. Res.*, **102**, 7431–7437.

Skinner, W.R., P.B. Hays and V.J. Abreu, 1987: Optimization of a triple etalon interferometer, *Appl. Opt.*, **26**, 2817–2827.

Slanger, T.G., D.L. Huestis, D.E. Osterbrock *et al.*, 1997: The isotopic oxygen nightglow as viewed from Mauna Kea, *Science*, **277**, 1485–1488.

Slanger, T.G. and D.E. Osterbrock, 2000: Investigation of potassium, lithium, and sodium emissions in the nightglow, and OH cross-calibration, *J. Geophys. Res.*, **105**, 1425–1429.

Slanger, T.G., P.C. Cosby, D.L. Huestis *et al.*, 2000: Vibrational level distribution of $O_2(b^1\Sigma_g^+, v = 0\text{–}15)$ in the mesosphere and lower thermosphere region, *J. Geophys. Res.*, **105**, 20 557–20 564.

Smith, R.W., K. Henriksen, C.S. Deehr, D. Rees, F.G. McCormac and G.G. Sivjee, 1985: Thermospheric winds in the cusp: dependence of the latitude of the cusp, *Planet. Space Sci.*, **33**, 305–313.

Solomon, S.C., P.B. Hays and V.J. Abreu, 1984: Tomographic inversion of satellite photometry, *Appl. Opt.*, **23**, 3409–3414.

Solomon, S.C., P.B. Hays and V.J. Abreu, 1985: Tomographic inversion of satellite photometry, Part 2, *Appl. Opt.*, **24**, 4134–4140.

Solomon, S.C., P.B. Hays and V.J. Abreu, 1988: The auroral 6300 Å emission: Observations and modeling, *J. Geophys. Res.*, **93**, 9867–9882.

Solomon, S., A.L. Schmeltekopf and R.W. Sanders, 1987: On the interpretation of zenith sky absorption measurements, *J. Geophys. Res.*, **92**, 8311–8319.

Steel, W.H., 1974: Field-widened Michelson interferometer with no moving parts: Comments, *Appl. Opt.*, **13**, 2189–2190.

Steel, W.H., 1983: *Interferometry*, Cambridge University Press, New York, 308 pp.

Stephens, G.L., 1994: *Remote Sensing of the Lower Atmosphere: an Introduction*, Oxford University Press, New York, 523 pp.

Strong, J., 1967: Balloon telescope optics, *Appl. Opt.*, **6**, 179–189.

Stoner, J.O., 1966: PEPSIOS: Purely interferometric high-resolution spectrometer. III. Calculation of interferometer characteristics by a method of optical transients, *J. Opt. Soc. Am.*, **56**, 370–376.

Strutt, G.R., 1964a: Robert John Strutt, fourth Baron Rayleigh, *Appl. Opt.*, **3**, 1105–1112.

Strutt, C.R. 1964b: The optics papers of Robert John Strutt, fourth Baron Rayleigh, *Appl. Opt.*, **3**, 1116–1119.

Suzuki, M., *et al.*, 1997: A feasibility study on solar occultation with a compact FTIR, *SPIE*, **3122**, 2–15.

Suzuki, M., K. Shibasaki, H. Shimoda and T. Ogawa, 1999: Overview of GCOM-A1 science program, *SPIE*, **3870**, 334–343.

Swenson, G.R., S.B. Mende and S.P. Geller, 1990: Fabry–Perot imaging observations of OH (8-3): Rotational temperatures and gravity waves, *J. Geophys. Res.*, **95**, 12 251–12 263.

Swenson, L.S., 1987: Michelson and measurement, *Phys. Today*, **40**, No. 5, 24–30.

Taylor, F.W., J.T. Houghton, G.D. Peskett, C.D. Rodgers and E.J. Williamson, 1972: Radiometer for remote sounding of the upper atmosphere, *Appl. Opt.*, **11**, 135–141.

Taylor, F.W., 1983: Pressure modulator radiometry, in *Spectrometric Techniques, Volume III* (Ed. G.A. Vanasse) Academic Press, New York and London.

Taylor, F.W., C.D. Rodgers, J.G. Whitner, S.T. Werrett, J.J. Barnett, G.D. Peskett, P. Venters, J. Ballard, C.W.P. Palmer, R.J. Knight, P. Morris, T. Nightingale, A. Dudhia, 1993: Remote sounding of atmospheric structure and composition by pressure modulated radiometry from space: The ISAMS experiment on UARS, *J. Geophys. Res.*, **98**, 10 799–10 814.

Thuillier, G. and G.G. Shepherd, 1985: Fully compensated Michelson interferometer of fixed path difference, *Appl. Opt.* **24**, 1599–1603.

Thuillier, G. and M. Hersé, 1988: Measurements of wind in the upper atmosphere: First results of the MICADO instrument, *Progress in Atmospheric Physics*, pp. 61–73, Kluwer Academic Publishers, Dordrecht, The Netherlands.

Thuillier, G., C. Lathuillere, M. Hersé, C. Senior, W. Kofman, M.-L. Duboin, D. Alcayde, F. Barlier and J. Fontanari, 1990: Coordinated EISCAT-MICADO interferometer measurements of neutral winds and temperatures in E- and F-regions, *J. Atmos. Terr. Phys.*, **52**, 625–636.

Thuillier, G. and M. Hersé, 1991: Thermally stable field compensated Michelson interferometer for measurement of temperature and wind of the planetary atmospheres, *Appl. Opt.*, **30**, 1210–1220.

Thuillier, G., V. Fauliot, M. Hersé, L. Bourg and G.G. Shepherd, 1996: The MICADO wind measurements from Observatoire de Haute-Provence for the validation of the WINDII green line data, *Geophys. Res.*, **101**, 10 431–10 440.

Thuillier, G., W. Gault, J.-F. Brun, M. Herse, W. Ward and C. Hersom, 1998: In-flight calibration of the Wind Imaging Interferometer (WINDII) on board the Upper Atmosphere Research Satellite, *Appl. Opt.* **37**, 1356–1369.

Tinsley, B.A., 1966: The circularly symmetric grille spectrometer, *Appl. Opt.*, **5**, 1139–1145.

Title, A.M. and H.E. Ramsey, 1980: Improvements in birefringent filters. 6: Analog birefringent elements, *Appl. Opt.*, **19**, 2046–2058.

Toth, R.A., 1993: Linestrengths (900–3600 cm^{-1}), self-broadened linewidths, and frequency-shifts (1800–2360 cm^{-1}) of N_2O, *Appl. Opt.*, **32**, 7326–7365.

Trondsen, T.S. and L.L. Cogger, 1997: High-resolution television observations of black aurora, *J. Geophys. Res.*, **102**, 363–378.

Trondsen, T.S. and L.L. Cogger, 1998: A survey of small-scale spatially periodic distortions of auroral forms, *J. Geophys. Res.*, **103**, 9405–9415.

Vallance Jones, A. and H.P. Gush, 1953: Spectrum of the night sky in the range 1.2–2 μ, *Nature*, **172**, 496–496.

Vallance Jones, A., 1974: *Aurora*, D. Reidel, Dordrecht-Holland, 301 pp.

Vanasse, G., 1977: *Spectrometric Techniques*, Vol. I, Academic Press, New York, 355 pp.

Vanasse, G., 1983: *Spectrometric Techniques*, Vol. III, Academic Press, New York, 334 pp.

van Rhijn, P.J., 1921: On the brightness of the sky at night and the total amount of starlight, *Publ. Ast. Lab., Groningen*, No. 31, 1–83.

Vest, C.M., 1974: Formation of images from projections: Radon and Abel transforms, *J. Opt. Soc. Am.*, **64**, 1215–1218.

Vincent, J.D., 1990: *Fundamentals of Infrared Detector Operation and Testing*, John Wiley and Sons, New York, 477 pp.

Vogt, S., 1994: HIRES: The HIgh Resolution Echelle Spectrometer on the Keck ten-meter telescope, *SPIE*, **2198**, 362–375.

Vountas, M., V.V. Rozanov and J.P. Burrows, 1998: Ring effect: Impact of rotational Raman scattering on radiative transfer in the Earth's atmosphere, *J. Quant. Spect. Rad. Transfer*, **60**, 943–961.

Wadsworth, F.L.O., 1896: The modern spectroscope XV. On the use and mounting of the concave grating as an analyzing or direct comparison spectroscope, *Astrophys. J.*, **3**, 47–62.

Wang, D.Y., S.P. Zhang, R.H. Wiens and G.G. Shepherd, 1993: Gravity waves from O_2 nightglow during the AIDA '89 Campaign III: Effects of gravity wave saturation, *J. Atmos. Terr. Phys.*, **55**, 325–3340.

Wang, S., G.G. Shepherd and W.E. Ward, 2000: Optimized reflective wide-angle Michelson phase-stepping interferometer, *Appl. Opt.*, **39**, 5147–5160.

Ward, W.E., Z. Pasturczyk, W.A. Gault and G.G. Shepherd, 1985: Multiple reflections in a wide-angle Michelson interferometer, *Appl. Opt.*, **24**, 1589–1598.

Ward, W.E., 1988: *The design and implementation of a wide-angle Michelson interferometer to observe thermospheric winds*, Ph.D. thesis, York University, Toronto.

Ward, W.E., D.Y. Wang, B.H. Solheim, and G.G. Shepherd, 1996: Observations of the two-day wave in WINDII data during January, 1993, *Geophys. Res. Lett.*, **23**, 2923–2926.

Ward, W.E., B. Solheim, G.G. Shepherd, 1997: Two day wave induced variations in the oxygen green line volume emission rate: WINDII observations, *J. Geophys. Res.*, **24**, 1127–1130.

Welford, W.T. and R. Winston, 1978: *The Optics of Nonimaging Concentrators: Light and Solar Energy*, Academic Press, New York, 200 pp.

Wiener, N., 1933: *The Fourier Integral and Certain of its Applications*, Dover Publications, New York, 201 pp.

Wiens, R.H., G.G. Shepherd, W.A. Gault and P.R. Kosteniuk, 1988: Optical measurements of winds in the lower thermosphere, *J. Geophys. Res.*, **93**, 5973–5980.

Wiens, R.H., S.P. Zhang, R.N. Peterson and G.G. Shepherd, 1991: MORTI: a Mesopause Oxygen Rotational Temperature Imager, *Planet. Space Sci.*, **39**, 1363–1375.

Wiens, R.H., S. Brown, S. Sargoytchev, R.N. Peterson, W.A. Gault, G.G. Shepherd, A. Moise, T. Ivanco and G. Fazekas, 1996: SATI – Spectral Airglow Temperature Imager, *Proc. SPIE*, **2830**, 341–344.

Wimperis, J.R. and S.F. Johnston, 1984: Optical cements for interferometric applications, *Appl. Opt.*, **23**, 1145–1147.

Wu, J., J. Wang and P.B. Hays, 1994: Performance of a circle-to-line optical system for a Fabry–Perot interferometer: a laboratory study, *Appl. Opt.*, **33**, 7823–7828.

Yee, J.-H., G.E. Cameron and D.Y. Kusnierkiewicz, 1999: An overview of TIMED, *Optical Spectroscopic Techniques and Instrumentation for Atmospheric and Space Research III* (Allen M. Larar, Ed.) *Proc. SPIE*, **3756**, 244–254.

Zhang, S.P., R.H. Wiens, R.N. Peterson and G.G. Shepherd, 1993a: Gravity waves from O_2 nightglow during the AIDA '89 Campaign I: Emission rate/temperature observations, *J. Atmos. Terr. Phys.*, **55**, 355–376.

Zhang, S.P., R.H. Wiens, R.N. Peterson and G.G. Shepherd, 1993b: Gravity waves from O_2 nightglow during the AIDA '89 Campaign II: Numerical modeling of the emission rate/temperature ratio, *J. Atmos. Terr. Phys.*, **55**, 377–396.

Zhang, S.P., R.G. Roble and G.G. Shepherd, 2001: Tidal influence on the oxygen and hydroxyl airglows: WINDII observations and TIME-GCM simulations, *J. Geophys. Res.* **106**, 21381–21393.

Zhang, S.P. and G.G. Shepherd, 2000: Neutral winds in the lower thermosphere observed by WINDII during the April 4–5th, 1993 storm, *Geophys. Res. Lett.*, **27**, 1855–1858.

Zwick, H.H. and G.G. Shepherd, 1971: Defocusing a wide-angle Michelson interferometer, *Appl. Opt.*, **10**, 2569–2571.

LIST OF SYMBOLS

a	effective filter width
A	area, also amplitude
A_1, A_2	apodizing functions
A'	fractional area
b	number of pixels in a bin
B	black body radiance
$B(\lambda), B(\sigma), B(\tau)$	black body monochromatic radiance
c	velocity of light
C	capacitance, also used as a constant
d	difference in Δ, such as between successive pulses
D	diameter
D^*	detectivity
$D(\sigma)$	etalon imperfection function
e	electronic charge, base of natural logarithms
E	integrated irradiance in energy units
$E_\lambda, E(\lambda), E(\sigma)$	monochromatic irradiance
E_p	irradiance in photon units
E_{photon}	energy of one photon
$\mathcal{E}$	electric field amplitude
f	focal length
$f(t), f(\Delta)$	input signal, varying in time or optical path difference
F_p, F_e	photon, photoelectron count rate
$F(\nu), F(\sigma)$	Spectrum, in frequency or wavenumber, also FPS aperture function
$\mathcal{F}$	finesse, $\mathcal{F}_A$, $\mathcal{F}_S$, $\mathcal{F}_G$, Airy, spherical and Gaussian finesse, respectively
g	sampling interval
$G(\Delta)$	interferogram filter function
h	Planck's constant
$h(\Delta), h(t)$	impulse response function in optical path difference or time
$H(\nu), H(\sigma)$	frequency response function in frequency or wavenumber
i	incident angle outside the dispersing system
I	integrated emission rate

I_D	dark current
$I(\Delta)$	interferogram
j	$\sqrt{-1}$
J_1, J_2, J_3	Fourier coefficients for a Doppler Michelson interferometer
k	Boltzmann's constant, also used as an index number, also mass absorption coefficient
K	Kelvins
L	integrated radiance
$L(\sigma), L(\lambda)$	monochromatic radiance
m	order of interference; also metres
M	fractional surface error in an etalon, as in $1/M$, also molecular weight
n	refractive index, n_e – effective refractive index
nm	nanometer
NEP	noise equivalent power
N	accumulated electron count, also used for number of pixels, or number of rulings
p	pixel size
P	pressure
$q, q(\lambda)$	quantum efficiency for photoelectron production
Q	stored charge, also Doppler Michelson interferometer constant
r	radial distance, also used as an index number
r_a	radiance reflectance
r_s	radius of curvature
rad	radians
R	responsivity, also used to denote the rayleigh unit (R)
$R(\nu), R(\sigma)$	real part of the spectrum
$\mathcal{R}$	resolving power
$R(\tau)$	autocorrelation function
s	seconds
sr	steradian
S_G, S_{FPP}, S_{FPS},	superiority of the grating spectrometer, planar and spherical Fabry–Perot spectrometers respectively
$S(\nu), S(\sigma)$	power spectral density
t	time, thickness
t_a	radiance transmittance
t_f	filter transmittance, t_{fMAX} – transmittance at the peak
t_s	system transmittance
T	temperature, also used for period, or time interval
T_E	temperature of the Earth's surface
u, v	index numbers
U	instrument visibility
V	volume emission rate, also voltage (V), also source visibility
V_{BG}	band gap energy (eV)

w	velocity
W	watts
$W(\sigma)$	instrumental response, or passband function
(x,y,z)	position in three-dimensional space
$X(\nu), X(\sigma)$	imaginary part of the spectrum
$y(t), y(\Delta)$	output signal in time or optical path difference
$Y(\nu), Y(\sigma)$	output spectrum
z	altitude
α	absorption coefficient, linear expansion coefficient, also species adjustment parameter
β	spectrometer angular slit height
δ	Dirac delta function
δ_R	phase shift on reflection
δ_T	phase shift on transmission
Δ	optical path difference (OPD), also used to denote increment, e.g. Δt
Δ_{eff}	effective optical path difference
$\Delta\sigma = \text{FSR}$	free spectral range in wavenumbers
$\Delta\lambda = \text{FSR}$	free spectral range in wavelength
ε	mirror or grating tilt angle, phase shift
ϕ	interferometer wind phase, also phase generally
Φ	interferometer instrument phase
ν	frequency
λ	wavelength
η	linewidth
$\eta_D, \eta_{ph}, \eta_R$	dark noise, photon noise and readout noise, respectively
μm	micrometre
θ	off-axis angle inside the dispersing system, also used as vector rotation angle $2\pi\nu t$
ρ	atmospheric density
σ_{SB}	Stefan–Boltzmann constant
σ	wavenumber
$\tau(z)$	atmospheric optical depth
τ	delay in time or optical path difference, as in the autocovariance function
Ω	solid angle
ω	angular frequency $= 2\pi\nu$
ξ	departure from the field-widened condition
Ψ	polarization rotation angle
ζ	zenith angle
$\star$ (open)	convolution
$\star$	cross-correlation

LIST OF ACRONYMS AND ABBREVIATIONS

ACE	Atmospheric Chemistry Experiment
A/D	Analog to Digital
ADEOS	ADvanced Earth Observing System
ADU	Analog Digital Units
AIS	Arizona Imager Spectrograph
ALIS	Auroral Large Imaging System
ASP	Auroral Scanning Photometer
ASTER	Advanced Spaceborne Thermal Emission and Reflection radiometer
ATMOS	Atmospheric Trace MOlecular Spectroscopy experiment
ATS	Application Technology Satellite
AU	Astronomical Unit
AVHRR	Advanced Very High Resolution Radiometer
CBR	Cosmic Background Radiation
CCD	Charge Coupled Device
CDB	Characterization Data Base
CDM	Capacitive Discharge Mode
CERES	Clouds and the Earth's Radiant Energy System
CID	Charge Injection Device
CLAES	Cryogenic Limb Atmospheric Etalon Spectrometer
CLIO	Circle to Line Interferometer Optical system
CMOS	Complementary Metal Oxide Semiconductor
CNES	Centre National d'Etudes Spatiales
CNRS	Centre National de la Recherche Scientifique
CODACON	Coded Anode Converter
COSPEC	COrrelation SPECtrometer
CRISTA	CRyogenic Infrared Spectrometers and Telescopes for the Atmosphere
CSA	Canadian Space Agency

CTE	Charge Transfer Efficiency
CTI	Charge Transfer Inefficiency
DE	Dynamics Explorer
DEM	Double Etalon Modulator
DFT	Discrete Fourier Transform
DGS	Diffraction Grating Spectrometer
DMI	Doppler Michelson Interferometer
DMSP	Defence Meteorological Satellite Program
DOAS	Differential Optical Absorption Spectroscopy
ENVISAT (also Envisat)	earth observing ENVIronmental SATellite
EUMETSAT	European Organisation for the Exploitation of Meteorological Satellites
EOS	Earth Observing System
EPIS	Etudes Polaires par Interférométrie Svalbard
ERBS	Earth Radiation Budget Satellite
ERWIN	E-Region Wind INterferometer
ESA	European Space Agency
EUV	Extreme UltraViolet
f/	*f*/number
FET	Field Effect Transistor
FFT	Fast Fourier Transform
FPI	Fabry–Perot Interferometer
FPS	Fabry–Perot Spectrometer
FSR	Free Spectral Range
FT	Fourier Transform
FTS	Fourier Transform Spectrometer
FUV	Far UltraViolet
FWHM	Full Width at Half Maximum
GCOM	Global Change Observing Mission
GGS	Global Geosphere System
GLO	Arizona Imager Spectrograph, also called AIS
GOES	Geostationary Operational Environmental Satellite
GOME	Global Ozone Monitoring Experiment
GOMOS	Global Ozone Monitoring by Occultation of Stars
GONG	Global Oscillations Network Group
GUVI	Global UltraViolet Imager
HALOE	HALogen Occultation Experiment
HRDI	High Resolution Doppler Imager
HIRES	High Resolution Echelle Spectrometer
HRE	High Resolution Etalon
IMG	Interferometric Measurement of Greenhouse gases
IMO	Inverted Mode Operation
InGaAs	Indium Gallium Arsenide
IRIS	InfraRed Interferometer Spectrometer

ISAMS	Improved Stratosphere and Mesosphere Sounder
ISAS	Institute for Space and Aeronautical Sciences (of Japan)
ISIS	International Satellites for Ionospheric Studies
ISTP	International Solar Terrestrial Program
LBH	Lyman Birge Hopfield bands of N_2
LRE	Low Resolution Etalon
LWIR	Long Wave InfraRed
MAESTRO	Measurement of Aerosol Extinction in the Stratosphere and Troposphere Retrieved by Occultation
MCP	Micro Channel Plate
MCT	Mercury Cadmium Telluride
MI	Michelson Interferometer
MICADO	Michelson Interferometer for Coordinated Auroral Doppler Observations
MIMI	Mesospheric Imaging Michelson Interferometer
MIPAS	Michelson Interferometer for Passive Atmospheric Sounding
MISR	Multi-angle Imaging Spectro-Radiometer
MKS	Metre Kilogram Second
MLS	Microwave Limb Sounder
MLTI	Mesosphere Lower Thermosphere Ionosphere
MODIS	MODerate Imaging Spectroradiometer
MOPITT	Measurement Of Pollution In The Troposphere
MORTI	Mesopause Oxygen Rotational Temperature Imager
MOS	Metal Oxide Semiconductor
MPP	Multi Phase Pinned
MRE	Medium Resolution Etalon
MTF	Modulation Transfer Function
MWIR	Medium Wave InfraRed
NASA	National Aeronautics and Space Administration
NASDA	National Aeronautics and Space Development Agency
NDRO	Non Destructive Read Out
NEP	Noise Equivalent Power
NICMOS	Near Infrared Camera and Multi Object Spectrometer
NIR	Near InfraRed
NOAA	National Oceanic and Atmospheric Administration
NRCC	National Research Council of Canada
ODUS	Ozone Dynamics Ultraviolet Spectrometer
OGO	Orbiting Geophysical Observatory
OLS	Operational Linescan System
OPD	Optical Path Difference
OSIRIS	Optical Spectrograph and InfraRed Imaging System
PAMI	Polarizing Atmospheric Michelson Interferometer
PBS	Polarizing Beam-Splitter
PMT	Photomultiplier

PMR	Pressure Modulator Radiometer
PPBV	Parts Per Billion by Volume
PQI	Phase Quadrature Interferometer
PRESTO	PRogrammable Etalon Spectrometer for Twilight Observations
QBO	Quasi Bienniel Oscillation
QWP	Quarter Wave Plate
RLP	Red Line Photometer
SABER	Sounding of the Atmosphere using Broadband Emission Radiometry
SAGE	Stratospheric Aerosol and Gas Experiment
SAMS	Stratosphere And Mesosphere Sounder
SAO	Semi Annual Oscillation
SATI	Spectral Airglow Temperature Imager
SCARI	Self-Compensating All Reflection Interferometric spectrometer
SCIAMACHY	SCanning Imaging Absorption spectroMeter for Atmospheric CHartographY
SCISAT	Scientific Satellite
SCR	Selective Chopper Radiometer
SEE	Solar Extreme ultraviolet Experiment
SEIS	Single Element Imaging Spectrograph
SHIMMER	Spatial Heterodyne IMager for MEsospheric Radicals
SHS	Spatial Heterodyne Spectroscopy
SISAM	Spectromètre Interférentiel a Sélection par l'Amplitude de Modulation
SNR	Signal-to-Noise Ratio
SOFIS	Solar Occultation FTS for Inclined Satellite
SOHO	SOlar and Heliospheric Observatory
SOI	Solar Oscillations Investigation
SPINR	Spectroscopic and Photometric Imaging with Numerical Reconstruction
SRI	SRI International
SSCC	Spin Scan Cloudcover Camera
SWAMI	Scanning Wide Angle Michelson Interferometer
SWIFT	Stratospheric Wind Interferometer For Transport studies
SWIR	Short Wave InfraRed
SI	System International
TDI	Time Delay and Integrate
TERRIERS	Tomographic Experiment using Radiative Recombination Ionospheric EUV and Radio Sources
TESS	Tomographic Extreme ultraviolet SpectrographS
TIDI	TImed Doppler Imager
TIMED	Thermosphere Ionosphere Mesophere Energetics and Dynamics

TIME-GCM	Thermosphere Ionosphere Mesosphere Electrodynamics-Global Circulation Model
TIROS	Television and InfraRed Observation Satellite
TOMS	Total Ozone Mapping Spectrometer
UARS	Upper Atmosphere Research Satellite
UV	UltraViolet
UVI	UltraViolet Imager
VAE	Visible Airglow Experiment
VLBI	Very Long Baseline Interferometry
WAMDII	Wide Angle Michelson Doppler Imaging Interferometer
WAMI	Wide Angle Michelson Interferometer, also WAves Michelson Interferometer
WINDII	WIND Imaging Interferometer
WINTERS	WINds and TEmperature by Remote Sensing
XUV	X-ray UltraViolet

AUTHOR INDEX

Names used in the text are included as well as cited references. For the latter, the name of the first author only is indexed.

Abel et al., 1970, 161
Abreu and Skinner, 1989, 122
Airy, 1877, 52
Akasofu, 1963, 92
Anger and Murphree, 1976, 74
Anger et al., 1973, 73, 74
Anger et al., 1987, 95
Aso et al., 2000, 94

Babcock, 1923, 102
Bahsoun-Hamade et al., 1989, 1994, 122
Baker, 1977, 147, 187
Baker et al., 1981, 147
Barringer, 1969, 160
Bates, Sir David, 171
Batten and Rees, 1990, 125
Baumert, 1964, 153
Beer and Marjaniemi, 1966, 138, 187
Benedict, 160
Bens et al., 1965, 118
Bernath, 249
Bernath, 2001, 250
Bernoulli, 31
Biondi et al., 1985, 115
Biondi et al., 1995, 125
Bird et al., 1995, 182
Blamont and Donahue, 1961, 165
Blamont and Luton, 1972, 111
Bluth et al., 1992, 160
Bose, 1995, 72
Bouchareine and Connes, 1963, 145, 175
Bovensmann et al., 1999, 271
Boyle and Smith, 1970, 80
Bracewell, 1986, 31, 38, 39
Bradley and Khun, 1948, 117
Brekke and Egeland, 1994, 4
Broadfoot and Sandel, 1992, 257, 258, 260
Broadfoot et al., 1977a, 258, 259
Broadfoot et al., 1977b, 260
Broadfoot et al., 1981, 260, 261
Broadfoot et al., 1992, 21, 261, 262
Broten, 1967, 140
Buijs, 139
Burkert et al., 1983, 249
Burnett and Burnett, 1983, 118
Burns, 1906, 4
Burnside et al., 1981, 125, 209

Chabbal, 112, 117
Chabbal, 1953, 102
Chabbal, 1958, 107, 117
Chakrabarti et al., 1994, 276
Chakrabarti et al., 1999, 263, 275, 277
Chakrabarti, 1998, 164
Chamberlain and Hunten, 1987, 10
Chamberlain, 1961, 4
Chamberlain, 1979, 39
Chapman, Sidney, 10
Clairmidi et al., 1985, 93
CNRS, 1958, 1967, 102
Cogger and Shepherd, 1969, 115
Conde and Smith, 1998, 125, 126

Connes and Gush, 1959, 1960, 139
Connes and Michel, 1975, 138
Connes, 1956, 145
Connes, 1956, 1958a, 109
Connes, 1957, 1958b, 156
Connes, 1958a, 110
Connes, Janine, 130, 134
Connes, Pierre, 130
Cormack, 1963, 16
Cotton et al., 1994, 262
Cotton et al., 2000, 262, 263
Cuisenier and Pinard, 1967, 187
Curtis et al., 1974, 161
Czerny-Turner, 258

Daley, 1991, 241
Davis, 1969, 93
Deans and Shepherd, 1978, 14, 113, 154
Decker, 1969, 1971a, 1971b, 1977, 153
Dick, 160
Dirac, 40
Dobrowolski et al., 1985, 180, 195
Dobrowolski, 1981, 186
Drummond et al., 1999, 19, 226
Drummond, 1980, 161
Drummond, 1989, 224
Duboin, 1974, 172
Dufour, 1951, 117
Dupoisot and Prat, 1979, 115

Eather, 1980, 4
Ebert-Fastie, 257
Einstein, 70
Elsworth et al., 1974, 147
Englert et al., 2000, 276
ESA, 2000, 251
Evans, 1949, 156, 181
Ezhevskaya and Sinitsa, 1978, 187

Fabry and Perot, 1896, 1897, 102
Fastie, 1952, 257
Fauliot et al., 1993, 212
Feldstein, 1964, 92
Fellgett, 130
Fellgett, 1951, 137, 138
Fischer and Oelhaf, 1996, 249, 252
Fisher et al., 1999, 127
Fisher et al., 2000, 210
Fourier, 30
Françon and Mallick, 1971, 181
Frank and Craven, 1988, 74
Frank et al., 1981, 74
Frank et al., 1986, 75
Frank et al., 1995, 98
Fraunhofer, 6, 257

Gault et al., 1983, 121
Gault et al., 1985, 181
Gault et al., 1996a, 200
Gault et al., 1996b, 207, 208, 210
Gault, et al., 2001, 237
Girard et al., 1988, 159
Girard, 1963, 158, 159, 160
Golay, 1951, 158
Goldman, 244
Goody, 1968, 160
Grainger and Ring, 1962, 164
Grassl et al., 1995, 213
Greet et al., 1989, 118
Greet et al., 1994, 117
Gunson et al., 1990, 21
Gush, 130
Gush and Buijs, 1964, 140, 141
Gush and Halpern, 1992, 141
Gush and Vallance Jones, 1955, 139
Gush et al., 1990, 5, 141, 143
Gush, 1988, 139

Hadamard, 152
Haidinger, 131
Hanel et al., 1970, 141
Hanel, 1983, 141
Hansen and Kinder, 1958, 145
Hansen, 1955, 145
Harang, 175
Hariharan and Sen, 1961, 119
Hariharan, 1987, 1989, 173
Hariharan, 1991, 181
Harlander et al., 1992, 274, 275, 276, 277
Harlander et al., 1994, 276
Harvey et al., 1988, 183
Hays and Anger, 1978, 74
Hays and Walker, 1966, 171
Hays et al., 1973, 55, 56
Hays et al., 1979, 119

Hays et al., 1981, 115
Hays et al., 1993, 213, 214
Hays, 1982, 213
Hays, 1990, 126
Hedin and Thuillier, 1988, 111
Hercher, 1968, 110
Hernandez and Mills, 1973, 120
Hernandez and Roble, 1979, 119
Hernandez and Smith, 1984, 117
Hernandez et al., 1990, 120
Hernandez, 1979, 1986b, 109
Hernandez, 1985, 111
Hernandez, 1986a, 48, 52, 65
Hernandez, 1986b, 102, 103, 110, 111, 119
Hernandez, 1987, 119
Hersom and Shepherd, 1995, 199
Hicks et al., 1974, 121
Hilliard and Shepherd, 1966a, 143, 144, 168, 170, 175, 178, 180
Hilliard and Shepherd, 1966b, 176
Hines and Tarasick, 1987, 26
Hines, 1974, 26
Hirschberg and Platz, 1965, 115
Hirschberg et al., 1971, 115
Hirshberg, 1974, 138, 187
Hirshberg and Cornwall, 1979, 187
Houghton et al., 1984, 11, 16
Houston, 1927, 117
Huffman, 1992, 11
Hunten et al., 1956, 9
Hunten et al., 1967, 156, 158
Hunten, 1953, 103
Hutley, 1982, 257, 258

Jacka, 1984, 117, 119
Jacquinot, 102, 117, 129, 151
Jacquinot and Dufour, 1948, 102
Jacquinot, 1948, 1954, 67
Jacquinot, 1950, 50
Jacquinot, 1954, 1960, 65
Jacquinot, 1958, 279
Jacquinot, 1960, 10, 62, 67, 155

Kanasewich, 1981, 31
Kaneda et al., 1981, 95
Kaneda and Yamamoto, 1991, 95
Katzenstein, 1965, 116
Keck, 264
Kerr et al., 1996, 117
Khun, 1950, 117
Killeen et al., 1983, 115
Killeen et al., 1999, 231, 232
Kirchhoff, 10
Kramer, 2002, 191

Laws, 1962, 159
Lissberger, 1968, 113
Llewellyn et al., 1997, 268, 269
Lorentz, 242
Lu et al., 1988a, 1988b, 147, 148
Lyot, 1944, 156

Mack et al., 1963, 117, 213
Mani, 2001, 247
Margottin-Maclou et al., 1996, 243
McClintock et al., 1982, 91
McDade et al., 1986, 10
McDade et al., 1991, 17
McElroy, 273
McLandress et al., 1996a, 28, 203
McLandress et al., 1996b, 218
McLeese and Margolis, 1983, 241
McLeese, 1992, 241
McLennan and McLeod, 1927, 102
McLeod, 1954, 116
McNutt, 1965, 118, 213
McWhirter et al., 1982, 91
Meaburn, 1976, 115
Meinel and Meinel, 1983, 4
Mende et al., 1977, 93
Menke, 1989, 14
Menzel and Purdom, 1994, 78
Mertz, 130
Mertz, 1959, 145
Mertz, 1965, 181
Michelson, 129
Michelson, 1927, 168, 169, 170, 175
Miller and Shepherd, 1968, 113
Morley, 129
Murcray et al., 1969, 141

Nagaoka, 1917, 117
Narayanan et al., 1989, 165
Naylor et al., 1981, 140, 142
Nelson and Fredman, 1970, 153

Neo and Shepherd, 1972, 154
Newcomb, 1901, 4
Newton, 30
Nilson and Shepherd, 1961, 112
Nobel, 70
Nossal et al., 1997, 124
Noxon, 1968, 164
Noxon, 1975, 268
Nyquist, 42

Offermann et al., 1999, 18, 270
Ogawa et al., 1994, 251
Oguti, 1990, 95
Okano et al., 1980, 115
Ortland et al., 1996, 216, 240
Osterbrock et al., 2000, 264, 265

Papoulis, 1962, 31
Pease, 1991, 76
Peck, 1948, 138
Peteherych et al., 1985, 115
Petersen and Kieffaber, 1973, 93
Philip et al., 1995, 89
Phillips, 1992, 3
Piotrowski McCall et al., 1989, 186
Planck, 2
Powell, 1986, 195

Rao, 1990, 78
Rayleigh, 1930, 9
Rayleigh, the fourth Lord, 4
Rayleigh, the third Lord, 3
Reber et al., 1993, 18, 192
Rees et al., 1982, 115
Rees et al., 1984, 125
Rees, 1989, 26
Regalia et al., 1995, 243
Richter et al., 1998, 271
Ring, 249
Ring and Schofield, 1972, 147
Rittenhouse, 257
Roach and Gordon, 1991, 4
Roach et al., 1958, 14
Roble and Ridley, 1994, 207
Roche et al., 1993, 7, 221, 222, 224
Rochon, 1999, 15, 201, 204
Rodgers, 1976, 1990, 14
Roesler and Mack, 1967, 117
Ronchi, 1964, 159
Rowland, 257, 262
Russell et al., 1993, 249

Salby, 1996, 10, 242
Sandercock, 1971, 119
Scherrer et al., 1995a, 1995b, 183, 184, 185
Schumann-Runge, 10
Scrimger and Hunten, 1957, 165
Seyrafi and Hovanessian, 1993, 6
Shannon, 42
Shepherd et al., 1965, 112
Shepherd et al., 1973, 73
Shepherd et al., 1978, 154
Shepherd et al., 1985, 173, 179, 181, 192
Shepherd et al., 1991, 130, 144, 181
Shepherd et al., 1993, 179, 192, 194, 201, 244
Shepherd et al., 1997, 201, 202
Shepherd, 1960, 112
Shepherd, 1974, 68
Shepherd, 1979, 75
Shepherd, M. et al., 1999, 204, 205
Siegmund et al., 1984, 1986, 91
Sipler et al., 1983, 119
Sivjee and Shen, 1997, 264
Skinner et al., 1987, 215
Slanger and Osterbrock, 2000, 264, 265
Slanger et al., 1997, 265
Slanger et al., 1997, 2000, 264
Slanger et al., 2000, 265
Smith et al., 1985, 119
Snell, 103
Solomon et al., 1984, 1985, 1988, 18
Solomon et al., 1987, 268
Steel, 1983, 181
Steel, 1974, 187
Stephens, 1994, 11
Stoner, 1966, 47
Strong, 1967, 160
Strutt, C.R., 1964b, 4
Strutt, G.R., 1964a, 4
Suzuki et al., 1997, 251
Suzuki et al., 1999, 248, 272, 273
Swenson, 240
Swenson et al., 1990, 123
Swenson, 1987, 129

Taylor et al., 1972, 161
Taylor et al., 1993, 162, 163
Taylor, 1983, 160, 161
Thuillier and Hersé, 1988, 172, 173
Thuillier and Hersé, 1991, 179, 211
Thuillier and Shepherd, 1985, 179, 211, 212
Thuillier et al., 1990, 212
Thuillier et al., 1996, 200, 212
Thuillier et al., 1998, 200
Tinsley, 1966, 158, 159
Title and Ramsey, 1980, 177, 178, 181
Toth, 1993, 243
Trondsen and Cogger, 1997, 94
Trondsen and Cogger, 1998, 93

Vallance Jones and Gush, 1953, 139
Vallance Jones, 1974, 26
van Rhijn, 1921, 4, 13
Vanasse, 1977, 1983, 137
Vest, 1974, 17
Vincent, 1990, 72
Vogt, 1994, 265
Voigt, 242
Vountas, 1998, 164

Wadsworth, 262
Wadsworth, 1896, 258
Wang et al., 1993, 123
Wang et al., 2000, 187, 188, 189
Ward et al., 1985, 183, 195
Ward et al., 1996, 1997, 205
Ward, 1988, 138, 176, 177
Welford and Winston, 1978, 63
Wiener, 1933, 31
Wiener-Khinchin, 45
Wiens et al., 1988, 180, 192
Wiens et al., 1991, 26, 122
Wiens et al., 1996, 123
Wimperis and Johnston, 1984, 180
Wu et al., 1994, 126

Yee et al., 1999, 230
Young, 30

Zhang and Shepherd, 2000, 207
Zhang et al., 1993a, 1993b, 123
Zhang et al., 2001, 207
Zwick and Shepherd, 1971, 180, 181

SUBJECT INDEX

The entries for instruments and missions are generally ordered according to their acronyms, rather than by their full names. A list of acronyms is available on page 300.

airglow, 4
  all-sky image of, 5
  altitude measurement, 14–17, 201
  dayglow, 24, 164–166, 202
  dynamical influence on, 205–207
  first spectrum of the near-infrared airglow, 139
  nightglow, 23
  non-thermal, 10, 170–171
  photolysis of O_3, 235
  rotational temperature, 26–27, 123, 239
  spectra from 120 to 900 nm, 21–26
  spectra from Keck astronomical facility, 264–266
  subtraction from astronomical spectra, 265
  temperatures from line widths, 175–176
  VAE (Visible Airglow Experiment), 55–56
  wind measurements from airglow, 115–116, 119–120, 124–126, 191–220
apodization, 50, 133
array detectors (*see also* CCD; detector; photomultiplier)
  charge injection device, 81, 90
  CODACON (Coded Anode Converter), 91–92
  extrinsic and intrinsic, 88
  frame transfer, 81, 85
  HgCdTe, 88, 237, 247
  infrared array detectors, 81, 88–89
  InGaAs, 88
  intensifiers, 90–91
  inter-line transfer, 81
  introduction to, 80–81
  MCP (Micro Channel Plate), 91
  multiplexer, 89
  NICMOS, 89
  position sensitive, 91–92
  resistive anode, 91
  silicon photodiode arrays, 89–90
  spectral response and materials, 87–88
  wedge and strip, 91–92
atmosphere, 11
  absorption spectrum observed by ATMOS, 22
  atmospheric holes, 75–76
  atmospheric radiation, 1–8
  Brewer-Dobson circulation, 240
  diurnal tide, 28, 202–203
  exosphere and geocorona, 124
  gravity waves, 5, 26–27
  quasi-biennial oscillation, 216–217
  semi-annual oscillation, 216–217
  structure and regions, 11–12
aurora, 4
  auroral oval, 97–98
  emission rate profiles, 15
  global image of, 76–77
  magnetospheric cusp, 73–75
  N_2^+ First Negative bands, 14
  narrow-field images, 94
  polar map in 630 nm emission, 75
  power manifested in, 78–80
  spectrum during magnetic cloud event, 264
  spectrum from 120 to 900 nm, 25

temperatures from line widths, 175–176
ultraviolet global image, 98

baffle
OSIRIS (Optical Spectrograph and InfraRed Imaging System) baffle, 269
VAE (Visible Airglow Experiment) baffle, 55–56
WINDII (Wind Imaging Interferometer) baffle, 197–198

Canadian Space Agency (CSA), 192
ACE (Atmospheric Chemistry Experiment), 248–251
MAESTRO (Measurements of Aerosol Extinction in the Stratosphere and Troposphere Retrieved by Occultation), 249, 273–274
MOPITT (Measurements Of Pollution In The Troposphere), 223–227
MIMI (Mesospheric Imaging Michelson Interferometer), 235–240
OSIRIS (Optical Spectrograph and Infra Red Imaging System), 267–269
SCISAT-1 mission, 239, 273
SWIFT (Stratospheric Wind Interferometer For Transport studies), 240–248
UVI (Ultra Violet Imager), 95–98
WINDII (Wind Imaging Interferometer), 192
CCD (Charge Coupled Device), 81–87 (*see also* array detectors; imagers, satellite)
back side illuminated, 88, 271
basic structure and charge shifting, 83
binning, 86
channel and channel stops, 82
characteristics, 86–87
clocking, 82–84
correlated double sampling, 85
frame transfer, 84
horizontal transfer register, 85
MOS capacitors, 82
readout, 84–85
signal-to-noise ratio, 87
used in a diffraction grating spectrometer, 154
CLAES (Cryogenic Limb Array Etalon Spectrometer), 220–225
chopper alternates views between Earth and a cold surface, 222–223
cutaway drawing, 222
examples of spectra obtained, 223–224
HNO_3 profile, 7
instrument design, 221–223
linear arrays of solid state Si/Ga photoconductors, 223
Lyot stop, 222
paddle wheel combining etalon selection and tilting, 221–222
constituents
atomic oxygen, 6
BrO and OClO to be measured with MAESTRO, 274
CF_2C_{12} measured with CLAES, 223,225
HNO_3, 7–8
ISAMS species, CO_2, CO, NH_4, NO, N_2O, NO_2 and H_2O, 162
minor, 7–8
MOPITT measurements of CO and CH_4, 224
ozone, 235, 240, 266
sodium, 117–118
sodium, potassium and lithium, 157
coolers, 163, 225, 247, 251
cryostats, 222, 269–270
Cosmic Background Radiation, 4–5
measured with a Michelson interferometer, 141–143
spectrum of, 6
cube corners, 138, 176–177, 249, 251

data assimilation, 241
detector (*see also* array detectors; CCD; photomultiplier)
background-thermal noise limited, 72
CLAES focal plane assembly, 222
intrinsic-thermal noise limited, 72
multiple ring anodes, 115
photon noise limited region, 71
shot noise, 71
diffraction grating, 47–51 (*see also* diffraction grating spectrographs; spectrometers)
aberration-corrected holographic concave grating, 261
analog Fourier series calculator, 50–51
classification of optical spectroscopic devices, 67
comparison of spectrometer and spectrograph, 258

diffraction grating (*continued*)
dispersion and resolving power, 64
grille spectrometer, 158–160
imaging and dispersion directions, 262
invention of, 257
linear dynamical system, 47–51
Littrow arrangement, 65
SISAM instrument, 154–156
superiority of, 65
used as a beamsplitter, 275
with imaging detectors, 154, 255
diffraction grating spectrographs
auroral spectra, 25, 264
curved grating and curved detector, 257–258
HIRES (High Resolution Echelle Spectrometer), 265
hyperspectral imaging, 255
linear CCD array for SAGE-III, 267
MAESTRO use of two spectrographs, 273
ODUS use of Ebert-Fastie type, 272–273
OSIRIS (Optical Spectrograph and Infra Red Imaging System), 267–269
SEIS (Single-Element Imaging Spectrograph), 262–263
SHS (Spatial Heterodyne Spectroscopy), 274–277
slit-less spectrograph, 259, 271
SPINR (Spectroscopic and Photometric Imaging with Numeric Reconstruction), 263
TESS (Tomographic Extreme-ultraviolet SpectrographS), 262–263
Voyager ultraviolet spectrograph, 260–261
with array detectors, 154
diffraction grating spectrometer, 255
CRISTA infrared spectrometer, 269–271
difference between the geometries of the FPS and DGS instruments, 256
Ebert-Fastie version, 257
Hadamard spectroscopy, 152–154
Mariner 10 ultraviolet airglow experiment, 258–259
mechanical collimator, 259
noise distribution across the spectrum, 152
TOMS (Total Ozone Mapping Spectrometer), 266–267
Dirac delta function, 39
Dirac comb, 40–41
Doppler Michelson Interferometry (*see also* wide angle Michelson interferometer)
achromatizing a field-widened Michelson Interferometer, 177–178
atmospheric temperature from visibility, 170
cube corner configuration, 176–177
defocusing a wide angle Michelson interferometer, 180–181
Doppler shift in wavenumber, 172
effective optical path difference, 173
four point algorithm, 174
Fourier coefficients, 174–175
fully compensated Michelson interferometer, 179
GONG (Global Oscillations Network Group) instruments, 183
Hilliard and Shepherd method, 175–176
measurement of Doppler temperature, 168–170
measurement of Doppler wind, 172–173
Michelson's observation of fringe visibility by eye, 168–170
phase quadrature Michelson interferometer, 185–187
phase-stepping interferometry, 173–175
polarizing Doppler Michelson interferometer, 181–183
reflective wide-angle Michelson interferometer, 187–189
solid interferometer, 178–180
thermal stabilization, 178–179
visibility for a single line, 169

emission rate, 8
apparent emission rate, 9, 12
integrated emission rate, 9, 14, 17
rayleigh, 9
volume emission rate, 8
ERWIN
component fringes nearly in phase, 209
optical path difference, 209
Resolute Bay wind observations, 210
suppression of F-region $O(^1S)$ emission, 209
étendue, 10
geometric factor, 68
Jacquinot's definition, 62
light grasp, 67–68

European Space Agency, 248
Envisat spacecraft, 498, 253
EUMETSAT, 79–80
GOME (Global Ozone Monitoring Experiment), 271
GOMOS and SCIAMACHY on Envisat, 271–272
MIPAS (Michelson Interferometer for Passive Atmospheric Sounding), 251–254

Fabry–Perot etalon, 51–52 (*see also* Fabry–Perot spectrometer; FPS applications)
Airy function, 52
depiction of elementary etalon, 103
free spectral range, 104–105
full width at half maximum, 106
imperfection function, 107–108
impulse response function, 51–52
instrumental passband function, 52, 105
linear dynamical system, 51–52
optical path difference, 103
photograph of Fabry–Perot fringes, 104
quadratic dispersion, 66–67
real etalon and imperfection function, 107–108
reflective finesse, 106
roughness imperfection function, 108
spherical Fabry–Perot etalon, 109–111
spherical imperfection function, 107–108
Fabry–Perot spectrometer (*see also* Fabry–Perot etalon; FPS applications)
aperture function, 108
axicon system, 116
capacitive servo controlled system, 121
CCD imager, 122
CLIO (Circle to Line Interferometer Optical system), 126–127
comparison of planar and spherical etalons, 111
configuration, 108–109
day airglow observations, 118
Double-Etalon Modulator, 119
fibre optic bundle, 115
low light level applications, 116–117
mechanical scanning, 112
multiple ring apertures, 114–115
multi-ring anode detector, 115
overall instrument passband function, 108–109
scanning methods, 112–114
stabilized Fabry–Perot spectrometers, 119–121
tandem Fabry–Perot spectrometers, 117–119, 213–220
Fabry–Perot spectrometer applications
Balmer-α observations, 124–125
CLAES (Cryogenic Limb Array Etalon Spectrometer), 221 (*see also* CLAES)
Dynamics Explorer Fabry–Perot interferometer, 116
ground-based measurement of winds, 119–121, 124–126
helium 1083 nm triplet, 116–117
HRDI (High Resolution Doppler Imager), 213–220 (*see also* HRDI)
Λ–doubling of OH lines, 117
MORTI (Mesopause Oxygen Rotational Temperature Imager), 122–123
Ogo-6 (Orbiting Geophysical Observatory) spherical Fabry–Perot spectrometer, 111
PEPSIOS (Purely Interferometric High Resolution Scanning Interferometer), 117–118
PRESTO (PRogrammable Etalon Spectrometer for Twilight Observations), 121–122
SATI (Spectral Airglow Temperature Imager), 123
sodium dayglow observations, 118
TIDI (TIMED Doppler Imager), 230–235 (*see also* TIDI)
wavelength of the auroral green line, 102
wind measurement, 119–121
Fourier spectrum, 31
autocorrelation function, 44–45
convolution and cross-correlation, 38–39
discrete Fourier transform, 41
fast Fourier transform, 44
folding frequency, 42–43
Fourier Transform, 33–35
impulse response function, 46
negative frequency, 32
power spectral density, 45
properties of, 35

Fourier spectrum (*continued*)
sampling interval, 42–43
Shannon's sampling theorem, 41–42
sinc function, 36
sine and cosine signals, 31–32
spectral element, 32

HRDI (High Resolution Doppler Imager), 213–220
comparison of HRDI and WINDII, 217–220
filters, 215
fully gimbaled telescope, 213–214
HRE, MRE and LRE etalons, 215
image plane detector (IPD), 216
optical system, 213–215
slit field converted to circular, 215
wind velocities in the stratosphere, mesosphere and lower thermosphere, 213, 216, 220

imagers, CCD satellite, 95–100 (*see also* imagers, scanning satellite)
AIS (Arizona Imager Spectrograph), 261
IMAGE satellite, 98–99
inverse Cassegrain optical system, 95–96
Polar VIS Imager, 98–99
Time Delay and Integrate, 97
UVI intensifier assembly, 96
Viking Ultra Violet Imager (UVI), 95–98
imagers, ground-based array detector, 92–100
all-sky camera, 5, 92, 94
ALIS (Auroral Large Imaging System), 94
étendue of a pixel, 93
SATI (Spectral Airglow Temperature Imager), 122–123
telecentric system, 93
imagers, scanning satellite, 72–73 (*see also* imagers, CCD satellite)
ASP (Auroral Scanning Photometer), 73–74
cross-track scanning, 73
DE (Dynamics Explorer) -1 imager, 74–77
DMSP (Defense Meteorological Satellite Program), 78–80
responsivity, 73
RLP (Red Line Photometer), 74–75
weather satellite imagers, 76–80
whisk broom or pendulum mode, 80

information domains, 12
geometrical space, 12
one-dimensional spatial information, 13–16
spectral information, 12–13, 20–26
temporal information, 26–28
three-dimensional information, 18–20
time domain, 13
two-dimensional information, 16–19
interference filters, 58, 113
cooled etalon filters, 247–248
equivalent width, 59
instrumental passband shape, 60, 113
inversion, 14–16
DOAS (Differential Optical Absorption Spectroscopy), 268
image and spectra recovered through image reconstruction, 263
onion peeling, 14–16
irradiance, 1, 12
integrated irradiance, 1
lunar irradiance, 3
monochromatic irradiance of sun and Earth, 2–3
radiant emittance, 6
radiant exitance, 6
isotropic, 12, 57

Lambertian, 57
light, 30
velocity of light, 30, 129
wave character of light, 30

measurement requirements, 13
Michelson interferometer, 129 (*see also* Doppler Michelson Interferometry; Michelson interferometer applications; wide angle Michelson interferometer)
apodizing functions, 133–134
cat's eye retro-reflector, 138–139, 187–189
consisting of three types of glass and a large air gap, 208–209
cubical beamsplitter, 175–176
depiction of elementary interferometer, 130
distinction between the SISAM and the Michelson interferometer, 155
dual pendulum interferometer, 249
dynamic alignment system, 139

elements fixed to a fused silica baseplate by molecular contact, 211
Fellgett (multiplex) and Jacquinot (superiority) advantages, 137
field widening, 142–148 (*see also* Wide Angle Michelson Interferometer)
field-widened versions, 130
Fourier transform of the spectrum, 132
fully compensated solid Doppler Michelson interferometer, 179
hexagonal beamsplitter, 180
high pressure scanning, 187
induced Doppler shift generated by reflected laser light, 212
interferogram, 132
its revival, 129–130
linear dynamical system, 133
Michelson-Morley experiment, 129
output radiances, 131–132
phase stepping, 173–175
phase quadrature Michelson interferometer, 185–187
photograph of fringes, 131
polarization states, 182
real interferometer, 135
sampling interval, 135–136
scanning methods, 137–139, 211, 249–250, 251–252
spectral resolution, 133–134
Stonehenge design, 184
superiority, 136–137
three glass system, 180
variation of optical path difference across the field of view, 134
visibility, 135
WAMI (Wide Angle Michelson Interferometer), 175–176
ZnSe beamsplitter/compensator, 245–246
Michelson interferometer applications (*see also* ERWIN; MIMI; MIPAS; SWIFT; WINDII)
ACE (Atmospheric Chemistry Experiment), 248–251
ATMOS (Atmospheric Trace MOlecule Spectroscopy), 21–22
atmospheric observations, 139–142
balloon borne observations from 1.2 to 2.5 μm, 140–141
balloon borne observations from 30 to 80 cm^{-1}, 140–142
Cosmic Background Radiation rocket measurement, 141–143
first satellite-borne Michelson interferometer, on Nimbus 3, 141
first Michelson interferometer spectrum of the near infrared airglow, 139
long baseline interferometry, 139–140
MICADO (Michelson Interferometer for Coordinated Auroral Doppler Observations), 211–213
solar absorption from 30 to 80 cm^{-1}, 140–142
with the EISCAT radar, 212
MIMI (Mesospheric Imaging Michelson Interferometer), 235–240
appearance of O_2 lines in the field of view, 239–240
filter selection, 239
four images recorded simultaneously, 236–237
general description, 235–237
optical path difference, 238
pointing mirror, 236
thermally compensated Michelson interferometer, 237–238
two images produced of each of two emissions, 237
MIPAS (Michelson Interferometer for Passive Atmospheric Sounding), 251–254
focal plane detector subsystem, 252
simplified optical layout, 252
modulators, 67, 154–166
Benedictine slits, 160
birefringent filter, 156–158
chopper, 162
circularly symmetric grille, 158–159
classification of optical spectroscopic devices, 67
correlation spectrometer, 160
COSPEC, 160
grille spectrometer, 158–160
ISAMS as a limb-viewing instrument, 162
ISAMS oscillating piston suspended on diaphragm springs, 163
molecular sieves for gas cells, 163

modulators (*continued*)
Noxon's grating spectrometer and chopper technique based on polarization, 164
ordinary and "split" birefringent photometers, 156–157
pressure modulation effects, 161
Pressure Modulator Radiometer, 160–164
resolution of the SISAM instrument, 155
Selective Chopper Radiometers, 161
SHS similarity to SISAM, 274
SISAM (Spectromètre Interférential a Sélection par l'Amplitude de Modulation), 154–156
specially designed chopper for dayglow observations, 164
superiority of the birefringent photometer, 158
Zeeman modulation with a sodium cell, 165
MOPITT (Measurements Of Pollution In The Troposphere), 223–227
carbon monoxide sources, 224
length-modulated radiometry, 223–224
Terra spacecraft, 226–227
multiplexers, 67, 152–154
classification of optical spectroscopic devices, 67
construction of Hadamard matrices, 153
Hadamard codes, 153
Hadamard spectrometer, 152–154
multiplex advantage, 151–152
noise distribution across the spectrum, 152
SHS (Spatial Heterodyne Spectroscopy), 274–277
visible region instrument, 151–152

National Aeronautics and Space Administration (NASA)
AE (Atmospheric Explorer)
ATMOS (Atmospheric Trace MOlecule Spectroscopy experiment), 21–22
EOS (Earth Observing System), 226
IRIS (Infra Red Interferometer Spectrometer), 141
SAGE (Stratospheric Aerosol and Gas Experiment), 267–268
SOHO SOI (Solar Oscillations Investigation), 183–185
TOMS (Total Ozone Mapping Spectrometer), 266–267
UARS (Upper Atmosphere Research Satellite)
National Aeronautics and Space Development Agency (NASDA), 248
ADEOS (ADvanced Earth Observing System), 248
GCOM (Global Change Observing Mission), 248
ODUS (Ozone Dynamics Ultraviolet Spectrometer), 272–273
SOFIS (Solar Occultation FTS for Inclined-orbit Satellite), 251
National Oceanic and Atmospheric Administration (NOAA), 76
GOES (Geostationary Operational Environmental Satellite), 78
TIROS (Television and InfraRed Observation Satellite), 76
nomenclature, 68
Nyquist folding frequency, 42–43
aliasing, 43

occultation
locations of solar occultations, 268
solar occultation mode for SAGE, 267
stellar occultation, 271
using either the sun or the moon with SCIAMACHY, 272
optical depth, 11, 16
optical path difference, 45
effective optical path difference, 173
ozone
Antarctic ozone hole, 266
GOMOS (Global Ozone Monitoring by Occultation of Stars), 271
ozone trends, 267
spatial resolution of 20 km for ODUS, 272
SWIFT (Stratospheric Wind Interferometer For Transport studies), 240–248
TOMS (Total Ozone Mapping Spectrometer), 266
total column amount of ozone, 266

photometer
birefringent photometer, 156–158
calibration, 59–61
elementary, 54–55
field lens, 55

multi-channel interference filter, 113–114
photon count rate, 54–55
photomultiplier, 70
dark current, 71
gain, 70
image disector, 73–74
photocathode, 70
photon, 3
emission and collection of atmospheric photons, 9
photon energy, 3, 30
photon noise, 71
photon rate, 9
work function, 70
Planck equation, 29
planetary missions
Mariner 10 ultraviolet airglow experiment, 258–259
Voyager ultraviolet spectrograph, 260–261
polarization
birefringent material, 182
birefringent photometer, 156–157
circularly polarized light, 182
polarizing beamsplitter, 182
quarter-wave plate, 182

quantum efficiency, 58, 71

radiance and monochromatic radiance, 5
radiation, 10–11
Earth radiation, 2, 3, 7
emissivity, 6, 10–11
mass absorption coefficient, 11
rayleigh, 9
remote sensing, 8, 11
ultraviolet remote sensing, 11
resolving power, 21
responsivity, 56–57
for line and continuum sources, 57–59
generalized definition, 61–62
luminosity, 68
luminosity resolving power product, 65
rocket experiments, 14–15
auroral emission rate profiles, 15
Cosmic Background Radiation measurement, 141–143
cryogenic wide angle Michelson interferometer covering 2 to 8 μm, 148
image reconstruction with SPINR method, 263
inverted volume emission rate, 17–19
multi-channel photometers, 113–114

satellite measurements, 14–16
albedo, 74
nadir viewing, 16, 19–21
orbits and tracks, 18–19, 27
scan pattern, 21
satellite missions
ACE (Atmospheric Chemistry Experiment), 248–251
ADEOS (Advanced Earth Observing System) and ADEOS-II, 248
AE (Atmospheric Explorer), 18
ATMOS (Atmospheric Trace MOlecule Spectroscopy experiment), 21–22
CRISTA (CRyogenic Infrared Spectrometers and Telescopes for the Atmosphere), 18–19, 269–271
DMSP (Defense Meteorological Satellite Program), 78–80
EOS (Earth Observing System) satellites Terra, Aqua and Aura, 226
EUMETSAT, 79–80
GOES (Geostationary Operational Environmental Satellite), 78
GOME (Global Ozone Monitoring Experiment), 271
GOMOS (Global Ozone Monitoring by Occultation of Stars), 271
HRDI (High Resolution Doppler Imager), 213–220
IRIS (Infra Red Interferometer Spectrometer), 141
ISAMS (Improved Stratosphere and Mesosphere Sounder), 161–164
MAESTRO (Measurement of Aerosol Extinction in Stratosphere and Troposphere Retrieved by Occultation), 273–274
MLS (Microwave Limb Sounder), 18
MOPITT (Measurements Of Pollution In The Troposphere), 19–21, 223–227
ODUS (Ozone Dynamics Ultraviolet Spectrometer), 272–273
OSIRIS (Optical Spectrograph and Infra Red Imaging System), 267–269

satellite missions (*continued*)
  SAGE (Stratospheric Aerosol and Gas Experiment), 267–268
  SAMS (Stratosphere And Mesosphere Sounder), 161
  SCIAMACHY (SCanning Imaging Absorption spectroMeter for Atmospheric CHartographY), 271–272
  SOHO SOI (Solar Oscillations Investigation), 183–185
  TIROS-1, 76
  TOMS (Total Ozone Mapping Spectrometer), 266–267
  weather satellite imagers, 76–80
  WINDII (WIND Imaging Interferometer), 14–17, 191–207
scattering and scattered light
  DOAS (Differential Optical Absorption Spectroscopy), 268
  minimizing the scattering of visible light into the UV region, 268
  molecular scattering, 3
  Rayleigh scattered background, 157–158
  Rayleigh scattered light in dayglow observations, 164
  Rayleigh scattering, 3
  resonance scattering in a sodium cell, 165–166
  Ring effect, 164
  zodiacal light, 4, 265
semiconductor, 81–82
SHS (Spatial Heterodyne Spectroscopy)
  all-reflective instrument, 277
  explanation of concept, 276
  schematic configuration, 274–275
  SCARI (Self Compensating All-Reflection Interferometric spectrometer), 275–277
solar, 3
  chromosphere, 6
  constant, 6
  corona, 3
  oscillations, 183–185
  spectrum, 2, 3
solid angle, 8
spectra
  as time-varying quantities, 30
  Chamberlain bands, 22–23
  first 16 vibrational levels of $O_2(b^1\Sigma_g^+)$, 265
  Gaussian line spectra, 169
  Herzberg bands, 22–23
  interstellar Ly-α, 260–261
  isotopic emission lines of O_2 A-band and OH emission, 265
  Jupiter system, 261
  Meinel bands, 22–23, 266
  N_2 First Positive, Second Positive and Vegard-Kaplan bands, 24–26
  N_2^+ First Negative and Meinel bands, 24–26
  numerical transformation, 129–132
  O_2 Atmospheric band, 22–23, 123, 213, 217
  space shuttle glow, 260–261
  spectral concepts, 31–33
spectral states and transitions
  allowed transition, 168
  Doppler shift of a line, 172
  forbidden transition, 10, 168
  interferogram for a single line, 169
  Maxwellian velocity distribution, 168–169
  $O(^1S)$ and $O(^1D)$, 168
  pressure broadening, 242
  radiating atoms in thermal equilibrium, 168
  relation of visibility to temperature, 170
  spectral lineshapes for radiation without collisions, 171
  spectral lineshapes for various gas pressures, 161
  thermal radiation, 7
  transition probability, 10
  Voigt profile, 242
Stefan-Boltzmann law, 1
superiority, 63–65
  birefringent photometer, 158
  comparison of diffraction grating and Fabry–Perot spectrometers, 66
  gain of the wide angle over the ordinary Michelson interferometer, 147
  grating spectrometer, 65
  influence of array detectors on, 279
  Jacquinot's Factor of Merit, 279–280
  Michelson Interferometer, 136–137
  planar Fabry–Perot Spectrometer, 66
  spherical Fabry–Perot Spectrometer, 111

SWIFT (Stratospheric Wind Interferometer For Transport studies), 240–248
choice of thermal emission line, 241–243
collisionally broadened line, 242
Doppler wind determination, 242–244
Fourier coefficients, 242
instrument description, 244–248
mechanical chopper replaced by rapid phase stepping, 244–245
Michelson interferometer, 245–246
ozone transport, 241
solid etalon filters, 247–248

telescope, 63
afocal, 63
anamorphotic design, 253
astronomical, 63
crossed cylindrical lenses, 261–262
non-imaging concentrators, 63
ten-meter Keck facility telescope, 265
three viewing telescopes as used on CRISTA, 269
thin film multilayers, 185–187
TIDI (TIMED Doppler Imager), 230–235
circle-to-line imaging optics (CLIO), 231
data coverage, 231
daytime measurement summary, 233–234
fiber-optically coupled telescopes, 231–232
five FP rings, each corresponding to one etalon order, 231–232
observed emissions, 230
TIMED (Thermosphere Ionosphere Mesosphere Energetics and Dynamics) mission, 230
GUVI (Global Ultra Violet Imager), 234–235
orbit inclination of 74°, 233
SEE (Solar Extreme ultraviolet Experiment), 235
SABER (Sounding of the Atmosphere using Broadband Emission Radiometry), 235
TIDI (TIMED Doppler Interferometer), 230–235
tomography, 16–19, 262, 268

UARS (Upper Atmosphere Research Satellite), 191–192
CLAES (Cryogenic Limb Array Etalon Spectrometer), 220–225
daily change of local time, 194
HALOE (Halogen Occultation Experiment), 249
HRDI (High Resolution Doppler Imager), 213–220
ISAMS (Improved Stratosphere And Mesosphere Sounder), 161–164
MLS (Microwave Limb Sounder), 18
orbit inclination, 193
WINDII (WIND Imaging Interferometer), 191–207
yaw maneuver, 194
ultraviolet, 11, 95, 234–235
Extreme, Far and X-ray Ultra Violet, 11

van Rhijn effect, 4

wavelength, wavenumber, 1
weighting function, 16, 46
wide angle Michelson interferometer (*see also* ERWIN; MIMI; SWIFT; WINDII; Doppler Michelson Interferometry)
Bouchareine and Connes, Mertz, configurations, 145
chromatic and thermal compensation, 177–179
derivation of off-axis optical path difference, 145–146
first spectral instruments, 145
fourth order dispersion, 146
Hilliard and Shepherd configuration, 144
introducing an optical path difference, 142–143
Lu configuration, 148
MICADO (Michelson Interferometer for Coordinated Auroral Doppler Observations), 211–213
parabolic primary mirrors, 187
phase variation across the field of view, 196
photographic comparison of the size of the fringes, 144
polarizing Doppler Michelson interferometer, 181–183
reflective cat's eye interferometer using spherical mirrors, 187–189
rocket measurement over a spectral range from 2 to 8 μm, 147–148
scanning methods, 147

wide angle Michelson interferometer
(*continued*)
SHS (Spatial Heterodyne Spectroscopy)
application of, 276
SOHO SOI and GONG instruments, 183–185
spherical surfaces, 187–189
superiority of, 147
superwidening, 181
WINDII (WIND Imaging Interferometer),
191–207
array detector, 197
calibration, 199–200
Centre National d'Etudes Spatiales, 192
characterization data base, 199
comparison of HRDI and WINDII, 217–220
daytime aperture, 198
Doppler temperature, 203–204
filter wheel, 195–197
frequent and infrequent calibration, 199
fully compensated, 195
inversion to true height profiles, 201
limb viewing in two different directions, 193
measurement procedure, 200–201
meridional wind, 203
Michelson interferometer, 195
multiple reflections, 195
optical system, 194–195
overall view, 198
Rayleigh scattering temperature, 204–205
standard deviation of the measurements,
201–202
tidal influence on atomic oxygen, 205–207
tidal wavelength, 203
validation, 200, 212
vertical profiles through imaging, 192
WAMDII (Wide Angle Michelson Doppler
Imaging Interferometer), 179–180, 192
WINTERS (WINds and TEmperatures by
Remote Sensing), 192
winds
imaged from a single ground station, 126
measured from Resolute Bay, 210
meridional winds, 28, 203, 220
South Pole observations, 120
stability of a wind-measurement system, 179
zonal winds, 28, 216

zenith angle, 13

International Geophysics Series

EDITED BY

RENATA DMOWSKA
Division of Engineering and Applied Science
Harvard University
Cambridge, MA 02138

JAMES R. HOLTON
Department of Atmospheric Sciences
University of Washington
Seattle, Washington

H. THOMAS ROSSBY
University of Rhode Island
Kingston, Rhode Island

Volume 1 BENO GUTENBERG. Physics of the Earth's Interior. 1959*

Volume 2 JOSEPH W. CHAMBERLAIN. Physics of the Aurora and Airglow. 1961*

Volume 3 S. K. RUNCORN (ed.) Continental Drift. 1962*

Volume 4 C. E. JUNGE. Air Chemistry and Radioactivity. 1963*

Volume 5 ROBERT G. FLEAGLE AND JOOST A. BUSINGER. An Introduction to Atmospheric Physics. 1963*

Volume 6 L. DUFOUR AND R. DEFAY. Thermodynamics of Clouds. 1963*

Volume 7 H. U. ROLL. Physics of the Marine Atmosphere. 1965*

Volume 8 RICHARD A. CRAIG. The Upper Atmosphere: Meteorology and Physics. 1965*

Volume 9 WILLIS L. WEBB. Structure of the Stratosphere and Mesosphere. 1966*

Volume 10 MICHELE CAPUTO. The Gravity Field of the Earth from Classical and Modern Methods. 1967*

Volume 11 S. MATSUSHITA AND WALLACE H. CAMPBELL (eds.) Physics of Geomagnetic Phenomena. (In two vols)

Volume 12 K. YA KONDRATYEV. Radiation in the Atmosphere. 1969*

*Out of Print

Volume 13 E. PALMÉN AND C. W. NEWTON. Atmospheric Circulation Systems: Their Structure and Physical Interpretation. 1969*

Volume 14 HENRY RISHBETH AND OWEN K. GARRIOTT. Introduction to Ionospheric Physics. 1969*

Volume 15 C. S. RAMAGE. Monsoon Meteorology. 1971*

Volume 16 JAMES R. HOLTON. An Introduction to Dynamic Meteorology. 1972*

Volume 17 K. C. YEH AND C. H. LIU. Theory of Ionospheric Waves. 1972*

Volume 18 M. I. BUDYKO. Climate and Life. 1974*

Volume 19 MELVIN E. STERN. Ocean Circulation Physics. 1975*

Volume 20 J. A. JACOBS. The Earth's Core. 1975*

Volume 21 DAVID H. MILLER. Water at the Surface of the Earth: An Introduction to Ecosystem Hydrodynamics. 1977

Volume 22 JOSEPH W. CHAMBERLAIN. Theory of Planetary Atmospheres: An Introduction to Their Physics and Chemistry. 1978*

Volume 23 JAMES R. HOLTON. An Introduction to Dynamic Meteorology, Second Edition. 1979*

Volume 24 ARNETT S. DENNIS. Weather Modification by Cloud Seeding. 1980*

Volume 25 ROBERT G. FLEAGLE AND JOOST A. BUSINGER. An Introduction to Atmospheric Physics, Second Edition. 1980

Volume 26 KUG-NAN LIOU. An Introduction to Atmospheric Radiation. 1980*

Volume 27 David H. Miller. Energy at the Surface of the Earth: An Introduction to the Energetics of Ecosystems. 1981*

Volume 28 HELMUT G. LANDSBERG. The Urban Climate. 1981

Volume 29 M. I. BUDYKO. The Earth's Climate: Past and Future. 1982*

Volume 30 ADRIAN E. GILL. Atmosphere-Ocean Dynamics. 1982

Volume 31 PAOLO LANZANO. Deformations of an Elastic Earth. 1982*

Volume 32 RONALD T. MERRILL AND MICHAEL W. MCELHINNY. The Earth's Magnetic Field. Its History, Origin, and Planetary Perspective. 1983*

Volume 33 JOHN S. LEWIS AND RONALD G. PRINN. Planets and Their Atmospheres: Origin and Evolution. 1983

Volume 34 ROLF MEISSNER. The Continental Crust: A Geophysical Approach. 1986

Volume 35 M. U. SAGITOV, B. BODKI, V. S. NAZARENKO, AND KH. G. TADZHIDINOV. Lunar Gravimetry. 1986*

Volume 36 JOSEPH W. CHAMBERLAIN AND DONALD M. HUNTEN. Theory of Planetary Atmospheres: An Introduction to Their Physics and Chemistry, Second Edition. 1987

Volume 37 J. A. JACOBS. The Earth's Core, Second Edition. 1987*

Volume 38 J. R. APEL. Principles of Ocean Physics. 1987

Volume 39 MARTIN A. UMAN. The Lightning Discharge. 1987*

Volume 40 DAVID G. ANDREWS, JAMES R. HOLTON, AND CONWAY B. LEOVY. Middle Atmosphere Dynamics. 1987

*Out of Print

INTERNATIONAL GEOPHYSICS SERIES

Volume 41 PETER WARNECK. Chemistry of the Natural Atmosphere. 1988

Volume 42 S. PAL ARYA. Introduction to Micrometeorology. 1988

Volume 43 MICHAEL C. KELLEY. The Earth's Ionosphere. 1989*

Volume 44 WILLIAM R. COTTON AND RICHARD A. ANTHES. Storm and Cloud Dynamics. 1989

Volume 45 WILLIAM MENKE. Geophysical Data Analysis: Discrete Inverse Theory, Revised Edition. 1989

Volume 46 S. GEORGE PHILANDER. EL NIÑO, LA NIÑA, and the Southern Oscillation. 1990

Volume 47 ROBERT A. BROWN. Fluid Mechanics of the Atmosphere. 1991

Volume 48 JAMES R. HOLTON. An Introduction to Dynamic Meteorology, Third Edition. 1992

Volume 49 ALEXANDER A. KAUFMAN. Geophysical Field Theory and Method.
Part A: Gravitational, Electric, and Magnetic Fields. 1992
Part B: Electromagnetic Fields I. 1994
Part C: Electromagnetic Fields II. 1994

Volume 50 SAMUEL S. BUTCHER, GORDON H. ORIANS, ROBERT J. CHARLSON, AND GORDON V. WOLFE. Global Biogeochemical Cycles. 1992*

Volume 51 BRIAN EVANS AND TENG-FONG WONG. Fault Mechanics and Transport Properties of Rocks. 1992

Volume 52 ROBERT E. HUFFMAN. Atmospheric Ultraviolet Remote Sensing. 1992

Volume 53 ROBERT A. HOUZE, JR. Cloud Dynamics. 1993

Volume 54 PETER V. HOBBS. Aerosol-Cloud-Climate Interactions. 1993

Volume 55 S. J. GIBOWICZ AND A. KIJKO. An Introduction to Mining Seismology. 1993

Volume 56 DENNIS L. HARTMANN. Global Physical Climatology. 1994

Volume 57 MICHAEL P. RYAN. Magmatic Systems. 1994

Volume 58 THORNE LAY AND TERRY C. WALLACE. Modern Global Seismology. 1995

Volume 59 DANIEL S. WILKS. Statistical Methods in the Atmospheric Sciences. 1995

Volume 60 FREDERIK NEBEKER. Calculating the Weather. 1995

Volume 61 MURRY L. SALBY. Fundamentals of Atmospheric Physics. 1996

Volume 62 JAMES P. MCCALPIN. Paleoseismology. 1996

Volume 63 RONALD T. MERRILL, MICHAEL W. MCELHINNY, AND PHILLIP L. MCFADDEN. The Magnetic Field of the Earth: Paleomagnetism, the Core, and the Deep Mantle. 1996

Volume 64 NEIL D. OPDYKE AND JAMES E. T. CHANNELL. Magnetic Stratigraphy. 1996

Volume 65 JUDITH A. CURRY AND PETER J. WEBSTER. Thermodynamics of Atmospheres and Oceans. 1998

Volume 66 LAKSHMI H. KANTHA AND CAROL ANNE CLAYSON. Numerical Models of Oceans and Oceanic Processes. 2000

Volume 67 LAKSHMI H. KANTHA AND CAROL ANNE CLAYSON. Small Scale Processes in Geophysical Fluid Flows. 2000

Volume 68 RAYMOND S. BRADLEY. Paleoclimatology, Second Edition. 1999

*Out of Print

Volume 69 LEE-LUENG FU AND ANNY CAZANAVE. Satellite Altimetry. 2000

Volume 70 DAVID A. RANDALL. General Circulation Model Development. 2000 (in press)

Volume 71 PETER WARNECK. Chemistry of the Natural Atmosphere, Second Edition. 2000

Volume 72 MICHAEL C. JACOBSON, ROBERT J. CHARLSON, HENNING RODHE AND GORDON H. ORIANS. Earth System Science: From Biogeochemical Cycles to Global Change. 2000

Volume 73 MICHAEL W. MCELHINNY AND PHILLIP L. MCFADDEN. Paleomagnetism: Continents and Oceans. 2000

Volume 74 ANDREW E. DESSLER. The Physics and Chemistry of Stratospheric Ozone. 2000

Volume 75 BRUCE DOUGLAS, MICHEAL KEARNEY, AND STEPHEN LEATHERMAN. Sea Level Rise: History and Consequences. 2000

Volume 76 ROMAN TEISSEYRE AND EUGENIUSZ MAJEWSKI. Earthquake Thermodynamics and Phase Transformations in the Earth's Interior. 2000

Volume 77 GEROLD SIEDLER, JOHN CHURCH AND JOHN GOULD. Ocean Circulation and Climate. 2001

Volume 78 ROGER PIELKE. Mesoscale. Meteorological Modeling, Second Edition. 2001

Volume 79 S. PAL ARYA. Introduction to Micrometeorology, Second Edition. 2001

Volume 80 SALTZMAN. Dynamical Paleoclimatology. 2001 NYP

Volume 81 LEE/KANAMORI/JENNINGS. International Handbook of Earthquake and Engineering Seismology. 2001 NYP

Volume 82 GORDON G. SHEPHERD. Spectral Imaging of the Atmosphere. 2002

Volume 83 R. P. PEARCE (ed.) Meteorology at the Millennium. 2002

*Out of Print

Printed and bound by CPI Group (UK) Ltd, Croydon, CR0 4YY
08/05/2025
01864902-0003